CW01332567

THE LANCHESTER LEGACY

A Trilogy of Lanchester Works

Volume One – 1895 to 1931

C. S. Clark

COVENTRY
UNIVERSITY

VOLUME ONE – 1895 TO 1931

PUBLISHED TO CELEBRATE THE CENTENARY OF THE FIRST LANCHESTER CAR OF 1895.

All rights reserved. No part of this book may be used or reproduced by recording, mechanical, electronic or any other means, including photocopying, without the express permission of the publisher:

Coventry University Enterprises Ltd.
City of Coventry
United Kingdom

ISBN 0 905 9493 07

ANY NEW INFORMATION OR AMENDMENTS RELATING TO THIS WORK WILL BE GRATEFULLY RECEIVED BY THE AUTHOR, VIA THE PUBLISHERS.

© C. S. CLARK 1995

This work is published with the assistance of the Michael Sedgwick Memorial Trust.

Founded in memory of the famous motoring researcher and author Michael Sedgwick (1926–1983), the Trust is a registered charity to encourage new research and the recording of motoring history.

Suggestions for future projects, and donations, should be sent to the Honorary Secretary of the Michael Sedgwick Memorial Trust, c/o John Montagu Building, Brockenhurst, Beaulieu, Hampshire SO42 7ZN, England.

Printed and bound in Great Britain by
Butler & Tanner Ltd, Frome and London
in January 1995

Dedicated to my wife, Lin, and my children,
Angie and Steve, without whose loyal support
this book would not have been accomplished.

ACKNOWLEDGEMENTS

My most grateful thanks go to my brother, Ray, my brother-in-law, Graham Hall, my parents-in-law, Betty and Leslie Hall, and my friend, Graham Pilmore-Bedford, for their patience in checking the script for errors.

For their assistance and encouragement over the years, my sincere gratitude extends to :

The late Mrs. George Lanchester, Frank Lanchester's two daughters: Mrs. Marjorie Bingeman and Mrs. G.A. Mobbs, Elaine Lanchester, John Bingeman and the whole Lanchester family.

Mrs. Francis Hutton-Stott and Mrs. Anthony Bird.

John Fletcher, Barry West, June Dawson and Roy Thomas.

Matt Baker, Robert Boden, Jim Cole, Dennis Corben, Fred Eite, Tom Grimmett, Alf Martin, Stan Powers, Eric Sisman, Bill Sweeney, George Uzzell, and other ex-employees of the Lanchester Works in Birmingham, as well as the descendants of their fellow workers.

In addition, I would like to thank:

Nick Baldwin, Derek Bonhomme, Peter Card, Jeremy Collins, Tony Falstein, David Fletcher, E. Roland Fox, Charles Howard, Peter Hull, Malcolm Jeal, Andrew Marfell, Bob Marshall, Rosie Shapland, Harvey Sharpe, Alen Warner.

The Lanchester Register, The Daimler & Lanchester Owners Club, The Veteran Car Club, The Vintage Sports Car Club, The Brooklands Society, The Royal Automobile Club, The Rolls-Royce Enthusiasts Club, The Sir Henry Royce Foundation, Historic Caravan Club, The National Museum of Science & Industry, Birmingham Museums & Art Gallery, The National Motor Museum at Beaulieu, The Tank Museum, Hull City Museums, The Yorkshire Car & Exhibition Company, The Behring Auto Museum of California, The Hulton Deutsch Collection, John Whybrow Photographers of Birmingham, Vince Ferry LBIPP, Neil Bruce: Photolibrary, Sotheby's, Christies, B.S.A. Tools Ltd, Jaguar-Daimler Heritage Trust, Society of Motor Manufacturers & Traders, Institute of Mechanical Engineers, Royal Institute of Naval Architects, Warwick University, The Autocar, The Motor.

Finally, to all the Lanchester owners and the scores of other people who have assisted me in my efforts but are too numerous to mention, thank you.

SPONSORS

The publishers wish to express their gratitude to the following specialist organisations, for their support and encouragement in producing this Lanchester history:—

Mr. Yoichi Takedo, Lanchester Systems K.K., Fukuoka, Japan.

Jaguar Cars Ltd., Registered owner of the Lanchester trade mark.
Browns Lane, Allesley, Coventry, West Midlands. U.K.

Manganese Bronze Components Ltd. Manufacturer of non-ferrous metal powders. Owners and occupiers of Armourer Mills, now known as The Lanchester Building.
Montgomery Street, Sparkbrook, Birmingham, West Midlands. U.K.

Premier Plating. Chrome, nickel and all plating for veteran, vintage and classic cars.
Lincoln Road, Cressex, High Wycombe, Buckinghamshire. U.K.

C.G. Jarvis and Son. Veteran and vintage coachbuilders.
Churchyard Farm, Neenton, Bridgenorth, Shropshire. U.K.

Gary W. Woodhead. Veteran, vintage and classic car mechanical restorations.
The Old Farmhouse, Standen Manor, Hungerford, Berkshire, U.K.

Richardson Hosken Group. Veteran, vintage and classic motor insurance.
Library House, New Road, Brentwood, Essex. U.K.

C.K. Motor Engineering. Engine machining and precision engineering.
Coronation Road, Cressex, High Wycombe, Buckinghamshire. U.K.

Leonard Reece and Company Ltd. Manufacture, repair and re-profiling of engine camshafts.
Clifton Road, Huntingdon, Cambridgeshire. U.K.

Vintage and Classic Shock Absorbers. New, secondhand, reconditioned and re-manufactured units.
203 Sanderstead Road, South Croydon, Surrey. U.K.

Auto Trimming Services. Vintage and classic hoods, carpets and seats renovated and renewed.
Unit 2, 3a Old Reading Road, Chineham, Basingstoke, Hampshire. U.K.

Stephen Langton Ltd. International dealers in veteran, vintage and historic cars.
Blindley Heath, Surrey. U.K.

Frenchay Garage. All electrical work, including rewiring, to the highest of standards.
Frenchay Common, Frenchay, Bristol, Avon. U.K.

Penrite Oil Company Ltd. Oils and lubricants for specialist and collectors' cars.
31 Dollman Street, Birmingham, West Midlands. U.K.

P.H. Donnelly (Rubber) Ltd. Mouldings and extrusions in natural and synthetic rubber.
68 Lower City Road, Tividale, Warley, West Midlands. U.K.

Vintage Tyre Supplies. Worldwide distributors of veteran, vintage and classic tyres.
National Motor Museum, Beaulieu, Hampshire. U.K.

Wefco Services. Manufacturer and supplier of leather spring gaiters since 1919.
90 Redland Road, Redland, Bristol, Avon. U.K.

Courtaulds Engineering Ltd. Project engineering and design.
P.O. Box 11, Foleshill Road, Coventry, West Midlands. U.K.

CONTENTS

	ABOUT THE AUTHOR	page vi
	FOREWORD	vii
Chapter One	THE VICTORIAN YEARS	1
Chapter Two	THE SINGLE-CYLINDER 5 H.P. STANHOPE PHAETON	11
Chapter Three	THE TWIN-CYLINDER 8 H.P. STANHOPE PHAETON	19
Chapter Four	THE TWIN-CYLINDER 8 H.P. GOLD MEDAL PHAETON	24
Chapter Five	THE TWIN-CYLINDER 8 H.P. MAIL PHAETON	30
Chapter Six	THE TWIN-CYLINDER 10 H.P.	38
Chapter Seven	THE TWIN-CYLINDER 16 H.P.	58
Chapter Eight	THE TWIN-CYLINDER 12 H.P.	64
Chapter Nine	THE TWIN-CYLINDER 18 H.P.	75
Chapter Ten	THE EDWARDIAN YEARS	80
Chapter Eleven	THE FOUR-CYLINDER 20 H.P.	90
Chapter Twelve	THE SIX-CYLINDER 28 H.P.	114
Chapter Thirteen	THE GEORGIAN YEARS	125
Chapter Fourteen	THE SIX-CYLINDER 38 H.P.	130
Chapter Fifteen	THE FOUR-CYLINDER 25 H.P.	140
Chapter Sixteen	THE SIX-CYLINDER SPORTING 40	147
Chapter Seventeen	THE FIRST WORLD WAR	153
Chapter Eighteen	THE SIX-CYLINDER 40 H.P.	163
Chapter Nineteen	THE VINTAGE YEARS	190
Chapter Twenty	THE BROOKLANDS RACERS	205
Chapter Twenty-One	THE SIX-CYLINDER 21 H.P.	217
Chapter Twenty-Two	DR. FREDERICK LANCHESTER	232
Chapter Twenty-Three	THE ARMOURED CARS	244
Chapter Twenty-Four	THE EIGHT-CYLINDER 30 H.P.	249
Chapter Twenty-Five	THE END OF THE ROAD	259
Appendix	SURVIVING LANCHESTERS	267
	TABLE OF ABBREVIATIONS AND COMPARISONS	272
	BIBLIOGRAPHY	272
	INDEX	273

ABOUT THE AUTHOR

Chris Clark's involvement with Lanchester cars began thirty years ago, when he was fourteen. Encouraged by his elder brother, he purchased a 1937 model for the princely sum of just £2.00, and from this humble beginning grew his love affair with the Lanchester. This first car was followed by a succession of Lanchesters from the post-1931 era, as well as other marques of earlier vintage. All these cars required considerable restoration work which Chris, a trained engineer, undertook himself.

Chris's first pre-1931 Lanchester was found abandoned as a bodyless garage breakdown truck and this, together with a further three 1920s Lanchesters which he has renovated, forms the basis of his present business of Vintage Car Hire.

The statistical information which was also required in order to carry out these restorations led to a burgeoning interest in the history of vintage cars in general, and Lanchesters in particular. He took on the responsibilities of Registrar and Historian of the Lanchester Register, a club which caters for all Lanchesters manufactured between 1895 and 1931, under the auspices of the Daimler and Lanchester Owners Club. He also organises the annual Lanchester Register Days, the only event on the social calendar when these extremely rare cars get together.

All these experiences combine to make Chris an expert in his chosen field and uniquely suited to write this history of the Lanchester Company and of the three Lanchester brothers who had such a profound influence on the British Motor Industry. This concise and definitive account, the result of many years of research, provides details of their incredible achievements and includes scores of previously unpublished photographs. The author's involvement with practically every surviving employee of the Lanchester Company, as well as members of the Lanchester family themselves, is shown in the anecdotes which abound throughout his work.

Michael Goldstein
Vice-Chancellor
Coventry University

FOREWORD

It has always been a source of sadness to me that in this country we seem to reflect so seldom on the achievements of our great engineers, and so I was delighted to see the production of this comprehensive description of the work of the Lanchester brothers. When I took over as Chief Executive of Jaguar & Daimler it was with a sense of pride stemming from not only what the Company stands for today but also the traditions and commitment to engineering excellence that have characterised those companies from which we have developed, and among these the name of Lanchester stands unequalled.

I was delighted also to learn that the description has been the work of Chris Clark, a lifelong Lanchester enthusiast who has had unique access to details of the brothers' work in his capacity as the Lanchester Register's official Historian. Being a trained engineer himself, he has combined this authority with a passion for his subject which makes this book so enjoyable to read.

Nick Scheele
Chairman and Chief Executive
Jaguar Cars Ltd

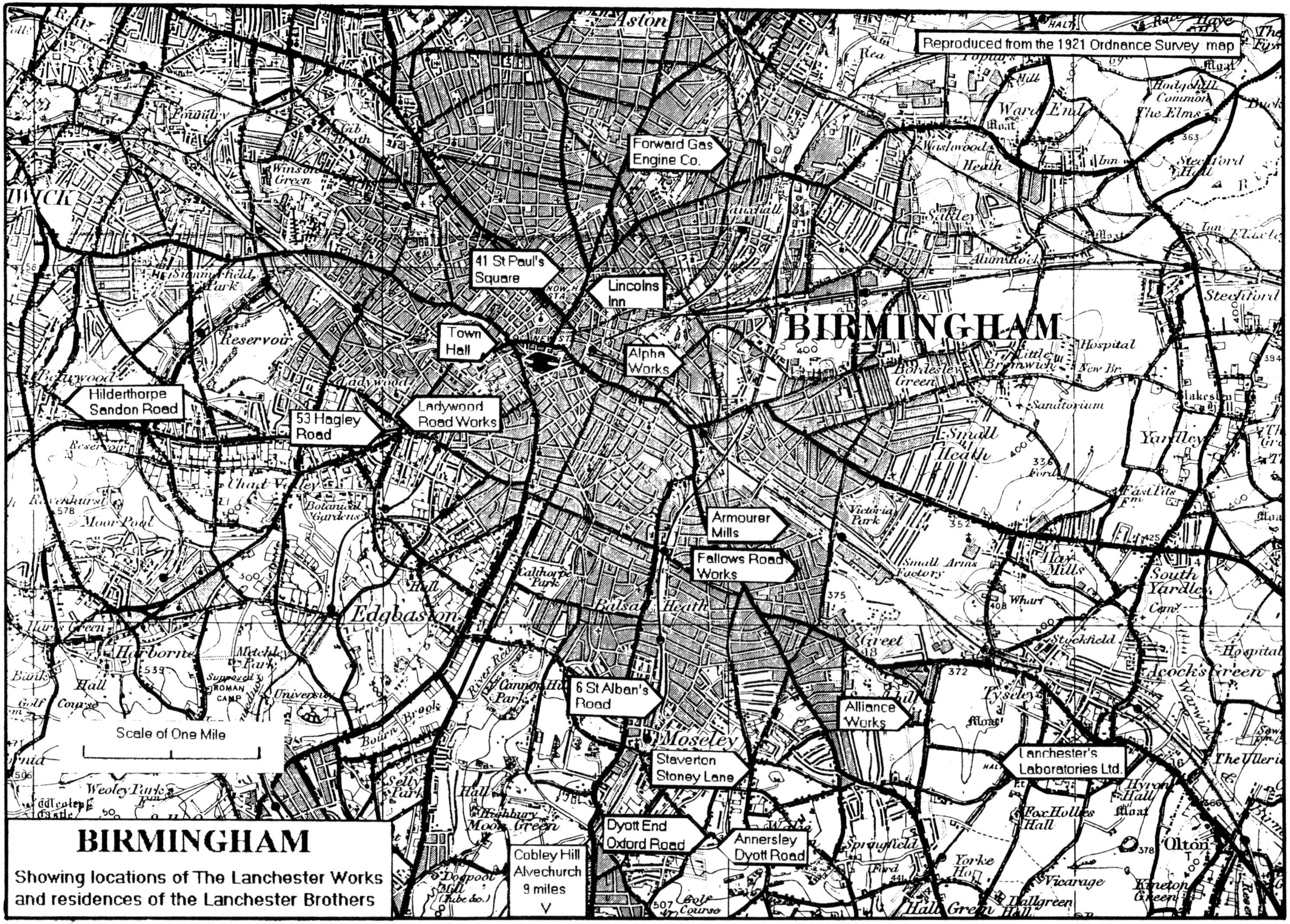

BIRMINGHAM

Showing locations of The Lanchester Works and residences of the Lanchester Brothers

Chapter One

THE VICTORIAN YEARS

The marriage in 1862 of Henry Jones Lanchester, architect, surveyor and part-time inventor, to Octavia Ward, teacher of mathematics and Latin, produced a family of eight children, all of whom were extremely gifted in their own right.

One of these children, Frederick William, was to become a prolific inventor, destined to achieve fame throughout the United Kingdom, and indeed the world, as the instigator of 426 patent applications and the creator of the first all-British, four-wheel, petrol car.

In this he was aided and abetted by two of his younger brothers, Francis Charles and George Herbert, and the three were to form a formidable trio which later became known as 'The Unholy Trinity'.

Frederick William, known as Fred, was the fourth child in the family, and was born on 23rd October 1868 at 4 Sandfield Terrace, Lewisham Road, in the village of Lewisham, then some distance from London. The Lanchester family moved to 1 St. John's Terrace, Hove, near Brighton (now 49 Church Road, Hove) in 1870, where Fred enjoyed an informative youth, attending preparatory school from the age of six and boarding school from about four years later. There he excelled at mathematics and science, finding both subjects a joy. He was accepted by the Hartley Institution in Southampton (later incorporated in the University of Southampton) although, as he was not then fourteen, he had to wait a year before attending; he spent this intervening period at an army preparatory school. When he reached the Hartley Institution Fred's natural ability soon began to show, and after two years of study he won a national scholarship to the combined Normal School of Science (now named the Royal College of Science) and the Royal School of Mines, in Kensington.

The scope of the study there proved fascinating to him and he revelled in both theoretical and practical subjects, even to the extent of deliberately failing exams in order to re-take a stimulating course encompassing precision engineering. For the third year of that course, however, chiefly to escape the laborious teaching methods of one of his main tutors, Fred opted to study mining instead of mechanics. To compensate for this, he enrolled at evening classes in mechanical engineering at Finsbury Technical College, where he became more and more interested in the engineering side of toolmaking, and correspondingly less interested in his daytime mining studies.

When his family moved in 1886 to 7 Balham Grove in south west London, he soon commandeered the basement area of the house and spent his meagre resources in setting up a small workshop there. With scope to expand his ideas, he invented his 'pendulum accelerometer', a device to record the acceleration and retardation of an object (later used in revised form on rail and road vehicles). He also invented his 'radial cursor', which assisted the old slide-rule method of quick calculations, and proved such an effective device that it was later produced by an instrument manufacturer. (He delivered a lecture on its design to the eminent Physical Society seven years later.) Fred's future looked set, but he felt he was now wasting precious time with routine studies and he left the School without even finishing the course.

He was then twenty years old with no qualifications and little money, but a great deal of confidence and a first-class brain. He took a lowly-paid job at just sixpence per hour in a Patent Office, which brought him new experiences and taught him the method of patenting an invention, knowledge which was to prove invaluable to him in the future. Shortly afterwards, in November 1888, he patented his invention of a draughtsman's drawing aid for 'Ruling Parallel Equidistant Lines'. He called this an 'Isometrograph', and later sold the patent for the princely sum of £25: one thousand times his hourly wage. The true potential of Fred's inventiveness was beginning to show, and was to be given a greater opportunity to shine shortly afterwards, when a relative introduced him to the owner of a small Birmingham engineering works: the Forward Gas Engine Company.

Francis Lanchester, known as Frank, was born on 22nd July 1870 and was the fifth child in the family. Frank had a happy and adventurous youth in Brighton, his education being at preparatory school and subsequently at boarding school until the age of fourteen, when lack of family funds prevented him from continuing further. He must surely have envied the opportunities for advanced education which had been accorded to Fred.

George Herbert Lanchester was the last-born, on 11th December 1874. The family's financial situation was even more strained than in Frank's time, but nevertheless George had a stimulating and sometimes mischievous youth. Like Frank, he adored his brother Fred, who was more than six years his senior, and he longed for stories of the man who was later to play such an important part in his life.

George's high spirits frequently earned him a caning from the Headmaster of his boarding school, and indeed he was once almost expelled when his liking for mechanics led him into serious trouble. He had devised a visually undetectable false bottom to his school tuck-box for smuggling in forbidden snacks, which was only discovered when an item was dropped into the box, causing the base to "resonate like a drum" thus exposing its secret.

George and Frank used to get up to all sorts of pranks together and were very good friends. They engineered their small toy brass cannon to take gunpowder, for instance, and used it to fire paper ammunition at passers-by.

They were both very much influenced by Fred, and watched and helped him when he was conducting his experiments in the basement workshop of the Lanchester family home in London. Fred, Frank and George lived together there for a few years, with young George at Clapham High School, Frank working in a merchant bank and Fred still studying in Kensington. In the latter part of 1888, Fred left London for Birmingham, but it would not be too long before the three were to join forces once more.

The Forward Gas Engine Company was a small firm of engineers at 67 Scholefield Street,

The Forward Gas Engine Company's factory.

Saltley, Birmingham. As well as producing gas engines, they also made hoists and printing and film machinery. The Company was run by Tom Birkett Barker, who must have been extremely impressed by Fred's initial interview for he offered the unqualified twenty-year-old the post of Assistant Works Manager. The wage offered offset this accolade somewhat, as it was only £1 per week. However, as Fred later wrote, "I raised no objection, but told him that I was entering his service for the experience and other advantages; evidently he appreciated the fact. On the very next page of the contract was the clause that any improvements I might make would be the property of the company, and to this I protested. T.B.Barker replied with a smile that was half a sneer, 'You don't suppose you are going to teach us how to make gas engines?' To this I replied, 'Evidently you don't, and therefore cannot regard the clause as of any value.' In saying this I drew my pen firmly through the offending clause in both copies of the contract."

This forthright manner was typical of Fred; the amended contract stood and he was duly appointed Assistant Works Manager.

At this time, gas engines were in common use throughout the country and were fuelled by coal-gas. They were actually internal combustion engines, but due to their excessive size, weight and need of a gas supply, could not in any way be classified as portable, although smaller ones could be trolley-mounted and pushed alongside a gas-producing machine.

In 1889, within six months of starting at the Forward Gas Engine Company, Fred had invented and patented his 'pendulum governor'. This ingenious yet simple piece of equipment, which controlled an engine's speed, was embodied in the valve gear of the engine, and proved so successful that it became general practice for it to be fitted to small engines, superseding the previously popular centrifugal governor. T.B.Barker, despite his earlier remarks, ended up paying a ten shillings royalty on each engine to which the new governor was fitted.

The Works Manager was the experienced Charles Linford, whose father had been designing engines since the middle of the century. By 1890 Linford was very ill, but endeavoured to show Fred as much as he could before he died. Fred was then duly appointed Works Manager and Designer over a workforce of fifty men, but was left with much work to do and no drawings, as Linford had kept his designs mostly to himself. Linford's widow asked for payment from the Forward Gas Engine Company in return for her late husband's notebooks, in which were kept all the details belonging to the firm's products.

The only particulars available at the Works were the drawings which Linford and Fred had prepared together previously. T.B.Barker was willing to pay for the return of the precious notebooks, but Fred confidently assured him that he could manage without them. He then proceeded to make engineering drawings of all the Forward Gas Engine Company products and parts, thus bringing the details right up-to-date, as well as designing several new models in the process.

During this year, Fred developed and patented the world's first successful gas engine starter. The effects of this invention were warmly welcomed as, prior to this, gas engines had to be started by pulling over the flywheel by hand. Every week, it seemed, a case would be reported of a man injured or even killed by being thrown over the wheel by an engine's backfire. Fred's patented design was a simple idea but very effective. It was, in effect, a cylindrical chamber through which some of the gas passed on its way to the engine cylinder. A small amount was allowed to mix with air and pass over a pilot gas-flame. The resulting explosion of this combustible mixture flashed back into the chamber and forced the engine to turn slightly. A series of these low-pressure impulses continued until the engine gained sufficient speed to overcome its compression and continue working. To many of the leading manufacturers, Fred's invention was hugely popular and was soon in use on many of their engines. It was patented by the patent agent, Dugald Clerk, and the transaction was the start of a lasting friendship between him and Fred, with the solutions to various engineering problems discussed endlessly between them. Dugald Clerk was fourteen years the elder, and Fred benefitted greatly from his wide experience and knowledge.

The success of Fred's invention led him to a meeting with the Crossley Brothers, one of the leading gas engine manufacturers at the time. He wrote, "I was shown straight into Mr. W.J.Crossley's office and things moved with breathtaking swiftness. W.J.Crossley asked me when I could let them have the necessary fittings (to enable Fred's starter to be installed) and to his astonishment, I produced them from my overcoat and laid them on his desk, at the same time handing him the necessary instructions." Fred offered to enter into an agreement which would give him a royalty of £3 per starter, with a minimum of £1,500 per year. He continued, "Two days later I visited them again and in the presence of the brothers Crossley and members of their staff, I started the engine, a 20 h.p., straight away without a hitch." W.J.Crossley agreed to Fred's terms, becoming in effect a Lanchester licensee. Payments were made to Fred for approximately two years, averaging £2,000 per year, until Crossleys purchased a newer engine starter patent by Dugald Clerk. This up-dated design had one high-pressure impulse, as opposed to Fred's sequence of low-pressure impulses, but there was a dispute for some time about priority between the young inventor and the established expert. This was resolved when Clerk granted Fred a half-share in his form of the device, which was then sold as the Clerk-Lanchester starter, yielding a fair income for both of them for many years to come.

One of the products from the Forward Gas Engine Company.

Whilst at the Forward Gas Engine Company, Fred was responsible for the design and construction of 6-stroke and 4-stroke engines ranging from 2 h.p. up to 60 h.p. and he gained much valuable experience from his time there. One of the largest engines he designed for T.B.Barker was the 'M' type, which had a bore size of 14.75ins. and stroke of 24ins.: a massive machine, which he designed "somewhere about 1895". Fred also gained widespread knowledge from servicing, maintaining and repairing all the older engines from other manufacturers, many of which were a source of great entertainment to him. One of them, an old Otto & Langham engine, had a piston which "shot upwards like a projectile from a gun", and on another occasion, when he had the job of restoring an old engine on an upper floor, he had to "place struts like mine props from floor to floor down to the solid earth, to take the recoil".

Due to Fred's expertise in ensuring that the Company's engines worked efficiently and used gas sparingly, the Company prospered, boasting higher sales than many larger and more famous companies such as the Crossley Brothers.

Fred's knowledge was constantly expanding, and in 1892 he designed, at T.B.Barker's expense, a successful single-cylinder gas engine, running at a speed of 600 r.p.m., incredibly high for those days. At these revs, the engine produced 3 b.h.p. and it was later coupled directly to a dynamo which Fred financed and produced himself in his own workshop. The combination was a great success and was probably the first direct-coupled, internal combustion engine-dynamo installation in the world. It was used for many years to light the offices and part of the Works but, in Fred's own disappointed words, "Owing to a lack of enterprise, it was not put into production and was kept for demonstration purposes only."

In October 1889, Fred sent for his younger brother, George, to join him at the Forward Gas Engine Company. Two months short of his fifteenth birthday, George therefore left his home in London to become an apprentice at the Company. Apprenticeships at that time had to be paid for, and he was sponsored by Fred for the three-year duration.

George's life at the Company began at completely the opposite end of the scale to Fred's. George recalled the early start of six o'clock, with latecomers being locked out until eight o'clock. His first jobs were fetching, carrying, cleaning and putting tools away, but George was undeterred. As he said, "It was with unalloyed pleasure I looked forward to the handling of tools and operating the machines." He was in no way given special treatment through his brother's position in the Company; he mixed well with everyone there and by the end of his apprenticeship he had become an accomplished engineer.

Towards the end of 1892, Fred visited the United States of America for the threefold purpose of trying to interest gas engine makers there in his patented starter, obtaining contracts for the Forward Gas Engine Company and also conducting some business for Dugald Clerk. The trip was not very beneficial on any of these counts, the market being completely wrong for the English style of gas engine, which was bigger than the Americans were used to. Whilst there, however, he did have some very interesting consultations with leading figures in the study of flight, a subject which greatly fascinated him.

Shortly after his return, he resigned his position as Works Manager, as there were other outside ventures he wished to explore, but he retained the post of Designer and Technical Adviser.

This situation created an incredible opportunity for George, who had by this time gained one year's experience in the drawing office, and also voluntarily attended evening classes to extend his general engineering knowledge. In the Autumn of 1893, George was appointed Works Manager, so following in his revered brother's footsteps. George had obviously proved himself a gifted engineer; he was in charge of a workforce of some sixty employees, and he was still only eighteen years old. As George recalled, "It was a great source of gratification to me that although I had worked under and alongside many of my fellow craftsmen, with two exceptions they were all loyal and accepted me as their boss." Both the management and the workforce had a great deal of admiration for George.

To ensure that he spent as little time as possible off-duty and would always be on call when the works were in operation, T.B.Barker made it a condition of employment that George should live in the worker's cottage adjoining the factory. A considerable amount of his duties consisted of engine testing and adjusting, which all had to be conducted after the main factory had closed down at eight or nine o'clock in the evening.

When Fred was away on business trips, either at home or abroad, George deputised for him in the role of Designer and Technical Adviser, giving advice to customers and attending to their complaints or concerns. As sales of the Company's products were worldwide, he also had the useful experience of making journeys abroad himself, travelling to Holland, Belgium, France, and other countries. He greatly enjoyed these trips, which broadened his knowledge tremendously.

Fred's mind was teeming with new ideas, and his latest project in a very crowded year was to start a business for the production and sale of his new invention: a novel design of pedals and wheel hubs for bicycles. For this, Fred rented a small factory in St. Paul's Square, Birmingham. At Fred's invitation, Frank Lanchester left his job at the merchant bank in London to take over the commercial and sales side of the new company as General Manager, and George assisted whenever and wherever time would allow. The name of the company was never in doubt: 'The Lanchester Brothers'. However, the high-quality designs specified by Fred were more expensive than the quantity-produced items of the established

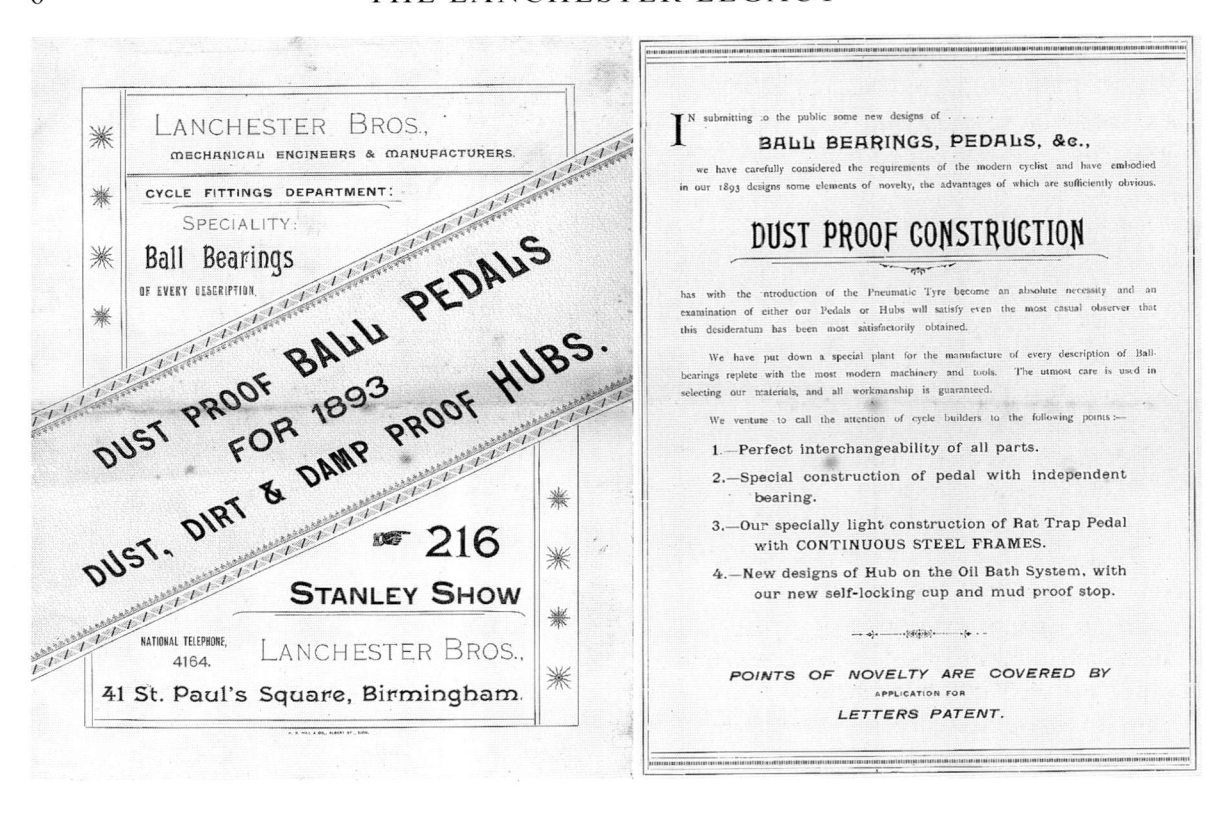

Fred Lanchester's cycle component company's brochure, detailing the advantages of his designs.

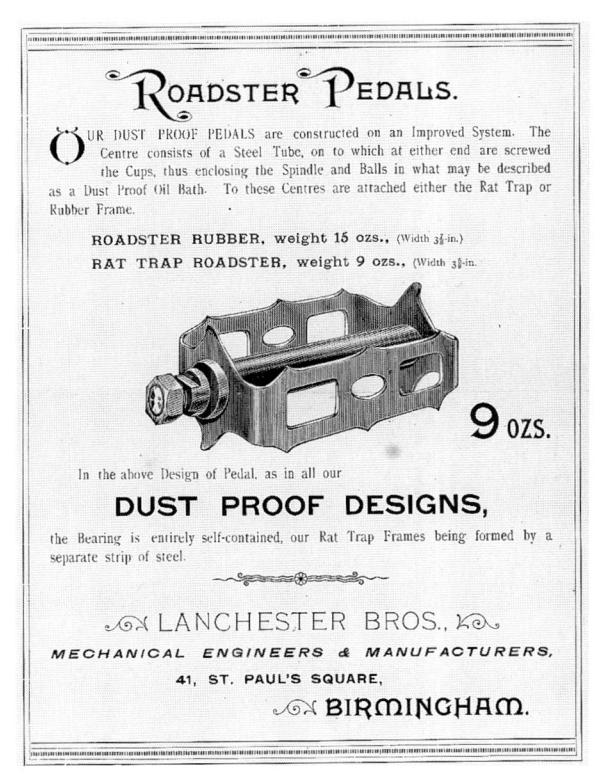

firms of component manufacturers. The business could not make sufficient headway so, rather than compromise his standards, Fred closed it down in late 1894. The bicycle components venture may seem a surprising diversion, but all three brothers were interested in cycling and were members of the Speedwell Cycle Club of Birmingham, Frank eventually becoming Secretary and Captain. Importantly, the business had brought the brothers together again, and this close association was to continue for many years.

It was at this time that Fred felt there had been an infringement of his first gas engine self-starter patent, as another company's product was underselling his. He asked Frank to investigate on his behalf, and George later recalled, "This task demanded quite an element of detective work and appealed to Frank's resourcefulness. The infringer was duly tracked down and the makers were served with an injunction to cease manufacture and pay royalties on all articles sold."

In 1895, Fred's starter patent was sold to Harry J.Lawson, the renowned and shrewd company promoter associated with the Great Horseless Carriage Company and other manufacturing concerns. The purchase was a way for Lawson to spread his influence even further over the rest of the motor industry, although George commented that "he only used the possession of the starter patents as a pawn in his company promoting schemes". Fred referred to Lawson and his directors as the "Coventry Company Promotion Gang". It was Frank who cleverly sold the starter patent on Fred's behalf; Fred was delighted, as by then he considered it somewhat outdated.

With these assignments finished, Frank was out of work, but a lucky break led to his employment at the Forward Gas Engine Company. T.B.Barker's cashier had absconded with a substantial sum of the firm's cash, and T.B.Barker offered the job to Frank at a salary of £1–10 shillings per week, plus commission on sales. Frank gladly accepted and, to Barker's amazement, his commission soon far outstripped his salary, such was his affable nature and flair for salesmanship. The three brothers were united again, although Fred was to resign his position as Designer and Technical Adviser within the following two years.

Fred Lanchester in the early 1890s, with one of his 'gliding machines'.

Since 1891, Fred had been studying the problems of heavier-than-air flight and devoted all the spare time he could to formulating the principles which promoted his 'sustentation and equilibrium' theories. His immediate aim was to establish the science of flight, and from the start he set up a framework of information in a methodical and professional manner. He researched ship design, thermo-dynamics, hydro-dynamics, the flight patterns of birds, gliding angles and indeed any aspect which could be used to enhance his theories. It became apparent to Fred that the future of flight did not lie with gliding machines, but instead could only be achieved successfully if powered by some sort of motor, much lighter than anything known of at that time. He therefore consulted his old friend, Dugald Clerk, with a view to raising money from institutions to finance the development of his proposed 'aero engine'. Dugald Clerk answered with the now famous words, "Lanchester, you may be right, but you are a young man and must consider your reputation. If you were to put forward such a proposition seriously, your reputation as a sane engineer would be ruined." Dugald Clerk was one of the few men whose opinion Fred respected and he reluctantly accepted this advice.

George Lanchester later recalled the prevailing attitude when he wrote, "The whole idea of flying in those days was popularly regarded as the wild dreams of crazy inventors." Fred continued, however, to carry out practical experiments in the theories of flight whenever time permitted. He cast aside his disappointment and re-channelled his energies into formulating designs for a motorised horseless carriage, knowing full well that the development of a high-speed, light, internal combustion engine for this purpose would also provide a solution to his dream of powered flight.

Some yards from the back door of George's cottage adjacent to the Forward Gas Engine Company was a wash-room, which Fred rented from T.B.Barker and converted into an experimental workshop. Here he designed a small high-speed vertical engine and used benzol as the fuel. This was the only light spirit then available and was a by-product from the manufacture of coal-gas, obtainable from the local gas works. The purpose of this engine was to gain experience in the problems of vibration at high speed, and was an important step along the path to his later vibration-free engines. The second high-speed engine he designed was a 2–3 h.p. single-cylinder vertical engine, with a 3.75ins. bore by 4.5ins. stroke, which ran at about 900 r.p.m. It was fitted with automatic, pressure-activated valve gear, a design taken from his patent of 1889. He devised a unique surface carburettor, which utilised a series of cotton wicks from which the benzol evaporated. As benzol was a low gravity spirit, the Lanchester wick carburettor proved to be an ideal vapouriser. The vapours emitted were fired by tube-ignition, the platinum tube of which was heated by a blow-lamp. The engine was funded and built by Fred and, after modification to take positively actuated exhaust and inlet valves, was subsequently fitted into a river launch.

Apart from being fascinated by all types of boats, Fred felt that a launch was the only practical option open to him for engine testing, as the stifling legal restrictions effectively denied him the use of the roads. The flat-bottomed launch had also been designed by Fred and was built by the three Lanchester brothers in the garden of his home: Fairview (now number 14), St. Bernard's Road, Olton, near Birmingham. Most of the work was done on Sundays, much to the disapproval of the local people passing by on their way to church. Due to this, the brothers were soon dubbed 'The Unholy Trinity', a soubriquet which was to stick with them for life.

The engine was connected to a stern-mounted paddle wheel of teak construction for its propulsion. When the boat was completed in the autumn of 1894, Fred and George transported it on a three-horse-drawn lorry to the River Thames at Oxford. It was launched from the slipway at Salters Boat Yard, thus becoming Britain's first motor boat. (It was used on the Thames for ten years and then on the River Wey for at least ten more.) However, some vibration was experienced within this experimental engine, and subsequent remedial work was carried out to rectify the problem.

The first all-British motor boat, launched by the Lanchester brothers in 1894.

Meanwhile, Fred's scientific research on his theories of flight was continuing. It was supplemented by the practical experimentation of catapulting model gliders from brackets fastened to the bedroom window frames of his house in Olton, to demonstrate his 'vortex theory' and to calculate the mean velocity of flight. Frank, who lived with him, and George, who stayed there at weekends, would act as 'spotters', estimating the height to which the models rose and marking the distance travelled. Their landlord, however, was more concerned with the structure of his house than world-shattering advances in flight, and these antics almost led to the brothers' eviction.

Fred was to leave the house shortly afterwards, in 1895, of his own accord, as his finances were extremely low. He moved to cheaper accommodation in a farmhouse at Cobley Hill, Alvechurch, some fourteen miles from Birmingham. He enjoyed living there, but in view of the long hours spent in his workshop, there must have been many times when he stayed overnight at George's cottage, just a stone's throw from his workshop. Finally, the distance between home and work became prohibitive and in 1898 Fred rented two rooms in a new residential building called Lincoln's Inn, in Corporation Street, Birmingham, where he worked on the many diverse challenges which interested him.

His life there was far from lonely as, shortly after he moved in, Frank suggested to George that they should pool their resources and rent a large room there also. As George was by then firmly established at the Forward Gas Engine Company and therefore able to dictate his mode of living, he readily agreed. A "colony of residents" soon grew around them, which included the pioneer motorist, Archie Millership, and the Pugh brothers, who owned the Whitworth Cycle Company and were to become business associates of the Lanchesters for over thirty years.

The many years that the Unholy Trinity lived there were happy ones, and the practical jokes they shared provided light relief from their heavy workloads. One such incident concerned a religious group who rented the rooms over theirs. Their weekly hymn meetings gradually became nightly, and the din of the congregation beating time with their boots proved too much for the Lanchester brothers, working throughout the evenings. Frank and George bought bugles, which neither could play, and "were able to make such noises as were a counterblast to the devotions". Fred's work must surely have ceased completely with this cacophony of sound. He was also singularly unimpressed by the dripping of water through his ceiling, caused by the adult baptisms held above in a large portable bath. George remembered, "On one occasion one of the bathers must have been hysterical with religious fervour and ejected a quantity of water over the side, which percolated through

Fred's ceiling and dripped onto his detailed drawings, completely ruining them." Fred's typically forceful and uncompromising letter of complaint addressed to "The High Priest of the Bathing Department", coupled with another attempt by his brothers to master the intricacies of bugling, resulted in the worship being conducted elsewhere.

Chapter Two

THE SINGLE-CYLINDER 5 H.P. STANHOPE PHAETON

Fred's annual overseas visits had shown him the progress made in eight years of automobile engineering design and development, and he was not at all impressed with what he saw. He felt that the designs were crude adaptations, either from the horse-drawn carriage or from the cycle industry. Fred's visit to the Paris exhibition in 1889 prompted his comment on the Daimler, "A vehicle of German origin having rather the appearance of two bicycles coupled side by side. A body with motor and transmission mechanism bridging the gap between them." He felt that none of the continental companies had attacked the problem from fundamental scientific principles, and was quite scathing in his comments on the designs of the Benz, Lenoir, and Peugeot; the popular Panhard he thought "a very crude affair".

Fred resolved to start anew with the design of his own horseless carriage, and in 1894 he began work in earnest. He operated from his small rented workshop, which he addressed as 12 Taylor Street, adjoining the Forward Gas Engine Company's main factory. T.B.Barker was to play an important part in the evolution of the Lanchester car when he introduced Fred to James and Allen Whitfield, so that they could discuss the provision of finance. Fred had already conceived general designs for a new petrol motor to supersede the one already in regular use on the River Thames. It was after one of these demonstration runs, obviously undertaken to assess the viability of the petrol motor, that the Whitfield brothers agreed to advance some money. A four-way syndicate, including a small share from T.B.Barker, was duly formed. T.B.Barker's part in all this was not done solely for altruistic reasons, however, as Fred was obliged to reward him for the introduction with the sum of £100.

The syndicate gave Fred the £700 he estimated would be necessary to build the car, although it was difficult to judge with no previous design principles or criteria to refer to. This was a great vote of confidence in Fred and no small risk by the syndicate partners, as they advanced the money at a time when all British legislation was stubbornly obstructive to the formation of a horseless-carriage industry.

The infamous Red Flag Act of 1865 had been slightly modified in 1878, making it unnecessary to carry the flag, but someone was still required to walk sixty yards in front of a moving vehicle to warn oncoming traffic and to help control restive horses. It also still required one person to steer the vehicle and another person to stoke the steam boiler. The three people with their carriage were allowed a maximum speed of 4 m.p.h. in the open country and 2 m.p.h. in towns; hardly conducive to encouraging a new industry which would transform the world. The Whitfield brothers and T.B.Barker were therefore to be commended for their forward thinking, which enabled Fred to begin this important venture.

Fred had patented many different systems of motor vehicle design for his own interest and evaluation, and had therefore formed definite opinions of the type of car he would design. The making of his first Lanchester car was typical of the man, inasmuch as he would not accept second best in any aspect of the enterprise. Consequently the components from the cycle and horse-drawn carriage industries, which could possibly have been adapted, were not used. Of the very few items which were bought from outside specialists, the main ones were the tyres. For the first time in motoring history, Fred's car was actually designed to run on air-filled tyres; no car made previously had used the flexibility and shock absorption characteristics of pneumatic tyres as an intrinsic feature of the overall design. The Dunlop Company, who had a virtual monopoly for tyres in the United Kingdom, hand-made a set of four tyres and tubes for the Lanchester car. These were the first motor vehicle tyres they ever produced, and were the beginning of a whole new industry. On the Continent, the Michelin brothers had converted a Peugeot to take pneumatic tyres. When it was entered for the Paris-Bordeaux Race, it incredibly used up their entire stock of twenty-two tyre tubes. They were consequently forced to retire after twenty hours of racing, causing the winner of the race, Levasser, to declare, "The air-filled tyre will obviously never be of the slightest use for motor cars."

The Lanchester rear wheels were 44ins. in diameter and 2.5ins. wide. They had a 6ins. wide malleable iron hub for the spokes, of which there were seventy-two in all, each of 0.125ins. diameter. The front wheels were of a different size from the rear, being 36ins. diameter with hubs of a 5.5ins. wide spoke base, incorporating sixty spokes. In a typical piece of original thinking, Fred discarded the normal practice of direct perpendicular spokes and set them tangentially, thus making them exceedingly strong and able to withstand far heavier loads.

The front axle consisted of a 2ins. diameter, three-piece, tubular structure, pivoted at its centre to support the main chassis frame. This gave vertical movement to each wheel to allow for uneven road surfaces, as no road springs were employed. The wheel track was 4ft.8.5ins., which was very wide, the same width in fact as the standard gauge of railway track.

The chassis design of Fred Lanchester's first car of 1895, showing its pivoting front axle.

The steering was of the 'castor' method, more commonly known as the Ackerman principle, with two wheels being mounted on separate stub axles and linked together to steer simultaneously, unlike a pivoted cart axle. Fred developed the system to give a wider and more accurate range of steering than that on other vehicles. The stub axle steering heads were machined to receive ball bearings for the swivel (king) pins and, unlike any other make, a swivel pin castor-angle was incorporated in the Lanchester design, thus setting a precedent for all manufacturers to follow. Fred's swivel pin inclination of 3 degrees was, however, later reduced to 1.5 degrees to ensure that the steering self-centred. Due to these sound design principles, the steering, by tiller, was dynamically stable. The steering column bearing was located on the centre pivot of the front-axle and was free to align itself with the axle's angular movement. The wheelbase distance from front to rear axle was 5ft.

In common with the vast majority of cars of the period, Fred used a chain-driven rear

THE SINGLE-CYLINDER 5 H.P. STANHOPE PHAETON

axle, the first and only instance on Lanchester cars. The chain sprockets were of 6:1 ratio and were mounted on the differential housing. Other manufacturers made the chain sprockets integral with the rear wheels, running on one continuous axle shaft and technically termed a 'dead axle'. This created difficulties on bends as the axle's wheels would travel at different speeds, resulting in one wheel jumping or skidding and thus losing traction. Lanchester's design was different. His 'live axle' set the pattern for the type accepted universally ever since, and consisted of two half-axles, or half-shafts, in conjunction with four bevel-type differential gears.

The whole rear-axle assembly was carried on four bearings, secured directly to the rear chassis in order to resist side-thrust when cornering. The near-side, rear, road wheel was in keyed connection with the near-side half-shaft. The off-side, rear, road wheel rotated on ball bearings around its half-shaft and was provided with a free-wheel roller-type clutch operated by a band-brake, whereby the half-shaft could be locked when desired, allowing that wheel to run free. The near-side shaft then became geared up in the ratio of 2:1 through the differential gear, and became the car's only driving wheel. George recalled that, "This high ratio 'overdrive' system was only suitable for accelerating on the level or slightly downgrade, which was not sufficiently down hill for coasting." It was the practice in those days to coast wherever possible.

Main braking was effected by first reducing the overdrive to normal drive. Fred's unique epicyclic gearbox, in its reverse selection, would then slow and stop the car, although the gear would have been disengaged before the car began to travel backwards. A second system of braking relied on brake blocks, called shoes, which actuated on the foremost part of both rear tyres. The gearbox was explained by George: "Two trains of epicyclic gears were given a low ratio of 3:1 and reverse 3.3:1. The gears were carried on a centre shaft extension from one crankshaft. The planet carriers of the low gear also comprised the female element of the top gear. In engagement the clutch locked the planets to the sun-wheel and thereby provided a direct drive, coupling the chain pinion direct to the drive shaft, thus rotating it at engine speed. Reverse was a simple train in which the planet pinions were carried by the brake drum (of 8ins. diameter) and therefore when the drum was arrested by the leather-lined band brake the ring, which was integral with its chain pinion, rotated in the reverse direction to that of the sun-wheel and driving shafts. The top gear clutch was mounted on the sleeve of the low gear planet carrier, in keyed connection thereto, but a sliding fit. It was operated by a fork engaging with quadrants running within the collars. The two sun-wheels were in solid keyed connection with the centre, or driving, shaft. The driving shaft was provided with a hollow cone coupling, engaging with the cone end of the crankshaft and secured by a long bolt passing through a drilled bore on the axis of the shaft. The outer extremity of the crankshaft extension was supported in a bearing carried on an outrig from the main frame." This 1:1 direct drive ratio he referred to was its first use in motoring history (although Renault later tried to claim the credit). Fred devised an ingenious system whereby a foot pedal was depressed to engage the gears (a method which was later used on Henry Ford's epicyclic-geared cars).

The vertical, air-cooled, single-cylinder engine was located centrally in the chassis, under the floor. It was of four-stroke design, with a bore of 5ins., a stroke of 5.25ins. and a maximum speed of 1,000 r.p.m.

The piston was of two-piece construction, the cast-iron upper part of which located three compression piston rings and was shaped to form a concave crown. In addition to these, an oil-scraper piston ring was used, again for the very first time in motoring history (although they are now accepted practice). Screwed and locked into the base of the crown was a malleable iron casting which housed the gudgeon pin. Incorporated in the piston's base were two guides, diametrically opposite each other, which located in cut-outs in the cylinder walls to resist alternating rotational forces given by the connecting rods. The whole piston, of advanced engineering design, was machined inside and out to reduce weight.

The aluminium crank-case housed two crankshafts and the connecting rod from each secured the single piston. The two connecting rods were fully machined to exact specifications to ensure that both were equal and dynamically balanced; their shanks were also drilled for weight saving. The two crankshafts were located one above the other and ran in opposite rotation, counteracting all reciprocating forces and ensuring perfect balance. Each had its own flywheel of equal diameter and weight to ensure equal moments of inertia, taking fully into account the one fitted with a centrifugal fan for cylinder cooling.

The Lanchester wick carburettor was adopted on this first car and was of similar design to that of the earlier river launch. It was installed between the seats within the petrol tank, the petrol being supplied by a hand-pump. Initially, ignition was by high-tension spark, but owing to electrical leakage the system was found to be none too reliable. Before taking the car on the road, the ignition was therefore altered to a unique Lanchester low-tension method; the source of current from an accumulator actuating a single-circuit inductance winding on an iron core. (In 1909, the Autocar was to state that "It is not generally known that the Lanchester was the first car in this country to fit a magneto as standard, or, in fact, to fit one at all.") Fred designed a special sparking plug called an 'igniter', which fired the fuel vapours allowed in by the mechanically operated inlet valve. The engine speed was governed by varying the tension on a spring connected to an inertia governor of the 'hit and miss' type, and was controlled by a sideways moving lever, adjacent to and operated by the driver's left knee.

The chassis design for Fred Lanchester's first car of 1895, showing the chain sprocket drive.

The chassis of the car was probably the first ever produced to be torsionally rigid and non-flexible. It employed longitudinal and transverse 2in. diameter steel tubes, thus making it both light and extremely strong. (This use of a tubular framework remains popular for vehicles of limited production.) During construction, the steel tubes were secured in the correct position by pegs and the whole framework had to be brazed together, as oxy-acetylene or electric welding had not then been invented. George recalled that, to his knowledge, no structure of such a size had ever been joined before by brazing and there was no brazing hearth large enough to take such an item. In view of this, the chassis was one of the few items which Fred could not construct within his small experimental workshop and so it was worked on in the Forward Gas Engine Company's factory. On this occasion, after the factory had closed down for the night, Fred and George assembled the chassis tubes and constructed the frame, avoiding distortion by heating and brazing each matching pair of joints simultaneously.

In view of the moderate travelling speeds anticipated at that time, spring suspension

THE SINGLE-CYLINDER 5 H.P. STANHOPE PHAETON

between the chassis and axles was considered unnecessary. However, for the comfort of passengers, the bodywork was given spring suspension, where a rare concession to traditional coach design allowed two 'C' shaped rear springs, consisting of five leaves, 1.75ins. wide by 0.25ins. thick. The rear ends of the springs were connected to the chassis brackets by shackles, and their forward ends were in sliding contact with the chassis stirrups. There was just one front spring which was placed transversely over the front-axle beam with its centre uppermost, and on this was mounted the front of the body.

The bodywork of the vehicle consisted of two side plates cut from a sheet of 0.125ins. thick mild steel, 9.5ft. long by 18ins. high, and then cut to shape. They were connected together by eight 2ins. diameter steel tubes, and on to this body skeleton were brazed the brackets for the suspension, control levers, petrol hand pump, and other fittings.

Despite the fact that this was the first experimental car, quality was not sacrificed in any way. In his later years, George recalled that, "The coachwork was built on to the body frame. Commencing from the rear end, the flat surface provided a luggage rack and the rear seat board was made of ash. The seat itself was in walnut, coach finished in natural colour and upholstered in a dark reddish-brown morocco leather, buttoned and pleated and made detachable. The seat, backsquab and seat arms were horse-hair stuffed. The centre well, the floor and toe board were ash covered with ribbed rubber matting and the seat apron framed and panelled in walnut, as were the two front bucket seats, also coach finished in natural colour and upholstered with leather, matching the rear seats. The deckboard between the seats was also coach finished. As for the rear seat, the seat apron was framed in panelled walnut and the floor board was ash covered with ribbed rubber-matting. The side plates were sheathed with 0.5ins. thick walnut, coach-finished in natural colour. The extreme cross-tubes in both rear and front were also sheathed in coach finished walnut, giving a cylindrical appearance and an attractive finish. Steps of the normal horse-drawn carriage type were incorporated, two on each side for the rear passengers and one each side for the driver and front passenger."

The warning sounder, or the 'alarm' as the Lanchesters called it, was a foot-operated

The first all-British, four-wheel, petrol car: the 1895 Lanchester 5 h.p. Stanhope Phaeton.

two-tone bell, an enchanting feature which was to be used for many years to come. Lamps for night driving were a pair of parabolic reflectors containing electric bulbs lit by an ordinary lead-plate battery. As first built, the five-seater, Stanhope Phaeton car had no weather protection except for a shallow dashboard fitted to the front footrail. To some extent this kept the driver's and front passenger's feet dry, as well as preventing the wind from blowing up the rug wrapped firmly around their legs and feet.

And so it was that Frederick Lanchester's first car was built, the greater part made in his own small workshop aided by three or four trusted craftsmen, with the remainder made in the Forward Gas Engine Company Works under Fred and George's supervision. George helped his brother whenever he could get away from his duties and, although it is not recorded, it seems highly unlikely that Frank Lanchester would have missed out on this exciting project.

The total cost of the finished car exceeded Fred's estimate of £700, by £400. However, Fred felt that as he had quoted the initial figure, it would have been unfair to have asked the syndicate for further money; he therefore bore the brunt of the extra cost himself, without even mentioning it to them. This expenditure, in addition to the £100 introduction payment made to T.B.Barker, must have been a heavy financial burden for Fred to bear.

Having never bothered with publicity, no contemporary press articles acclaimed the completion of Fred Lanchester's first car, but later written reports redressed this omission. George Lanchester wrote, "The machine, an entirely original conception, was built in 1895." The car was "completed by the end of 1895", according to Dr. P.W.Kingsford in his biography of Fred Lanchester, the text of which was agreed by George Lanchester and also Fred's widow, Dorothea. The 1895 completion date was also recorded by the great motoring correspondent, Sammy Davis, who was acquainted with all three brothers. The motoring historians, Anthony Bird and Francis Hutton-Stott, both knew Frank and George and frequently wrote of this date, as did many other accomplished authors, such as St.John C. Nixon. Indeed, the esteemed Veteran Car Club, referring to an earlier debate about the choice of car to decorate their Club tie, wrote, ".... it was felt that a British car would be more appropriate, and the first Lanchester of 1895 was selected."

Positive confirmation of the Lanchester's completion date is also made by two sons of Mr. Hubert Chippendale Smith. They recall his vivid account of how, as a nineteen-year-old employed by the Forward Gas Engine Company, he witnessed the Lanchester being wheeled out of the workshop for its initial road trials to begin: the date was definitely late in 1895.

This 1895 date not only referred to the first Lanchester car, but also laid claim to a more exalted title. There had been many other makes of vehicle constructed previously of different classifications, such as Richard Trevithick's 1802 three-seater steam carriage, Samuel Brown's 1824 coal-gas vehicle, Arthur Bateman's 1881 steam tricycle, and Walter Bersey and Ratcliffe Ward's 1888 electric powered vehicles. Similarly, the list of experimental cars powered by petroleum spirit is wide and varied. Some cars were advertised but were never constructed, others were constructed but with uncorroborated dates. The majority were based on foreign designs, most were powered by foreign engines, and to avoid steering complications many were made as three-wheel cars. From 1889 the construction dates of cars made by Edward Butler, James Roots, Frederick Bremer, Thomas and Walter Santler, John Knight (assisted by George Parfitt), Percival Petter, the Roots and Venables partnership, Herbert Austin for the Wolseley Sheep Shearing Machine Company, and others, have all been taken into account and compared with the Lanchester. According to all available information, the 5 h.p. Lanchester, designed from 1894 and finished in 1895, was indubitably the first all-British, four-wheel, petrol car. More importantly though, the Lanchester was the first car to be designed as a stepping-stone to full production, and was the first British car to influence the development of motor vehicle design in general. Percy Kidner, when President of the Institution of Automobile Engineers, wrote, "It would not be an exaggeration to call Frederick Lanchester the designer of the first genuine automobile."

THE SINGLE-CYLINDER 5 H.P. STANHOPE PHAETON

After thorough testing of the Lanchester, and the installation of all the luxurious "fittings and furniture", the historic first road run took place before daybreak towards the end of February, 1896. In preference to using benzol, George recalled obtaining enough benzoline petroleum spirit for the test-run from the apothecary-chemist's shop, where "only medicinal quantities were available for sale". Sitting in the car with the engine running and the workshop doors open, Fred and George speculated on how far they should drive, and whether or not they would have to be towed home by a horse! The tension and excitement they felt must have been heightened by the knowledge that they were breaking the law, as they had no preceding footman "to give assistance to drivers of horses and riders to proceed past the machine".

They set off from the workshop but had only gone two or three miles when they experienced ignition trouble. Whilst trying to adjust the low-tension igniter in the dark, they must have thought their luck had run out when the local constabulary arrived. No doubt astounded by the sight of a horseless carriage, the policeman illuminated the engine by the light of his bullseye lantern. The adjustment was then quickly completed and the obliging policeman went on his way, rewarded with a piece of silver. As George recalled, "He had been very interested in the novel vehicle, the like of which he had never seen or

A later Lanchester advertisement.

dreamed of." They completed the run of about ten miles' duration and both were eager and excited about the future potential of mechanical road transport. George related, "It was a great thrill and sufficiently successful to instil in me an enthusiasm beyond expectation for the new mode of locomotion."

However, it was soon discovered that the 1,689 c.c. 5 h.p. engine was inadequate for the weight of a large five-seater car, which unladen was between 17 and 18 cwt. It had to be pushed up some of the hills, and it was obvious that redevelopment work was necessary. The brothers were very disappointed, especially Fred who had devoted so much time to the project. He recalled later that, "I think I may say without fear of contradiction that all the pioneers underestimated the speeds at which road vehicles could be driven and were correspondingly at fault in the matter of the power required. We believed in those days that the motor car would replace the horse vehicle, not the railway train, and we were thinking of speeds about 15 m.p.h. I think also that many of us did not realise that a horse in harness will often do work at the rate of 10 h.p. as defined. Of these matters we must plead ignorance and taking both into consideration, it will account for a great deal."

More trips were made in the car and to gain protection from the elements, a 'Victoria Hood' of the orthodox horse-drawn carriage design was fitted.

With the car in constant use, Fred pondered on the options open to him. Wanting to keep the car capable of conveying five adults and their hand luggage, he considered adding an extra gear to assist the car's hill-climbing capabilities, or alternatively installing a more powerful engine. By mid-summer he had decided on the modifications needed for improved performance.

More money was needed for this next phase of experimental work, so the syndicate was again consulted. They declared themselves willing to carry on, but during the lengthy discussions it emerged that in addition to the £100 introductory commission received from Fred, T.B.Barker had also accepted a similar sum for the same purpose from the Whitfield brothers. Whilst recognising that this might be good business practice, Fred was bitterly disappointed that the friend with whom he had shared so much had not been more obliging. Therefore, although Fred retained his rented workshop at the factory, T.B.Barker dropped out of the syndicate, leaving the others to carry on without him.

Chapter Three

THE TWIN-CYLINDER 8 H.P. STANHOPE PHAETON

By late July 1896, Fred had reached the decision that the original Stanhope Phaeton should be modified to incorporate, amongst other items, a more powerful engine. The 1895 car was stripped down for investigation, new designs were developed and reconstruction work began.

The new engine was a horizontally opposed twin-cylinder of 5ins. bore and 5.375ins. stroke, giving a capacity of 3,459 c.c. Two parallel contra-rotating crankshafts in vertical alignment were provided, one 3ins. above the centre line of the engine and one 3ins. below. The crankshafts themselves were fitted with weights whose function was to balance the rotational and the reciprocating components of the engine, as well as the resulting torque reactional forces. Each crankshaft was fitted with flywheels of identically-balanced weights and these neutralised the recoil impulses when the cylinder gases exploded, so that no shock or vibration would be transmitted to the chassis frame. Fred explained, "With the two crankshafts in opposite rotation, the reciprocating movement of the combined centre of gravity of the balanced weights was in a straight line along the axial line of the cylinders and pistons." It was also always half-way between the two crank pins.

Attached to each piston were three connecting rods. On each side of the main connecting rod were two smaller connecting rods which ran to the other crankshaft, sandwiching the main connecting rod attached to the other piston. Thus when the pistons were at mid-stroke a diamond shaped linkage was formed. The movement of the reciprocating parts, as well as the recoil from the ignition explosion, were both neutralised by virtue of this novel arrangement, which rendered the engine practically vibrationless.

It developed some 10 b.h.p. when run at 1,250 r.p.m., an extremely high speed in those days. Other makes of engine ran at much lower speeds, except for one small De Dion Bouton where the internal forces generated were correspondingly too small to be of any consequence.

Fred's forward planning required that all modifications had to be finished before the

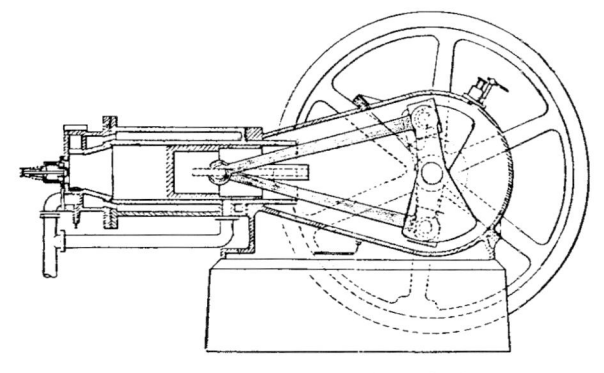

The first of Fred Lanchester's single-cylinder gas engines, which incorporated two contra-rotating crankshafts connected to a single piston. Patented in August 1895.

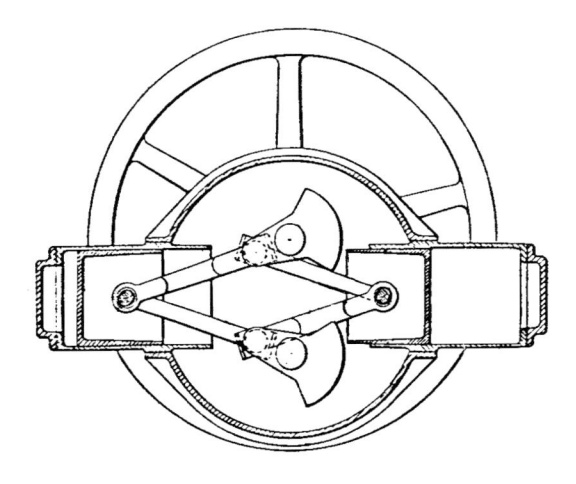

The first of Fred Lanchester's twin-cylinder gas engines, which incorporated two contra-rotating crankshafts connected by four rods to the two pistons. Patented in June 1896.

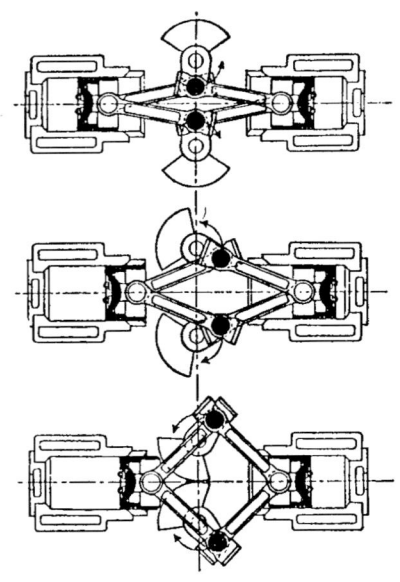

The three phases of crankshaft, connecting-rod and piston movement of Fred Lanchester's twin-cylinder car engines.
Top = left piston on power stroke.
Centre = right piston at top dead-centre after its compression stroke.
Bottom = left piston at bottom dead-centre.

middle of 1897, a tall order for a man with limited finances. Work progressed, but it soon became obvious that the facilities in his experimental workshop were not sufficient, and new premises were sought. In early 1897, Fred rented a plot of land at 33 Ladywood Road, five hundred yards from Five Ways, Edgbaston, in Birmingham, and also purchased its existing galvanised building. All the plant and equipment were transferred and duly installed in this, together with a substantial amount of new machinery which was required. The cost of this operation was between £1,300 and £1,400, for which Fred made himself personally responsible.

George resigned his post as Works Manager at the Forward Gas Engine Company in order to join Fred as his personal assistant. His new responsibilities included the supervision and control of all work in progress, managing the craftsmen, ordering materials, storekeeping, general accounting of materials used, checking of information and components produced and also that materials delivered were correct to specified requirements. George recalled that, "Though hard work it yielded a wealth of enjoyment, but in many respects it was certainly no sinecure." His work relieved Fred of many routine tasks, thus enabling him to carry on with new designs and specifications, as well as freeing a little time for his beloved aerodynamic studies.

A team of eight craftsmen were employed. Such was Fred's trust in these workers that, many decades later, he could still remember their names. Hopton, who had transferred from the earlier Cycle Works, was mechanic and chief toolmaker, and received 34 shillings per week. Chapman, the lathehand, received 30 shillings, and Chapman Junior received 26 shillings. Richard Cliff, the turner, who had also worked for Fred previously, received 32 shillings. Trikket, who worked the capstan lathe, received 20 shillings, as did Hayward, the smith and handyman. A labourer received 18 shillings, and the young apprentice, Albert Lee, five shillings per week. (Albert was to rise through the ranks to become manager of the coachbuilding works in the 1920s.) The weekly wage bill was £9-5 shillings, but

THE TWIN-CYLINDER 8 H.P. STANHOPE PHAETON

occasionally extra help would also be hired. George's salary was decided at £3 per week, making the total bill £650 per annum. The outgoings for outside work, materials and other items were to total £100 per week on average, a tremendous amount of money in those days. W.F.Thomas, a clever draughtsman and engineer, was later employed, and he prepared the drawings from Fred's layouts and original sketches, for Fred to check and dimension later.

The first commitment in the new Ladywood Road Works was the conversion of the 1895 car and the installation of the original gearbox coupled to its new two-cylinder, air-cooled, 8 h.p. engine and its new rear-axle transmission. The engine was carried on a tubular steel frame built out behind the rear axle. Its two horizontally-opposed cylinders of 5ins. bore and 5.375ins. stroke ran across the original chassis frame. The wick carburettor, or vapouriser as it was known, was transferred to the chassis. This allowed room for an extra passenger on the front seat, making the car a six-seater phaeton.

The reconstructed Stanhope Phaeton, complete with 8 h.p. twin-cylinder engine and worm drive, with Fred Lanchester at the tiller.

The reconstructed car was finished and ready for its road tests by the Autumn of 1897. However, not everything went according to plan for the 8 h.p. engine on its road tests of several hundred miles. Fred recalled, "At first it ran excellently, then it took to vibrating, not in the ordinary way but in a manner that would wake the dead; I could hardly sit on the car. This would happen quite intermittently without any apparent cause." The engine was stripped down and the problem was found. Without any gear connection between the two crankshafts there had been excessive lateral pressures acting on the pistons. To cure this, a train of three bevel gears was employed to connect the crankshafts, and produced a smooth, silent, fully-balanced engine. Due to this, the problems found in other components during these initial tests were not catastrophic. As Fred related, "We have had all sorts of things happen to these engines and they have gone on just the same; we have had most of the brasses melt out on one occasion and a piston broke and practically disappeared on another, and yet the engine remained serviceable. The difficulty was to get everything in line in the workshop, because any want of accuracy in alignment was fatal to smooth running."

The epicyclic gearing incorporated internal fibre-block band brakes, which replaced the leather-lined type, and gave one low-speed gear and one reverse gear. Gear actuation was altered to hand-lever operation.

A high-speed, direct drive was provided via a leather-faced, cone clutch. A new braking system was installed on this model. When the hand lever for actuation of the top gear was pulled backwards past its 'neutral' gear position, the reverse side of the top gear cone clutch engaged with a fixed braking cone of a similar taper. The system provided an efficient and

Fred Lanchester at the tiller of his 8 h.p. Stanhope Phaeton, accompanied by James Whitfield, in the rear seat, and his family.

powerful transmission brake. The drive-chain for the 'live' rear-axle was abandoned in favour of another unique feature: the Lanchester underhung worm gear drive. This device was perfectly quiet and efficient in operation, never needed adjustment, could not come apart and was fully enclosed, thus avoiding harmful dust and dirt erosion. The half-shafts also received the benefit of Fred's inspiration as, once more setting a precedent for other manufacturers, he designed and made the machinery to produce splined shafts for his cars. These innovations, along with other features such as the low-tension ignition system, inertia governor, wick-carburettor, centrifugal fan extractors and valve gear, were all novel features which were to be used on successive cars.

Another extremely important item was Fred's invention for simultaneously operating the governor on each cylinder. Unlike the 5 h.p. Phaeton, a foot pedal, which Fred christened the 'accelerator', was fitted in the front floorboard.

One obvious difference from the original design was the position of the steering tiller, sometimes called a side-lever. Due to the re-location of the new clutch, gear and brake levers, it was taken from its mid-seat position, where it had been operated by the driver's left arm, and re-positioned on the off-side of the car to be operated by the driver's right arm. (Later on, probably in 1898, wheel steering replaced this side-lever-tiller. In order to compare the advantages of both systems, Fred designed and fitted a 'rack and pinion' steering layout, anticipating their common use by some fifty years. He deemed his tiller steering system to be superior and employed it on all subsequent models, although it was never re-introduced on this particular car.)

The actual conversion of the 1895 Stanhope Phaeton to take the 8 h.p. engine, new transmission and ancillaries took a few months to complete, but it proved worthy of the time and money spent. It could easily undertake long journeys and was known to complete a hundred miles in one day, on many occasions. It could climb any hill, was capable of a

The 8 h.p. Stanhope Phaeton, modified for wheel-steering to compare with its original side-tiller.

maximum speed of 18 m.p.h., could easily maintain 12 m.p.h. as an average road speed, and would give 23 m.p.g. It was, in fact, the ideal vehicle to demonstrate to potential sponsors, and the speeds attained must have been exhilarating for them.

From the 14th November 1896, due to the repeal of the earlier, archaic laws which had held back Britain's car industry for so long, 12 m.p.h. was allowed on country roads. The famous Emancipation Run had taken place to commemorate this, although even if the Lanchester car had been in running order at the time, the brothers would certainly not have participated. They felt that the Run was nothing more than a clever publicity stunt for the roguish organiser, Harry Lawson, and his many business enterprises. However, Fred later acknowledged that the Emancipation Run had produced benefits to motoring in general, when he commented, "I do not myself consider Lawson was better than any other swindler, but I do say that owing to the stupidity of past legislators, the net result of his operations was to the national good."

Chapter Four

THE TWIN-CYLINDER 8 H.P. GOLD MEDAL PHAETON

In 1897, concurrently with the conversion of the original chassis to take the 8 h.p. engine and new worm gear transmission, a start was made on the 'Spirit Passenger Phaeton', officially designated Lanchester Car Number Two.

The chassis construction of this second car was similar to that of the first in both its original and revised form, in that it utilised 2ins. diameter tubes to make the whole structure torsionally rigid and very strong; the body frame was similarly built with 1in. diameter steel tubes. These seamless, drawn steel tubes were shaped and brazed to their locating lugs.

The body framework was built as a separate entity from the chassis and was suspended independently on half cantilever elliptics, both front and rear. The bodywork was detachable at the extremities of the front springs, and it was able to be hinged upwards from its pivot points at the rear of the spring eyes. This allowed the whole of the bodywork to be easily lifted up, single-handedly, for convenient access to the engine and mechanics.

Orthodox coach building methods were dismissed, as one would expect from the innovative Lanchester brothers. The tubular framework was allowed to remain visible, and inset panels for the coachwork were added. The distinctive Lanchester leather folding-dashboard made its first appearance, pivoted from the top edge of the angled footboard. Stowage drawers were located underneath the hide-covered seat squab and a woven container basket ran across the width of the car, behind the seat. A pair of large driving lamps were secured to brackets protruding from the body frame, above the front wheels of the car. It was originally

The chassis and body frames of the 8 h.p. Spirit Passenger Phaeton. The solid tyres used as a comparison with the earlier Lanchester were soon discarded in favour of pneumatic tyres.

THE TWIN-CYLINDER 8 H.P. GOLD MEDAL PHAETON

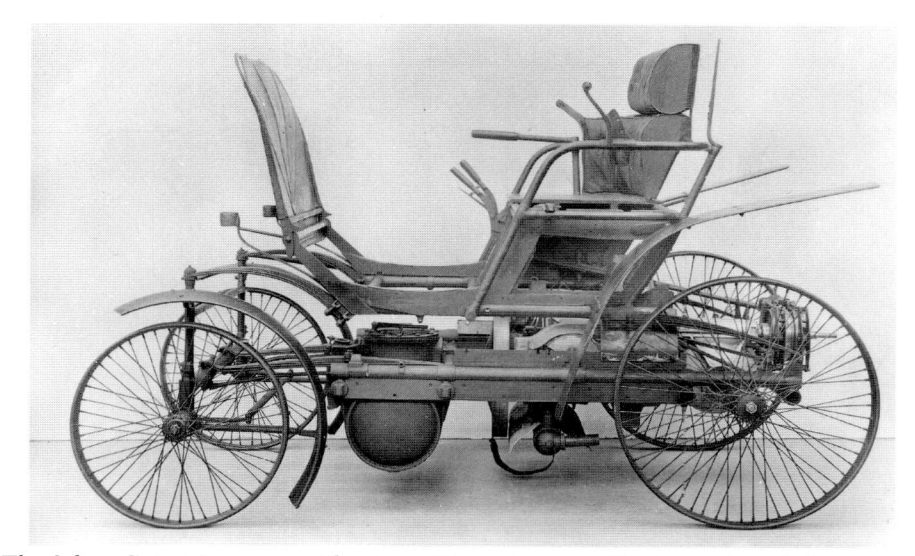

The 8 h.p. Spirit Passenger Phaeton with no tyres and little coachwork in position.

called a three-seater car, but a comfortable two-seater was a more realistic claim and the car was later described as such.

Set in each corner with no body or chassis overhang, the 2.5ins. wide wire wheels were again of tangentially spoked design and were of unequal size between the front and rear, the front wheels being 32ins. diameter and the rear 38ins. As a comparison to the earlier car, they were experimentally fitted with solid rubber tyres, but after initial road testing and assessment these were quickly converted to pneumatics. Both types of tyre were of smooth tread, as specified under the Light Locomotive Act, which came into force on 14th November 1896.

The wheelbase of the car was 5ft.10ins. and the wheel track was 3ft.10ins. The worm transmission was identical to the 8 h.p. Stanhope Phaeton, although the 'live' rear axle was given ball bearing thrust races and cageless roller bearings. As the bearing industry declared itself unable to consider Fred's idea of producing precision roller bearings, he decided to manufacture them himself. Fred therefore designed and produced the world's first centreless grinding machinery for their manufacture. (This was used until 1905, when the bearing industry caught up and was able to offer accurate, mass-produced rollers at a cheaper price.)

Fred's design was for a 'cheese' type, where the length of the roller was equal to its diameter, and was another 'first' in engineering history. Harry R.Ricardo, the famous engineer, later wrote of these roller bearings that Fred "succeeded in making them himself to a uniform accuracy of plus or minus two ten-thousandths of an inch, a remarkable achievement for that date".

Also remarkable for their time were the Lanchester brakes. The popular Punch magazine epitomised the general state of braking systems by its cartoon showing a horse and cart at full gallop alongside a speeding motor car. The farmer shouts to the motorist, "Pull up, you fool, the mare's bolting!" "So's the car", replies the terror-stricken motorist. This problem did not occur on the Lanchester, which had three methods of braking. The epicyclic and main braking systems remained the same as before, but the large section shoes acting on both rear tyres as a parking brake were of a different design from the Stanhope Phaeton. In contrast, many other cars of the period merely employed the basic method of allowing a metal rod, called a 'sprag', to stick in the road surface so that the vehicle would not roll backwards. However, the 'sprag', actuated by a cord from inside the car, was not always successful. A 'double-sprag' was then designed by some manufacturers to avoid the danger of the 'single-sprag' sliding over smooth surfaces, or of the car using it as a pole-vault!

The completed 8 h.p. Spirit Passenger Phaeton, ready for the road.

The 8 h.p. engine was carried on the steel tubular frame placed centrally beneath the body work. For the first time ever in motoring history, cantilever springs were employed. Their complementary parallel motion linkwork connected the chassis and front axle, whereas the rear axle, as on the first car, was unsprung.

The Lanchester wick carburettor, complete with its cotton wick air cleaners, the tiller-steering and the accelerator pedal were all retained from the previous vehicle. The epicyclic gearbox was identical to the 8 h.p. Stanhope Phaeton, but had its actuation of top gear and brake by one hand lever, and the low gear and reverse by another; a leather to metal cone clutch was employed.

As with all Lanchester cars, Fred used aluminium extensively. Cooling fans, cylinder jackets, crankcase covers, crankshaft phasing gear cases, worm gear boxes, carburettor, fuel pump and several body and chassis components were all produced in light alloys, which resulted in the unladen weight of the Spirit Passenger Phaeton being only 14 cwt.

The car was completed in 1898 and road testing began, with Fred driving and George acting as riding-mechanic. For the initial road tests they circled the neighbouring streets. "How the residents must have cursed us", recalled George, "hurtling round and round this course until some trouble made it necessary for us to return to the works!" The route must have become equally boring for Fred and George as, during yet another trip, they spontaneously decided to break away from this routine. George later related, "I well remember the joy of our first non-stop run of 68 miles. Fred and I had been testing the car round and round a block of houses in Edgbaston and were so fed up with just sitting that we set off, regardless of the fact that we were hatless and coatless, to Pershore, Evesham and back to the works without mishap." Thoughts of the 12 m.p.h. speed limit were ignored, as on that journey they touched 28 m.p.h. on level ground. It seems, in fact, that most of it was travelled at that speed, and much faster on the downhill stretches, as George logged the average speed as 26 m.p.h. Furthermore, the journey was accomplished with no fuss, no vibration, and no noise. Flushed with success, the brothers carried out more extensive testing throughout the remainder of 1898 and many trips of a similar length were undertaken.

The 8 h.p. Lanchester was surely worthy of their time, attention and hard work. At the Automobile Club Exhibition and Trials in Richmond, Surrey, it greatly impressed the public and motoring journalists alike. Indeed, the Motor Car Journal of 23rd June 1899 stated: "A car at the exhibition which comprises probably more novel points than any other, is that exhibited by Mr. F.W.Lanchester of 59 Lincoln's Inn, Corporation Street, Birmingham." The Motor Car Journal went on to describe the car in the most favourable of terms.

The tests were many and varied, involving hill climbing, the noise of which was generally described as "appalling", a static display of the car and its accessories, and fifty-mile road runs. The Lanchester behaved impeccably and the simple summary from the impressed judges of the rally stated, "Good on road and on hill." Of the eighteen cars competing, none received a better report. The Lanchester Spirit Passenger Phaeton received the Gold Medal Award for its 'Excellence of Design and Performance'. The Times newspaper wrote of the Lanchester in glowing terms, and summarised, "The most salient features of novelty occur in the motor, the principal departure from previous practice existing in the means taken to avoid vibration, one of the chief objections to ordinary motor cars driven by oil engines. The result of Mr. Lanchester's invention has been to produce an exceptionally easy-running carriage, equal in fact to one electrically driven, whilst the steering and manoeuvring properties are quite remarkable."

The journal 'Engineering' gave a damning opinion on the other motor cars at the event, but particularly picked out the Lanchester for special praise: "The display was, however, sufficient to make plain what we think most persons have recognised, namely that the average oil motor car is not the kind of vehicle most persons would select for a pleasure

George Lanchester, with Fred at the tiller of the 8 h.p. Spirit Passenger Phaeton, both wrapped up on a winter's day.

trip in the country. The rattle, jar, dust and stench are as opposed as anything can well be to the peace and quiet of the country. Some of the carriages underway were simply terrors from the rattle and noise they made. If we make an exception to this statement, it is in favour of the Lanchester car which appeared to us the smoothest running and pleasantest vehicle on the road." The unique silence of the Lanchester maybe prevented the driver of a Lifu steam vehicle being aware of its presence when Fred was giving trial runs, as the two vehicles collided head-on in the entrance gate. Neither was seriously damaged, and both were running again shortly after.

The 8 h.p. Spirit Passenger Phaeton on test at Petersham Hill in Richmond, on its way to a Gold Medal Award and a subsequent change of name.

All this attention and publicity were quite a novelty to the Lanchester brothers as, apart from an article in their local paper, nothing had ever been written about them. Whereas other company promoters sought publicity, even if they had nothing much to publicise, the Lanchester brothers had insisted on waiting until they were ready to manufacture cars for sale to the public.

After the well-deserved praise at this first public showing of a Lanchester, Fred and George must have driven back to Birmingham in the car feeling very proud indeed. A Punch magazine poem of the time stated:

> If you desire to travel fast,
> a Motor Car is unsurpassed;
> should you desire to travel far,
> trust not too much a Motor Car.

Fred and George's confident, six-and-a-half-hour return journey to Birmingham surely cast doubt on this sentiment! On their return, the jubilant syndicate suitably re-named the car the Gold Medal Phaeton.

Meanwhile, a variation of the car's 8 h.p. twin-cylinder engine had been developed, which encompassed another phase in Fred's long experimental programme. It was of similar form to those made previously, but this time was adapted to be cooled by water instead of air.

The engine was built to power a river launch, and James Whitfield gained approval from the syndicate for the construction of a 28ft. craft. Fred somehow found time from his car-manufacturing programme, and the constant expansion of his theories on flight, to design a beautifully-contoured boat in which to install this engine, instead of merely settling for a floating test bed. A cross-braced section provided two wells for the seating of passengers and, unlike his earlier boat launched in 1894, was tiller-steered from the rear. Surprisingly, he allowed a boat-building company, Bathurst of Tewkesbury, to construct the hull to his design. The Whitfield brothers no doubt had a big influence in this decision, as they were financing this section of the experimental programme. The boat made its maiden voyage at Christmas 1898 and was an obvious success, with the water-cooled engine proving most satisfactory in driving Lanchester's reversible propeller. (George later pointed out that this "bore a remarkable resemblance to one claiming to be a 'new' design exhibited at the Engineering and Marine Exhibition in 1947".)

Fred Lanchester's second boat of 1898, used as a test bed for his 8 h.p. twin-cylinder water-cooled engine, with James Whitfield at the helm.

The launch was able to cruise at 9 m.p.h. and was kept at James Whitfield's country home near Tewkesbury, fronting the River Severn. Apart from his regular use of the boat, it was no doubt also used for demonstration purposes, as Lanchester engines were shortly afterward advertised for sale as suitable for river craft. (This boat was in use at least until the Second World War, complete with its reversible propeller, although by then a Lanchester 20 h.p. engine had replaced the original.)

Chapter Five

THE TWIN-CYLINDER 8 H.P. MAIL PHAETON

The construction of the Mail Phaeton, officially classified as Lanchester Car Number Three, was started in 1897, at the same time as the Gold Medal Phaeton. The chassis and all the mechanics were identical but, much to Fred's abhorrence, the Whitfield brothers insisted on having the coachwork built by an outside company, as was common practice at that time. They contracted the famous coachbuilding firm of Mulliner and Son, but the end result was scorned by the Lanchester brothers. As George related, "The design was a four-seater, hooded Mail Phaeton with a relatively comfortable front seat and a sparsely upholstered dickey aft. It was very high and unwieldy and looked precisely like a horse vehicle without the horse." He added, "Not only was it made inordinately heavy by the unscientific application of coach iron-work, but it was incessantly in trouble, owing to the cracking of panels and loosening or parting of joints."

The Whitfields had told the Lanchester brothers: "You boys may be clever engineers, but you don't know anything about coachbuilding." How wrong they were proved to be, as the car was destined to be a source of embarrassment and a stain on the good reputation of Lanchester.

The outmoded and unsuitable bodywork of the contemporary coachbuilding companies was not the only problem that Fred had to grapple with in his quest to perfect the design of a usable motor car. Even though, compared to others, the Lanchester engines were vibration-free, many problems resulted through nuts, bolts and machine screws coming undone. Cycle threads, pipe threads, Whitworth threads and indeed some home-made threads were all in general use. The main one, the Whitworth, was regarded by Fred as partially unusable, as under 0.75ins. diameter securing nuts tend to 'back off' due to their thread pitch becoming proportionally coarser as their diameter decreases. In 1897, Fred therefore abandoned the Whitworth system for sizes less than 0.75ins. and engineered a new set of finer threads. These he called Lanchester 'M' Threads, as the diameter sizes below 0.3125ins. were given in metric sizes; although below Fred's 4mm. size, he employed the Imperial Standard Wire Gauge threads commencing with number 10. The Lanchester 'M' Threads were highly successful and it can only have been lack of publicity which prevented their use by other motor manufacturers and engineering concerns. (Some ten years later, the extremely similar British Standard Fine (B.S.F.) threads came into being, although no credit was ever accorded to Fred. His own threads were used on all Lanchester engineering projects until the end of the First World War, by which time B.S.F. had become fully established.) Considering every eventuality, Fred adopted special short-handled spanners to prevent the mechanics over-stressing the small-diameter studs and bolts.

The sheer volume of work necessary to prepare for full vehicle production meant that Fred had less time for other issues and he was increasingly forced to delegate. Frank Lanchester must have been called in many times to assist Fred and George and by the end of 1899 had, as George termed it, "successfully graduated as a demonstrator". He was much in evidence, promoting the cars, talking to people and gathering together influential friends who might be interested in the new business of motor manufacture. One such source of contacts was the Speedwell Cycle Club, of which he had become Captain.

Frank Lanchester, alongside Archie Millership and behind George, in Archie's 1900 4 h.p. Mors car.

The three brothers, George in particular, also became firm friends with Archie Millership. They had known Archie since 1898 when, as a pioneering motorist, he had sought their advice on his many imported cars. The Lanchester brothers were always eager to gain experience of other car designs and specifications, and so would often mend vehicles for their baffled owners, one of whom was Archie. It was George who saved Archie's Locomobile steam car from destruction when "a monumental pillar of flame enveloped it". There was no water supply available, but George, "that man of infinite resource and sagacity", had observed a quantity of milk churns awaiting collection nearby. Seizing one of these, he ran with it to the car and doused the flames with ten gallons of best Jersey. George recalled that, "Archie eventually steamed away, proud and happy but smelling unspeakably of burnt milk."

This was the start of an enduring friendship, which led to Archie being employed in the drawing office at the rate of £1 per week, and six months later becoming their vehicle tester and sales demonstrator. Another car owned by Archie was a Mors and, pleased with its recent overhaul, he spontaneously challenged George to a race. George picked up the gauntlet and sped off in the Lanchester. He recalled, "Believe it or not, we did that mile and a bit in three minutes, threading our way through the horse traffic." Although they were not stopped by the police, they were noticed. From then the police were on the lookout, and when the innocent Fred next drove that car he was stopped and issued with a summons. George wrote, "It cost him the maximum fine of £2 plus costs. What he said to Archie and me is unprintable."

Although Fred had to pay for his brother's guilt, at least an offence had been committed, which was not always the case. George related that, "In those days, police constables were invariably not so courteous nor so well-educated as they are today. The bobbies took their cue from the magistrates and were ever-alert to get a conviction against the motorist. The magistrates and the general public were horse-minded and hated the motor car and all motorists. The police had no difficulty in finding 'witnesses' from amongst their friends or from frequenters of their favourite pub. We early makers were regarded by an unenlightened public as cranks or emissaries of the Devil, or a combination of both, according to the individual's mental make-up. The vast majority of people were either horse owners, users of horse-drawn vehicles, or engaged with some trade connected with horses, their equipment or their food. The motorist was fair game for conviction on false evidence, or even a doubtful car. It is not surprising, therefore, that motorists should adopt unconventional strategy to defeat the police." One such occasion was related by John L.Milligan, who later worked with Fred. "Fred told me of the days when there were only about half-a-dozen cars in Birmingham; the police did not know one from another. The speed limit was 12 m.p.h. and a policeman had seen a car travelling fast along the Bristol Road. When Fred was travelling along that road in the reverse direction, a policeman stopped him and charged him with fast driving on the outward journey. Now, Fred had not gone that way, but he knew a Minerva had, and that was probably the one the officer had seen. There were no number plates then, as this was before 1904. Fred knew that the police were quite prepared for perjury, and would swear that he was the man they saw driving the defaulting car. So the three brothers, Frank, George and Fred all wore the same kind of collar and tie and had their hair cut the same way. George went up when the defendant was called, and was

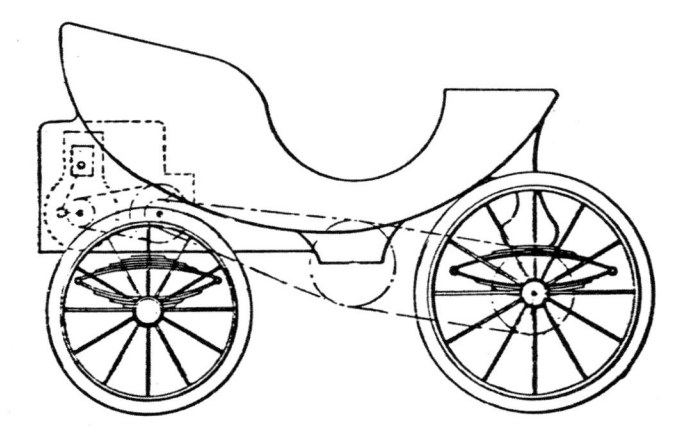

Fred Lanchester's 1896 patented design for chain-driven cars, in which an extra chain wheel was added to the system to allow movement over rough ground without the usual loosening and tightening of the drive chain which caused breakage. This system was not used on Lanchester cars, which had already adopted worm gearing.

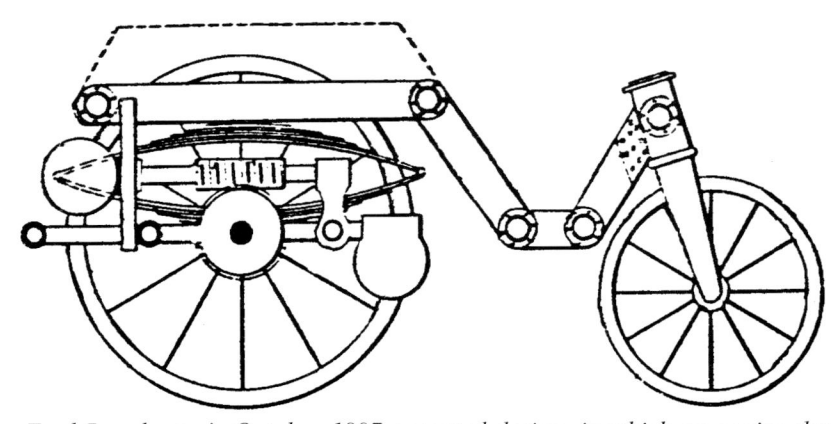

Fred Lanchester's October 1897 patented design, in which an engine drove directly onto the driving axle and where all the mechanics were isolated from the vehicle's body, ensuring great comfort. Not used on Lanchester cars.

identified, all witnesses having been ordered out of Court, then several came forward from the works who swore that George, not mentioned by name, was in the works at the alleged time of the offence. Frank, more like Fred than George was, then went in the box and swore that George was in the works, and the man who 'saw' and stopped Fred was now doubtful about the identity of the defendant. When Fred went in the box and gave evidence of George's alibi, the police witness, who was now scared, changed his mind again, and the solicitor said, 'Would the prosecution like to see the other two brothers?'"

On a more serious note, Fred was averaging ten patents per year throughout the last decade of the nineteenth century, on such diverse topics as gas engines, book marks, velocipedes, power hammers, aerial machines, and colour photography, as well as various car designs not destined to be used on his own cars. He had already received some publicity in the United States of America when a report in 'The Horseless Age' of 27th September 1899 stated: "F.W.Lanchester, of Birmingham, England, inventor of the Lanchester oil motor, which is to be manufactured at Pittsburg, Pa., is now on his way to this country." (No records remain to show whether anything came of this.)

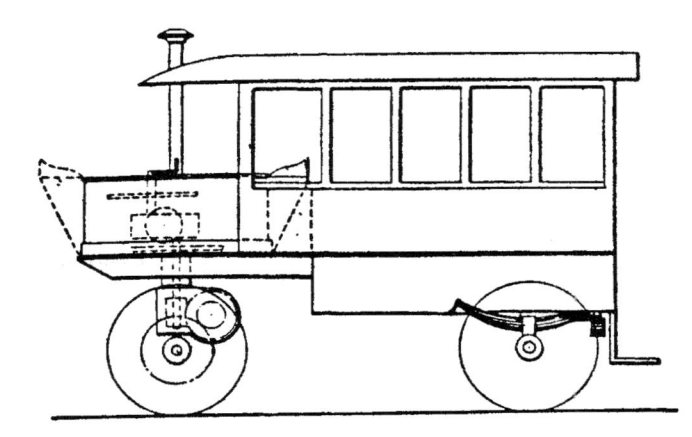

Fred Lanchester's patented design of April 1897, which detailed a front-wheel drive with easy access to all the mechanics, simple control and an unencumbered floor space.

Fred had also successfully navigated his syndicate through uncharted waters, although the experimental stages were now drawing to a close and the imminent need for the formation of a new company to promote and sell Lanchester cars necessitated the winding-up of the original, successful syndicate.

Such had been their financial arrangements that the Whitfield brothers had provided, after T.B.Barker's commission, £600 towards the first Lanchester car along with a further £2,800 to finance its reconstruction, the two new cars, the water-cooled engine and other experimental works, totalling £3,400 in all.

Fred's expenditure had been £300 for the first boat of 1894, £600 for the first Lanchester car, £1,400 for purchasing the building and the plant at Ladywood Road and £1,500 for five years' previously unpaid work, so totalling £3,800.

The new company were to repay Fred his £1,400 expenditure for Ladywood Road and also pay him a "modest" £1,000 for all his patents and improvements relating to motor vehicles. The final position for the new syndicate was that Fred Lanchester and the Whitfield brothers had, in effect, each contributed £3,400. In order to clear the way for the formation of the new company, the original syndicate was enlarged shortly before the actual flotation. And so it was that on 30th November 1899, the Lanchester Engine Company Limited was formed. The Company was privately subscribed and there was no flotation scandal as there had been in so many other firms.

The first directors, all Birmingham men, were Charles Vernon Pugh as Chairman, his brother, John Pugh, Joseph Taylor, Hamilton Barnsley and James Whitfield. (The first three mentioned had previously established the Whitworth Cycle Company.) Frank Lanchester, who had done so much hard work getting the Company off the ground, was appointed Secretary at a salary of £200 per year, to start on the 1st January 1900, "he giving his whole time and services to the Company". Frederick Lanchester was appointed General Manager, Works Manager and Chief Designer at the salary of £350 per year. George Lanchester was appointed his assistant, whose tasks were to relieve Fred of much routine work and enable him to deal with the many major problems peculiar to the new enterprise.

In financial terms, the Company's paid up capital was £45,525, ominously some £14,475 below the authorised capital of £60,000. Patents and assets of the original syndicate amounted to £20,525, so leaving the amount subscribed in cash at only £25,000. This dire lack of working capital, which was to bedevil the Lanchester Company throughout its existence, was in evidence from day one.

In April 1900, a sales catalogue was produced for public circulation. The document offered the specification for the Gold Medal Phaeton and described it as "A two-seater vehicle of best construction fitted with an 8 h.p. motor." With the painting and style of finish to suit the purchaser's requirements the price was £525, with the option of a hood in best quality patent calf leather for an extra £25. The Lanchester Engine Company assured the public that they had "no hesitation in claiming for our carriages that they are unsurpassed for safety, comfort, elegance and, in short, all the many points which go to make a satisfactory and efficient vehicle". The brochure highlighted the mechanical qualities of the car and also stated: "All our cars unless otherwise ordered are fitted with our safety tiller steering, which we believe to be the best for high speed cars up to the present devised. For those, however, who are accustomed to use the 'wheel' we are prepared to fit wheel steering when so ordered without extra charge."

A six-seater landau design for the 8 h.p. Mail Phaeton chassis.

An alternative six-seater landau design was also offered, but no mention was made of the Mail Phaeton. The brochure for Lanchester Motor Carriages proudly proclaimed the new location of the Lanchester Engine Company Limited as Armourer Mills, Birmingham. However, there are no records to indicate that any 8 h.p. motor carriages were built, other than Lanchester cars one, two and three, and indeed it is not known just how seriously the Company pushed the sales of their 8 h.p. car. Fred himself certainly regarded it as an experimental stepping-stone to full production of a new model, and would have been completely against customer sales. The publication of the brochure was, it seems, merely a way of drawing attention to the new Company and to the fact that it would be attending the week-long Royal Agricultural Hall Exhibition in London from 14th April.

The two cars exhibited were the Gold Medal and Mail Phaetons. Both were advertised for sale, although the price was only shown on the former, at £525. Each proudly displayed signs declaring 'Entered for the 1000 Mile Trial' which was to take place on 23rd April, starting from Hyde Park.

The event was organised by the Automobile Club of Great Britain and Ireland (later re-named the Royal Automobile Club) and was routed from London, via Reading, to Bristol, Birmingham, Manchester, Kendal, Carlisle, Edinburgh, Newcastle, Leeds, Sheffield, Nottingham and back into London. Among the sixty-five vehicles which departed for Bristol, each with an A.C.G.B.I. official on board, only the Lanchester and Wolseley cars were classified as wholly British.

The Lanchester Engine Company Limited stand at London's Royal Agricultural Hall in April 1900, showing the Mail Phaeton, in the foreground, and the Gold Medal Phaeton, prior to the 1,000 Mile Trial.

The Lanchester Mail Phaeton, which had only just been finished and was not properly tested, was driven by George Lanchester. It was soon in trouble when an ignition tweaker came loose and was unable to be found; sadly no spare was carried. With only one cylinder providing power, the car had to be pushed up the steeper hills. It eventually arrived in Bristol and was checked in as a 'late arrival'. The Motor Car Journal reported, "A Lanchester car also returned to the ranks having been taken off the route for some adjustment." The Gold Medal Phaeton driven by Archie Millership enjoyed a trouble-free run and was one of just twenty-six cars on the official list of vehicles completing the journey in the correct time. Overnight, George, with Archie's assistance no doubt, had made a temporary replacement for his missing tweaker and this proved sufficient to get to Birmingham the following day. Indeed, both Lanchesters were checked in as having completed the journey at speeds of "up to the legal limit of 12 m.p.h." A visit to Armourer Mills enabled George's car to be prepared with a new tweaker for the next stage of the run but, again, trouble was in store for the Mail Phaeton. The reporter allocated to accompany George and his official observer was, as George recalled, "a giant in stature weighing eighteen stone", and he sat in the rear-facing dickey-seat. George continued, "Somewhere near Derby, where we were going fairly fast over a pretty bad road, there was an ominous cracking noise and had not the aforesaid reporter deftly seized the back of the front seat, he, complete with dickey, would have been left in the road. It was a hopeless fracture and clearly showed that the coachbuilders didn't yet know much about motors." The coachwork referred to was that by Mulliner and Son, which the Lanchester brothers so despised. The considerable bad publicity was unfortunately directed at the Lanchester Engine Company.

Undaunted, George carried on and reached the destination of Manchester, albeit without the reporter. On checking the car over with Archie, "We noted a peculiar knock and vibration in the engine and, to our dismay, found one of the flywheels had cracked because of a casting stress. To have continued to run the engine might have caused a disaster, so we had to withdraw the ill-fated car." The castings were one of the few things not made in the Lanchester works and the fracture was another undeserved blow to the Lanchester reputation.

It was therefore left for the Gold Medal Phaeton, number 22, to carry the honour of the Lanchester Engine Company, and this it did admirably, right up to Edinburgh and then back down to Bradford. It was here that Archie realised he had some engine trouble. George, who had been following on a motor tricycle since he had lost his own car, worked with Archie to make the necessary repairs. George later described, "We soon found that a choked oil pipe had caused the cams to seize up solid in their bearings. In a couple of hours we had removed all the damaged parts and taken them to a local engineering shop. The boiler was being cleaned so there was no power available, but the owner let us have the run of his lathe and tools. Archie provided the power by pulling the belt round while I turned new parts. The tools were very primitive and I was only able to make a 'jury rig', but we telegraphed to the works for new parts to be delivered to us at Leicester. We had a trouble-free run there and stripped down again and completed the job properly." As the Motor Car Journal reported from Leicester, "The repair occupied some considerable time, but eventually the car was got in order and, although not reaching London until ten minutes after Saturday midnight, the return journey was made in good style." In fact, a non-stop run was made from Northampton averaging well over the legal limit. George and Archie drove for two hours alternately to ease their fatigue and, on arrival at London, left the car in the street until the next morning. Archie drove the car into the Crystal Palace well before the nine o'clock deadline, where an exhibition was staged as the concluding event of the Trial.

George Lanchester, lighting his pipe, next to the 8 h.p. Gold Medal Phaeton at Reading control on the 1,000 Mile Trial.

In overall terms, the two Lanchester cars had not received the best of press reports. Even though the Mail Phaeton had reached Bristol, Birmingham and Manchester, it was not mentioned in the final statistical report, and the Gold Medal Phaeton had had its impressive run spoilt on the final leg by the consequences of the oil pipe blockage. More misfortune was to occur. The Daily Mail reward of £10, accompanied by the important press coverage for each of the finishing cars, was not awarded to the Lanchester due to the A.C.G.B.I.'s incorrect statistics. Along with five other finishing vehicles, their cash reward was given a few days later, with negligible publicity. The exhibition and rally, however, had proved a great success with the public, and was certainly one of the major factors in gaining acceptability for the motor car.

The Gold Medal Phaeton was also influential in the design of the motor cycle. Joseph Barter, the Douglas designer visiting the Crystal Palace Show, was visibly impressed by the Lanchester car exhibit and was especially interested in its twin-cylinder engine. The centrepiece of the Lanchester display was a glass of water that rested on one of the car's wings. With its engine running, scarcely a ripple could be detected on the surface of the water, a vivid and effective demonstration of smooth operation; and this after just completing the 1,000 Mile Trial. Joseph Barter's attempts to emulate the design led directly to the construction of the twin-cylinder Douglas motor cycle.

Shortly afterwards, the Lanchester Engine Company took a stand at the Agricultural Show and displayed both 8 h.p. cars. This was to be the last showing of the ill-fated Mail Phaeton, for its mechanical components were gradually used as spare parts for its sister car. The newly-designed production car was not available for sale at that time, possibly due to the reluctance of Birmingham's financiers to support motor car development, but more probably because Fred's logical, assured approach did not allow for short cuts in the programme. With hindsight, this may be deemed as an error of judgement, for there is no doubt that the Lanchester was technically, scientifically, and practically, superior to the contemporary 'horseless carriages' from abroad and the few in this country. The delay was to prove costly to the future of car sales as, instead of allowing the Lanchester design to influence public perception of 'the right way to advance', the continental ideas were allowed to continue setting the fashion and, increasingly over future years, Lanchester cars were the ones to appear unconventional. Fred later acknowledged that, despite his efforts, the general design layout of the established type of car must be attributed to Panhard and Levassor, although he did not think highly of their product. Certainly though, Fred's systematic method of motor carriage development was the right way to progress. He realised that the full interchangeability of all parts produced was the key to correct car design, and accepted that its implementation would delay the debut of his production car for at least a year.

Chapter Six

THE TWIN-CYLINDER 10 H.P.

Fred Lanchester, confident of the future prospects for full production of his cars, had already inspected many large industrial buildings in Birmingham which were for sale, even before the new Company had been formed. He had purchased one of these, a 5,400 sq.ft. site in Sparkbrook, and the race to make cars available for sale was on. The year of 1900 was one of hectic activity, as all efforts were made to make the factory, Armourer Mills, usable. It was a red-brick, Victorian building of 1861 construction, originally used as a steel rolling-mill and tube-drawing factory for the Birmingham Small Arms group to produce armaments for the American Civil War, hence the building's name. The main entrance was in Montgomery Street and the building covered 70 yards to the rear boundary wall, which formed one side of the Warwick and Birmingham Canal (now re-named the Grand Union Canal). Much structural work had to be undertaken and existing buildings had to be altered to suit the required purpose. For this work, power was supplied to the factory from gas engines, still the most convenient form of power for small factories until the era of electricity.

The dedicated work force, as well as all the equipment, machinery and car components, were transferred from Ladywood Road. New plant and machinery also had to be bought and installed to suit Fred's overall design-layout in preparation for full production. A

Armourer Mills, Montgomery Street, Sparkbrook, Birmingham.

multitude of jigs, gauges and patterns had to be made as practically all the new cars' components were to be produced in the works. Design, production and sales schedules needed to be calculated, and a myriad of other items had to be carefully considered.

One ideal which Fred felt essential for car production was his system for ensuring the complete interchangeability of parts. Fred wrote: "Out of over a thousand different components listed, approximately 250 could be enumerated as major parts, of special design, requiring gauge limits to be prescribed and of which the manufacture involved many processes and needed more or less tool equipment for production on an interchangeable basis. This may give some idea of the magnitude of the task we undertook." He had actually realised the benefits of such a system in 1895, and by 1897 had utilised it exclusively on all the components he produced. He termed it his 'Unilateral System of Gauge Limits', and it showed clearly the tolerances allowed for any given measurement.

For an item of, say, 3.142 inches with a maximum finished oversize tolerance of 2 thousandths of an inch, it could be written as 3.142"/3.144" or 3.143" + or - 0.001". For draughtsman, machinist and inspector alike, the ease of understanding this tolerance is perfectly clear with Fred's obvious and simple figure of 3.142" +2. (To many, this proved superior to the bi-lateral system adopted by Newall and the British Standards Institute some ten years later.) Detailing another important advantage, Fred wrote, "The Unilateral System shows clearly whether a dimension relates to an internal or an external gauging depending upon whether the tolerance index has a positive or a negative sign."

He developed the system of 'go' and 'no-go' gauges, which he termed 'Yes' and 'No' and so brought the method of checking intricate work to the unskilled, without the need for a craftsman's measuring equipment. He remarked of his system, "The gauge limits are in many cases finer than are employed in the gun trade which has hitherto been considered the acme of high class workmanship." (The system was successfully used by the Lanchester Company until 1931, and portrayed the far-sightedness which Fred brought to all aspects of his work.)

Fred's agile brain was also eager to meet short-term challenges. John Milligan recalled stories about him: "Early in the century, Birmingham Corporation supplied electricity for power purposes at a cheap rate, by 3-phase current at 25 cycles per second; a periodicity which made lighting almost impossible, owing to the flicker, specially where machinery was being watched. Some of the lights at the Lanchester works were on that system, and he noticed that when the place had been freshly whitewashed, the lamps were much steadier. He analysed the cause; the canal behind the works was thick with sulphurated hydrogen, and that acted on the fresh whitewash to make calcium sulphide, a substance which can absorb light and re-emit it when the light becomes dimmer or extinct. This acted as a smoother, so he had all the reflectors coated with calcium sulphide and thus got all the lights onto the cheap current."

Another detail not missed by Fred's eagle eye concerned the water supply. He noted, "The canal company charged for water not by meter, but by the bore of the supply pipe, which had to maintain that bore for a certain length. By using a polished bronze pipe of small diameter, instead of the usual cast-iron pipe, the required supply was obtained at a lower rate."

The local telephone company also became involved in Fred's insistence on fair play. John Milligan remembered that "Fred found that the telephone company had overcharged him but they did not agree, so he had a coin box fitted. From the patent library he found out how the coin box worked; the pennies fell onto a spring and the noise was recognised by the operator. He had a tuning fork made to that sound, and used it to get calls until he recovered the overcharge."

Perhaps these measures went a small way to easing the pressure on the finances of the Company, which at this period were described by Archie Millership as "somewhat strained".

It had been decided initially to build six of the new production 10 h.p. cars, and these were completed in the summer of 1901, some eighteen months after the inception of The

Lanchester Engine Company. The cars were an engineering masterpiece, and Harry R. Ricardo wrote that they "created somewhat of a sensation". Showing the results of Fred's innovative experimental programme, the Lanchester was different from any other vehicle made in the world in that it was conceived as one entity. A.B.Filson Young's contemporary book stated, "A feature of the Lanchester car is that originality in one part of the design has evolved originality in another, and that each part of the car has had to be logically thought out in order to justify and make good what has gone before."

The 10 h.p. was designed for a production run of five or six years without the need for change, and made well enough to withstand a minimum of 5,000 miles' continual service without the need for attention. George later recalled, "One owner bought a car direct from the Motor Show and after three lessons of instruction in driving went on tour and returned six months later with a log book showing a mileage of over 10,000 miles; he had worn out more than one complete set of tyres, and the only other attention beyond the usual filling of petrol and lubricating oil was the adjustment of sparking plugs."

The chassis of the 10 h.p. car differed greatly from the 8 h.p. models. In place of the tubular framework, which would have restricted the design of the coachwork, the chassis frame consisted of two girder sections of steel, riveted to the top and bottom of a 0.125ins. thick aluminium plate, totalling a massive 18ins. deep. The top section provided a level surface which was used to secure the coachwork, and the bottom steel section supported the transmission and suspension brackets. The 12ins.-diameter steel cylindrical petrol tank, the bed plate for the engine, the oak floorboards and the oak lamp-board which carried the lamps, front wings and pivoting dashboard, all provided torsional stiffness and transverse cross-bracing to the chassis frame, providing great rigidity in proportion to its weight. The chassis spring-mounting points were secured well away from the frame ends, so concentrating all the forces created on its strongest sections. It was connected to the front and rear axles by exceptionally long cantilever springs, half of each spring being enclosed within the chassis frame.

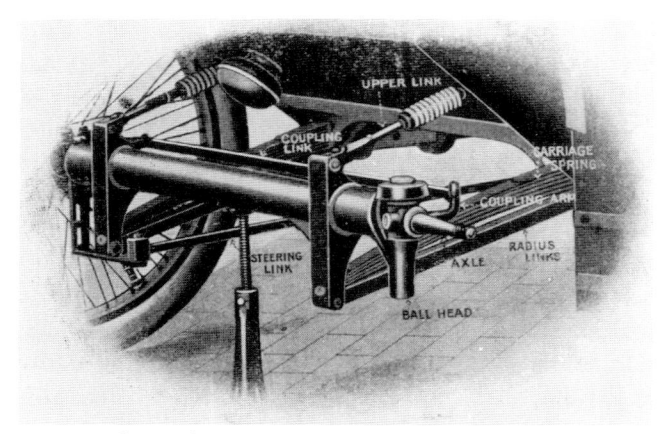

The 10 h.p.'s unique system of suspension.

The suspension allowed for some 7ins. movement at the front of the car and 5.5ins. at the rear, much more than was the case with all other cars, made possible by the rigid chassis. In fact, Lanchesters used this to good advertising effect, with pictures and details of a car with one wheel raised 2ft. from the ground, with no chassis or body distortion at all. Later, in 1904, Archie Millership dramatically made the most of this arrangement when he took a Lanchester to Crystal Palace for a demonstration. In front of the press and cine-film projectors, incidentally making this the first motoring film ever, he proceeded to drive up all the steps to the top with no problems whatsoever, although the leading edges of a few steps were chipped off! In a repeat of this spectacular performance, Archie drove up the steps of Lincoln Cathedral but, on finding that there was no turning space, was

THE TWIN-CYLINDER 10 H.P.

The chassis and layout of the twin-cylinder 10 h.p. Lanchester.

compelled to come all the way down again backwards, a manoeuvre which the suspension readily allowed.

The rate of oscillation of the road springs related to a person's own natural motion when walking, as Fred felt that drivers would then be better able to adjust themselves to this new form of travel. After experimenting with different rates, the ideal balance between passenger comfort and excessive roll when cornering was found to be between 85 and 90 vibrations per minute.

The innermost end of the springs rode on sliding 'slipper' plates, while rollers were used to take the weight of the vehicle at the axle end of the springs. These were free to function purely as springs, independent of any forces acting on the axle. The axles were located by Lanchester's system of 'parallel-motion links' which were placed above and below the springs, the lower ones being called radius links. In what was surely the very first such application of rubber, the upper links were provided with rubber buffers to act as shock absorbers, "......in order to intercept irregularities in the road and propulsion impulses," George explained. The parallel links restricted the movement of the axles to a scientifically correct vertical plane, instead of allowing them to travel in a curve, thus making for better road-holding performance. The 'live' rear axle incorporated the Lanchester under-hung worm and wheel drive contained within an aluminium housing. Although later the 10 h.p. was offered with different axle ratios, the first one employed a ratio of 7:36 (5.143:1). Axle casings of 3ins. diameter were provided for the high-tensile steel half-shafts. These incorporated splines to locate inside the differential assembly and ran in thrust ball bearings and roller bearings. At the wheel hubs the half-shafts were supported by similar roller bearings of equal 0.75ins. diameter and length.

The 34ins. detachable wheels of the proven 'tangent' wire design, took a tyre size of 870mm. x 90mm. and were pronounced by The Autocar as being the first rear wheels able to be removed whilst leaving any brake gear intact.

The wheelbase was 7ft.9ins, and was probably the longest of any car produced at the time. The front wheels were carried on the stub-axle shafts by adjustable cup and cone

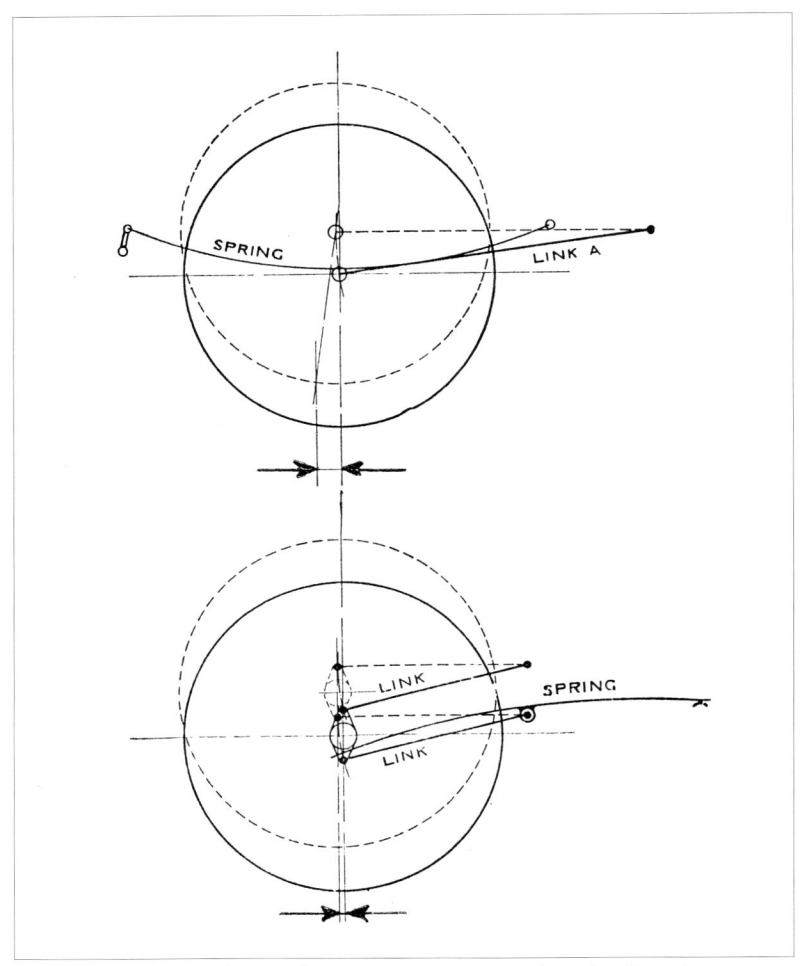

The ordinary form of suspension compared to the Lanchester parallel-link suspension (bottom). The distance between the base arrows shows the amount of skid with each system and the dotted lines indicate positions of wheel and links when the axle bounces. The practically-non-existent movement with Lanchester's system provided safer travel, more comfort and less tyre wear.

bearings, while bronze bushes were used for the axle swivel pins with thrust ball bearings taking the load from the top of the stub axles. These were connected to the 3ins. diameter, tubular front axle which was, one report suggested, initially designed to give a 4ft.4ins. track, although it would seem that this narrow track would have allowed only limited steering movement.

The method of steering was of the Ackerman principle and was actuated by a track rod set behind the front axle. The operating end of the steering assembly was the Lanchester side tiller, similar to the earlier cars. This near vertical shaft was set at right angles to the drag link and terminated in the horizontal tiller, covered with a leather handle. The lever, on which the driver rested his lower arm, was pushed in the direction the car was desired to travel and was very easy to use, dynamically stable and self-correcting: qualities missing in the vast majority of contemporary vehicles. Fred wrote in one of his handbooks: "Experience has shown that it is one of the best forms of steering for encountering unexpected obstacles, for the same reason that the bath-chair tiller proved to be one of the worst. (This latter design actually resulted in the car designer, M. Levassor, losing his life.) A car fitted with our side-lever steering tends always to steer in smooth curves, and to oversteer is almost a physical impossibility owing to the centrifugal force acting on the driver's body, tending always to moderate his steering efforts." However, as Fred conceded,

THE TWIN-CYLINDER 10 H.P.

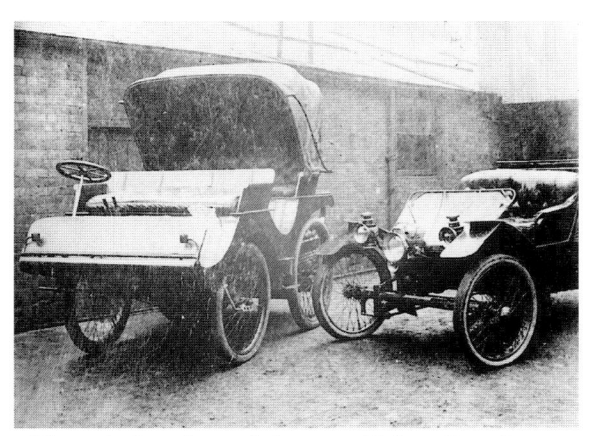

The relative sizes of the new 10 h.p. Lanchester and the old 8 h.p. Stanhope Phaeton.

"The system at first was not perfection. We did experience certain troubles; in particular the road irregularities, especially pot-holes, gave rise to disagreeable vibration of the steering tiller, at times very tiring to the driver." This obviously did not bother the Lanchester customers however, as, perhaps comparing it to the steering methods of other contemporary cars, each was full of praise for the Lanchester system.

Fred continued, "The side-lever vibration did not become a source of serious complaint until the Lanchester Company was turning out "hotted-up" cars and it was only at these speeds that the vibration became really disagreeable." The problem was solved in a very simple and effective way, as from then on the side-lever handle was made of solid lead weighing some 5 or 6lbs. Fred elaborated, "Since the total weight on the tiller restraining the vibration now amounted to about ten times the weight of the driver's hand, the speed of vibration was reduced about 3:1, and it became a harmless wobble of quite small amplitude; moreover, nine-tenths of the forces involved were borne by the lead mass, and only about one-tenth by the driver." Having claimed special status for their side-lever, it has to be said that the Lanchester Engine Company did allow for a steering wheel to be fitted to their cars if the customer so desired, although it is not known if this conversion was ever requested. Indeed, one satisfied customer declared, "I would as soon use a steam hammer to crack a nut, as have wheel steering fitted to my Lanchester!"

Although the 10 h.p. engine was based on the earlier successful 8 h.p. model, it was still radically different from any other make of engine in the world. Its two opposing cylinders of 5.25ins. bore and 5.687ins. stroke gave 4,029 c.c. When run at its normal speed of 850 r.p.m. the engine gave 10 h.p., the figure from which the car derived its name, and it developed 15.5 b.h.p. at 1,250 r.p.m. Slightly inclined downwards in the chassis towards the rear of the car, the engine looked awesome and complicated to the uninitiated observer.

Twin-cylinder engine parts, including one piston and its three connecting rods.

It was laid out in the horizontal plane, filling the space between the two longitudinal chassis frame members, and was positioned centrally in the wheelbase between the front and rear seating areas. The engine incorporated the twin-cylinder arrangement of the 8 h.p. with one crankshaft above and geared to the other.

Whereas most vehicles used both an inlet and exhaust valve in each cylinder, the Lanchester patent valve gear had just one. Also, instead of the conventional inlet valve operated by atmospheric pressure alone, this Lanchester main valve was mechanically operated, actuated by a lever off the camshaft and closed by its incorporated recoil spring. Owing to the fact that heat-resisting metals were not available in this era, the device was designed so that the main valve was cooled by the cold induction charge of petrol vapour passing around it, immediately after it had been heated by the outgoing exhaust. The temperature of the valve was therefore kept down by this method, so preventing corrosion of the valve seat. Whereas in other makes of car the grinding-in of valves was a regular chore, with the Lanchester system it was a rare occurrence, for the cars would run for many thousands of miles before attention was needed.

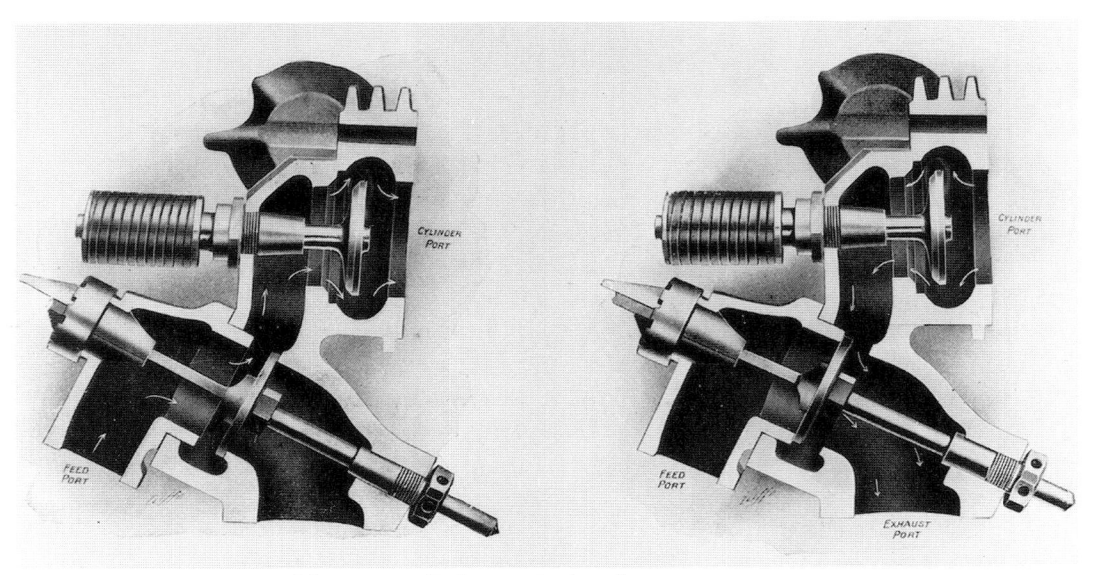

The twin-cylinder engine's valve arrangement.

Within the same assembly, but below the main valve and cylinder, was a double-seated disc-shaped valve, which was designed to communicate alternately with the inlet and exhaust pipes, sealing one off as it opened the other. This lower valve, or feed valve as it was known, had a separate actuating lever operated by a governor of the 'hit and miss' inertia type. The system depended on the governor blade making contact with the valve stem, so that with no contact the engine cylinder merely took in and then expelled exhaust gases. The two governors, one for each cylinder, therefore cut out the mixture when the speed was greater than that allowed by the accelerator pedal, and re-admitted it when the speed had dropped below that set by the driver's accelerator. The levers could also be individually used to test each cylinder separately. The governors prevented the inlet being opened when the tick-over speed rose above its usual 250 r.p.m., but the engine quickly responded when it was required, by the action of the accelerator foot pedal.

The Lanchester wick carburettor employed copper-gauze air filters which acted as air cleaners, air silencers and also flame-baffles in the event of a back-fire in the induction pipe.

The petrol tank incorporated yet two more novel features. The first was the large-diameter filler inlet underneath the driver's seat cushion, the shape of the inlet being that of a cone plug which, when the tank was not being filled, cut off the inlet's opening. The second was the patented level-gauge whose stem, graduated in gallons, had at its base a conical-shaped saucer, which was normally in the sunken position. When the amount of the tank's contents was required the stem, complete with saucer, was lifted clear of the petrol level, spun between the fingers to eject the contents by centrifugal force and then allowed to rest on the surface of the petrol so that the graduations on the stem could be counted. As there were no petrol pipes, joints, connectors, or valves anywhere on the system, there was no possibility of leakage or risk of fire. It was also safe, reliable and totally enclosed but, sadly, it seems to have been perplexing to anyone but a Lanchester owner.

The firing spark was provided by the low-tension ignition system, as used on the 8 h.p. engines, and fed the special design of Lanchester's sparking plugs, called 'ignitors'.

The method of lubricating the various components was automatic in operation when the engine was turning over. A large oil reservoir tank, situated above the uppermost part of the engine, was supported on two tubular posts. The tank carried enough oil for at least 200 miles of travel before refilling. Its oil was metered to each point by a transverse rotating spindle, indirectly gear-driven from the camshaft. The spindle had blind holes of different capacity, which filled with oil in their highest position and released it at their lowest position into twelve small bore brass tubes, which gravity fed to points requiring lubrication. Four additional, larger pipes allowed oil to pass to each of the crankshaft's four main bearings. Drillings in the crankshaft permitted the oil to flow to the big ends, initially by gravity to the crankshaft web and then by centrifugal force. Metered oil also ran through the oil tank posts to the cylinders, pistons and gudgeon pins. When used, the oil fell to the centre of the engine bed where it passed to the gearbox pump. Mixing with used oil from the countershaft main bearing, which had its own feed, the oil was pumped to the centre shaft of the gearbox and to the epicyclic gears. Metered oil fed the clutch-brake double ball bearing. A short undertray was used to keep the engine free from dust.

Cooling of the 10 h.p. engine was by circulation of air. As on his earlier engines, Fred had favoured this instead of the water-cooling method, which at that time he thought more involved and unreliable. In opting for his method of forced air-cooling by the implementation of two large-diameter centrifugal fans, he went against the run of fashionable thinking. However, by doing so he eliminated all the problems of water levels, draining and refilling in icy weather, water pumps, pipes, connections and so on. His system worked well, and indeed, "The engines do not lose power up a long hill, and they work at their best when an advocate of a water-cooled engine would be considering it awfully hot", quoted one customer.

As on the 8 h.p. car, the first and reverse speeds were of the epicyclic gear system, but Fred's newly-patented 'compounding' system was also employed to give a second, intermediate gear. The design of the actuating control meant that this was the first ever pre-selector system in the world (beating that of De Dion by two years). The top gear, direct drive for high-speed travelling was effected by a cone clutch, of slightly different design from the earlier model.

1901, the year in which Queen Victoria died and Edward VII acceded to the throne, was also notable as the year in which the first production Lanchester took to the road. In the August, as soon as the first 10 h.p. car was finished, Archie Millership was asked to test it by driving on a twenty-mile circular route. There must have been great excitement as the car left Armourer Mills, but before he had reached half-way there were problems. As Archie later recalled, "The clutch was slipping badly and finally gave up altogether. The car was towed to the Works and the clutch, which was leather-lined, was found to be completely burnt out. This was a serious matter, as we were hoping to deliver several of these new machines to customers."

An immediate assessment of the situation revealed that the problem was caused by a smaller clutch being fitted than previously, made necessary by a re-arrangement of the transmission system. Although the first and second gears were not affected as they were on the epicyclic system, Archie had called for the car to be towed back to the Works, presumably to avoid any possible further damage. As he recalled, "When the inquest was over, we endeavoured to persuade Fred to come home, but this he refused to do. He worked at the drawing board all night and when we arrived the next morning, he had produced a completely new design, substituting the leather for a metal-faced clutch, all surfaces being of cast-iron. In an equally impressive time, the construction of the new components was rushed through the works, and within a couple of days two cars had been converted. The deliveries were made possible, and all clutch troubles were completely eliminated for the future." With enormous admiration for Fred Lanchester, Archie continued, "This was typical of the man. He was never at a loss when difficulties were encountered and no detail was so small as to escape his notice."

No sooner had two of the cars been modified than they were driven to Scotland for the famous Glasgow Trial. Earlier, in the June of 1901, the Lanchester Engine Company had optimistically entered three vehicles to participate in the Trial: chassis numbers 7, 8 and 9. With the completion dates of the first cars uncertain, they felt obliged to withdraw from the list of official participants. This was unfortunate, as two of the cars they had originally entered were now, albeit at the very last moment, ready.

George Lanchester and Archie Millership nevertheless decided to attend the Trial, as the organisers had decided to allow the Lanchesters to be used for the conveyance of various officials and the press. They drove the cars from the Birmingham Works with two potential customers, Charles and James Dixon, as passengers. The third Lanchester, driven by Fred, accompanied them, but in view of the previous lack of time to modify his 10 h.p. as well as his reluctance to be away from the Works, he only intended to travel a short distance. There can be no doubt, however, that he was eager to try out his new car. On reaching Stafford, his car not unexpectedly suffered a similar fate to that of Archie's previously, and Fred drove it slowly home using first and intermediate gears only.

Even though the two Lanchesters had departed for Scotland immediately their conversion work had been completed, it was late at night by the time they arrived in Glasgow and they had already missed the first day's run.

One correspondent wrote, "On Tuesday my allotted place was on the Lanchester carriage, but I found that, owing to having arrived in Glasgow only the previous night at a late

George Lanchester, relegated to the tonneau compartment, with his mother and father. The 1902 10 h.p. car shows the new imitation cane-work panels.

hour, Mr. Lanchester was not proposing to run. I was disappointed, of course, specially as ever since the Petersham Hill Climb, now long ago, experts have been looking forward with high anticipation to the finished products of Mr. Lanchester's genius." Brightening, his report continued, "It was, however, arranged for me to have a short drive in the afternoon. The trial, short as it was, demonstrated one of the chief features of the Lanchester construction, that is to say, its suspension on rubber buffers as well as springs. It is little exaggeration to say that one travels over the roughest pavement without feeling it. In the open country, for the sake of showing this point, we drove repeatedly on to and off a sidewalk, perhaps 4ins. high. The control of the car, which of course is peculiar to itself, makes it very easy and pleasant to handle in traffic."

The two Lanchesters travelled with the convoy and undertook all the trials in their unofficial capacity without any problem, and it was a great shame that hardly any advertising was forthcoming from this. However, they were mentioned on the Friday's event by the Motor Car Journal, which reported, "The organisers did well to put Friday's route on the last day of the trials, for it was the most beautiful and the most exciting. Mr. Lanchester took me on board. The carriage we were upon was geared rather low and consequently not fast on the flat, but it was with delight that we skimmed along over surface bad and good indifferently, uphill and down, almost unvaryingly at a comfortable but by no means contemptible speed. Through Dumbarton and along to Gareloch the road is more and more charming at every turn. Between Gareloch Head and Whistlefield is the great climb, so severe that at the last moment cars were excused by the committee from having their performances officially chronicled, unless they should so signify their wish. We wanted to see the cars ascend, so waited on the steepest section. The Lanchester went up, meanwhile, unchecked." In the same journal, it was later reported that both the Lanchesters actually went up Whistlefield Hill with a full number of passengers on board. The reporter had drawn attention to the fact that although the Lanchester was not the fastest car on the flat, it was indeed faster than many on an average time, no doubt due to its suspension, brakes, steering and general layout, giving an all-round, confidence-inspiring performance. This ease of driving was picturesquely described in The Motor Journal by an article written about the Lanchesters' homeward trip: "On the descent of Shap Fell, George Lanchester gave a demonstration of the car's stability and road-holding by just letting her go, coasting on the steeper slopes at well over 60 m.p.h." This greatly excited his passenger, Charles Dixon, and on joining the others at the hotel in Kendal, he said to his brother, "Coming down Shap was glorious: the milestones went by fip, fip, fip, FIP - I thought we were driving through a cemetery!"

However, not all was rosy, as The Motor Car Journal reported that, "In the Lanchester car in which I was driven, two defects were apparent: the cooling was insufficient and the metal faces of the clutch squeaked, but these are to be remedied at once." No doubt he was the other rear-seat passenger with the official observer, who wrote of the trip in the Automotor Journal. He similarly praised the car, but remarked upon, "......occasional evidences of overheating and a disagreeable squealing every time the high-speed clutch was engaged." The noise was cured on the car's return to Armourer Mills by the occasional application of white lead and graphite paste, subsequently called 'Lanchester Clutch Dressing', to grooves cut in the mating surfaces, with access via drilled holes.

The reports of insufficient cooling and evidence of overheating were most unfortunate, as the cause of the problem was merely occasional drops of surplus oil dripping from engine components on to the hot exhaust silencer pipes. The inaccurate reports were believed by all, adding another nail in the coffin for air-cooled engines, already unpopular in public opinion. Shields were later fitted to protect the silencer system from oil drops, as well as ventilation air-scoop boxes on the near-side and off-side of each car to provide a greater flow of air around the engine. Surprisingly, these modifications had still not been done by January 1902, when Charles Dixon, the aforesaid passenger on the Glasgow Trial, bought his own 10 h.p. Lanchester. In his subsequent article, published in 'Car Illustrated', he remarked, "When the car has been running for some distance, a certain amount of smoke and smell is given off." It seems ridiculous that this problem should have been left as a cause of complaint, however minor, when such painstaking work had been undertaken on every other aspect of the car, with major modifications already completed, such as the early increase in the front wheel track by 6ins. to 4ft.10ins. in order to achieve the better steering lock of 39ft.

The first 10 h.p. cars produced brought about a dispute between Fred and the directors of the Lanchester Engine Company. The last experiment of using outside coachbuilders, on the Mail Phaeton, had been detrimental to the reputation of the Company, but it seems that this lesson had not been learnt. The Directors apparently thought that they should again follow the lead of other motor companies and, despite Fred's pleas, they decreed that the first twelve 10 h.p. Lanchesters built were to have bodywork provided by an independent coachbuilder of the highest reputation. As Fred explained, "The difficulty of getting bodywork made on an interchangeable basis was due to the fact that the bodybuilders would not work to instructions. In those days, when a bodybuilder was asked to work to drawings, gauges or templates he gave a sullen look, such as one might expect from a Royal Academician if asked to colour an engineering drawing. When the bodywork came it would not fit the chassis and we had to cut the chassis about and adapt the bodywork in each individual case." How frustrating this must have been for Fred's orderly mind, as he saw his identically-produced cars altered so that each was a complete "one-off", with correspondingly fewer interchangeable parts. The quality of the coachbuilder's workmanship was described by George as "shocking" and the situation as a whole as "a disaster". When Fred was able to show that nearly £10,000 worth of cars and stock could not be delivered, as they were all awaiting adjustments to bodies, chassis or both, the Directors swallowed their pride and allowed Fred to produce his own coachwork from then on.

In typical style he chose the long-term approach, based on a sound technical footing. He designed his coachwork to be built on special jigs, which he made so that each section would be identical to that of another, regardless of the date of manufacture. This, of course, matched the principles in use for the production of the cars' mechanical items (systems of

The Armourer Mills production line.

inter-changeability which long preceded that of De Dion Bouton, Cadillac or Ford, although they are often awarded the credit for the idea).

Although they would fit any style of coachwork desired, the Company "..... strongly recommend purchasers to choose the popular 'tonneau' design." This consisted of a front seat for two, the section over the engine compartment with its luggage deck and wind-scoops, and the 'tub-shaped' tonneau compartment itself, complete with folding picnic table. The tonneau seated two passengers comfortably and was built in right and left-handed sections. As a concession to contemporary coachbuilding, an ash body frame was constructed, but instead of the customary wood or papier-mache for the body panels and mudguards, aluminium sheet was used. This was a typical Lanchester innovation and yet another first in the motoring world (preceding the French coachbuilders, Rothschild et Cie, who introduced a similar method in 1902).

The coachwork consisted of a number of separate sections, each individually removable in a matter of minutes. In fact, with one person on each side of the car, it took less than five minutes to entirely remove all the bodywork, and only a little longer to replace it.

It will be recalled that on the formation of the Lanchester Engine Company Limited, Fred had accepted the position of General Manager and Designer. He had not wanted the post of Managing Director, as he realised he would then be partly responsible for the finance of the Company. This he was wholeheartedly against, as he felt that the Company was already under-financed. As General Manager, his responsibility was limited to seeing through the production programme sanctioned by the Board, which Fred trusted would coincide with his own ideas. For the most part, this was indeed the case, as Fred related, "My authority was practically the same as though I had taken the position of Managing Director, but my presence was only called for at Board meetings to present and discuss my own reports and to receive instructions and advice from time to time. A nod from the Chairman signified my release." He added, "Nevertheless, individual members of the Board were a worry to me at times, but that, I believe, is usually so."

As well as being General Manager, Fred had taken on the additional role of Works Manager. The enormous pressure of this dual responsibility should have been eased in the January of 1901, some twelve months after the formation of the Company, by the Board's decision to appoint a new Works Manager. Fred was given the power of appointment, but was only allowed to offer £150 per annum to fill the post. The £3 per week offered naturally did not entice a flood of applications from experienced engineers and the post was eventually granted to Mr. Max Lawrence, who unfortunately had no previous experience in this capacity and, in effect, had to be trained by the Lanchester brothers. Fred related, "When Max Lawrence took charge, a great deal of the production material was already well forward and a start was being made on group assembly."

The first six cars produced were allocated to the Directors for personal and demonstration runs, and a sales campaign was put into effect. It was, however, a difficult time for the Lanchester Engine Company, and for the fledgling British motor industry as a whole.

The long-standing Boer War had created an economic depression in Britain, the Company had been floated on far too small a capital, and the established continental car companies were enjoying massive sales. Indeed, De Dion Bouton had, up until the period to April 1901, already sold 1,500 of their cars. Although many motorists considered the Lanchester to be near perfection, it had sadly arrived too late in the day to greatly influence the general public's perception of a motor car's ideal appearance.

A massive rearguard action was needed for any British company to break into the French-dominated motoring scene, and this was especially so for Lanchester, with their radical approach to vehicle design. Fred later remembered the problems: "At the same time, the French automobile industry was booming. Few now understand what that meant; their financial resources enabled the French houses to advertise over our heads, and we British were acutely conscious of the extent to which this affected other forms of publicity." This view was supported by the respected Motor Car Journal, who wrote at the time, "In

LANCHESTER CARS

AS SUPPLIED TO WAR OFFICE TRANSPORT COMMITTEE.

Successful Air Cooling by Forced Draught.

Unfailing Magnetic Ignition.

Vibrationless Balanced Engine.

Silent Worm Drive Of Unequalled Efficiency.

Perfect independent Governing to Each Cylinder.

Luxurious Suspension.

FITTED WITH HOOD ARRANGED AS DUST SCREEN

**THE LANCHESTER ENGINE Coy., Ltd.,
Sparkbrook, BIRMINGHAM.**

Agents for Lanchester Cars and Official Repairers by Arpointment.
THE LONDON MOTOR GARAGE Coy., Ltd., 20, Regent Street, LONDON, W.

An advertisement taking advantage of the 1902 War Office order, showing George Lanchester behind the driver, and Frank.

many circles, the continued importation of French cars is occasioning some concern in view of the efforts of English makers to secure the market in their own country." Fred bitterly pointed out another problem when he continued, "Added to all this, the wealthy patrons of the automobile movement in this country purchased, ran, and praised, everything French." Indeed, the Lanchester brothers and other British pioneers must have been totally dismayed by the unequivocable statement by no less a person than the Honourable C.S. Rolls: "Nothing can assist English constructors more than the importation of French cars. When our British firms are able to build cars as good as the Panhard, for instance, it will then be time for them to begin and work out ideas of their own. Till then, experiments will be only a waste of time and money. We have first to realise that other countries are very much ahead of us." Incredibly, he continued, "If our makers had been wise, they would, three years ago, have bought modern French cars, and copied them exactly, instead of spending money and time in unnecessary experiments." Fred Lanchester countered this in the most forthright terms and emphasised that it was all down to image. Making a similar pronouncement, S.F. Edge of the Napier Company wrote, "Why do we buy French cars? I think that it is owing to the start the French have had; they have been able to a very great extent to set the fashion in automobile construction."

Against these difficult times then, the Lanchester Engine Company struggled to make headway. The whole team put great effort into the sales policy organised by Frank Lanchester. Max Lawrence, who "did as well as might be expected" as an inexperienced Works Manager, gained praise from Fred for his abilities as a sales representative. "He had an engaging manner and a wide social connection. He was able to influence sufficient business to have justified his salary twice over. Max Lawrence had a style and way with him, and many were the orders that came to us, as a result of his week-end efforts." Indeed, a perfect example of this was when he drove a demonstration 10 h.p. car to visit his sisters, who had recently founded Roedean School. They introduced him to a near neighbour, that

Cars from the Lanchester works in St. James's Park, London, in April 1902, for their press-publicity-run to Worthing.

most eloquent of authors, Rudyard Kipling, whose transport at that time was a Locomobile steam car. Rudyard Kipling was indeed impressed with Lawrence's Lanchester, and duly ordered a 10 h.p. which George delivered, chassis number 16, in early 1902. His two-hundred-mile journey from Birmingham was surely worthy of a Kipling tale in itself, as the pneumatic tyres on the car, which was christened Amelia, were a commodity which had not then been perfected by the embryo tyre industry and suffered no fewer than twenty-one punctures or bursts on the way!

Many other sales followed, with orders for Lanchester cars being taken from the Duke of Portland, the Marquess of Zetland, and many other dignitaries. An order from the War Office was taken in the July of 1902, following their exhaustive trials of the standard 10 h.p. car. This had been tested, with a full complement of passengers, over runs of 150 miles a day which included the steepest of hills, all of which the car easily accomplished. Much impressed by this performance, the War Office put in a multiple order, which was fulfilled by the end of the year.

This contract was welcome news and assisted the Company's financial position. As Fred reported, "The subscribed capital had been increased and the total liabilities, including issued capital, loan on debentures and sundry creditors amounted to £77,900. Of this, £16,000 stood for patents and goodwill, shares to that value having been issued to the vendors."

One of the Lanchester cars, built to the order of Lord Milner, was probably their first export order, as it was dispatched to South Africa in the November. Many more were to follow, partly due to the good offices of their agent, Rudge-Whitworth. Charles Pugh was the Chairman of this firm, which was the product of the amalgamation of his Whitworth Cycle Company with the Rudge Cycle Company. (Rudge-Whitworth were later to assist Lanchesters as sales distributors of their oils and lubricants throughout the United Kingdom.)

By the end of 1902, people were taking more notice of British cars, and not before time. The Motor Car Journal, in October, wrote, "If we might make some suggestions with regard to future editions, we think much more space might be devoted to a more lengthy account of the latest types of British car, more particularly, those which differ materially from vehicles of continental design, such as the Lanchester, Wolseley, Brooke and others." Sales of the Lanchester were increasing month by month, in spite of its unusual design, though they did not yet match those of the more conventional Wolseley, whose output from their works had actually reached the dizzy heights of ten cars per week.

A special-bodied Lanchester, showing the difficulty of side-entrance coachwork with the high chassis line of the 10 h.p.

Charles Dickson's report on the ownership of his 10 h.p. car summarised the technical success of the Lanchester production cars. "I would like to say that I have never seen any piece of complicated mechanism, such as a motor is, better designed, or with better material and workmanship than the Lanchester. It is a great pleasure to me, as an engineer, to see attention paid to the working-out of details to the same careful extent as is given to the more essential parts of the work."

With funds at last permitting, every outlet was sought for advertising and publicity, with even the Chairman driving one of the three Lanchester 10 h.p. cars entered for the earlier Southsea Anniversary Run. Three similar cars, standard in every detail, achieved another publicity stunt when they ascended the Worcestershire Beacon in the Malvern Hills; not up a track, but up the donkey path. Shortly after this, Archie Millership received useful publicity for the Company, when he drove to the summits of Porlock, Lynton and Parracombe Hills in Devon in a 10 h.p., the first car ever to do so; it was also recorded that he took with him two passengers and a full load. The press related, "On arriving at Porlock Village and enquiring the way to Ilfracombe, they were informed that they had no road and that it was impossible for the car to go up. However, they drove on, and the climb was accomplished in 28 minutes. There were two 'V' corners on the road, and it was more like a river bed than a high road. Mr. Millership then ran down into Lynmouth and enquired the way and was told straight on up the hill. The gradient of that hill is, if anything, worse than Porlock, and after one fearful turn at the start, this hill was ascended easily, doing the half mile in five minutes exact."

Although not strictly correct in all its detail, this was the kind of publicity that possibly the Lanchester Engine Company should have been promoting, as other firms were not averse to small, or large, reporting discrepancies in their favour.

One event which may have produced bad publicity, but for evasive action, was recalled by Archie. "In 1902 the Automobile Club of Great Britain and Ireland organised a trial at Eastbourne and district, and it was going to take place over a very hilly route. I suggested to Fred that we put more tension on the fans to avoid slip, or 'dope' the flywheel with some leather belt dressing. He did not take kindly to either suggestion but in a few moments he took a piece of paper and drew a dimensional sketch of grooves to be cut in the surface of the flywheel, and said to give that to the Machine Shop Foreman and tell him to 'get busy'. There was no time to test the car as it was being trailed to London where the Trial started. During the Trial the fans made a shocking noise, but we managed to complete the

THE TWIN-CYLINDER 10 H.P.

first day's run without any loss of marks. The cars were put into a large marquee and no drivers were allowed to remain in the tent after they had driven the car into its position. One of our fitters by name of Simpson was a passenger and I said to him, 'We must see what the trouble is with the fans as it is serious. You go and talk to the gatekeeper while I see what has happened.' The position of the car had been carefully marked and I dived underneath the canvas of the tent and removed the flywheel and fans. The fan bosses were made of paper discs tightly compressed, and looked like ivory. It was found the spiral groove in the rim had stripped every scrap of paper from the fan pulleys. This looked hopeless so a telegram was sent to the Works, where Fred immediately secured two new fans and a new flywheel, and sent them with a driver on a test car. The driver arrived in the small hours of the morning. Now the difficulty was - how to get the flywheel and fans back on the engine without anyone seeing us. Simpson the fitter was a born humorist so I evolved the following plan. Simpson was to keep the gatekeeper amused while the driver who had come down from Birmingham was to crawl under the canvas and collect the flywheel and fans, which I had already removed, from the back of the car and leave them under the hedge in the field well away from the marquee, to be picked up later. He was then to drive his car to the marquee with the new flywheel and fans and push them under the tent at the back of the car, and in half an hour I changed the flywheel and fans, and no one was any the wiser. It would have been impossible to do this with any other car, but in a Lanchester the rear flywheel which was very heavy was keyed in to the shaft by a long taper key, but thanks to Fred's genius all these keys had a screwed thread on the end, and a special nut withdrew the key without a sound. The new flywheel was left with the key pushed in place and two or three shrewd blows with a hammer in the morning before starting put everything right, and the Trial was completed satisfactorily with no trouble at all."

The publicity from these rallies helped the Company considerably and the 10 h.p. car was now selling well, but the price had become another cause of controversy between Fred

The brougham coachwork Lanchester.

Which way do we go? A 10 h.p. car with a hood each end.

and his Directors. Ever astute, he knew that the sale figure had to pay not just for the car but for the future investment of the Company. He considered all the details and decided on a figure of £600, which he put to the Chairman. Charles Pugh thought this excessive, and instead named a figure of £450. Fred informed him that the car could not even be produced at that figure without loss, notwithstanding any future monetary requirements of the Company. The Chairman finally agreed to compromise and the four-seater carced at £525, although Fred remained convinced that ".....the Lanchester Engine Company would not have sold one car less if the price had been £600, and the difference would have been sufficient to provide some badly needed capital."

Another design 'first' for the indefatigable Fred was a detachable hard-top for the Lanchester tonneau. The Guinness book 'Car Facts and Feats' records this as introduced in 1901, although other evidence suggests 1902, but either way the Lanchester 'brougham' clearly anticipated the versions by both Wolseley and Cadillac in 1903.

The brougham-top was, in effect, a secondary piece of coachwork which could be raised or lowered onto the tonneau bodywork as required, by means of a suitable pulley arrangement in the roof structure of one's 'motor house'. Due to its luxurious appointments and high sides, the brougham weighed 20 cwt. and a 2 m.p.h. loss in performance was conceded. With this detachable top removed, the remainder of the car was advertised as "a most comfortable touring car", instead of as a tonneau. This terminology relates to the fact that the Company were offering the brougham with extra wide tonneau bodywork, probably as a result of customer demand. In January 1903, this five-seater was for sale at £600, and without the brougham-top the price for the car with the extra wide tonneau coachwork was priced at £550.

An alternative arrangement offered was the Victoria hood, made from "best quality, patent calf leather", which could be fitted to the front seat, known as a chair, as an optional extra, at a cost of £25. A similar, but smaller, hood was centrally pivoted over the tonneau compartment so that it could be used as a dust hood with the back resting on the rear of the tonneau, or as a storm hood when resting on the forward section.

The Stanley Automobile Exhibition, in January 1903, demonstrated that, at last, the Directors appeared to realise the importance of advertising, for on stand number one, twelve of their 10 h.p. cars were shown, with the London Motor Garage Company advertising its role as agents. A complete chassis showing full details of design was proudly exhibited, along with a "10 h.p. touring car with standard tonneau; painted white throughout and relieved with crimson line, upholstered in scarlet buffalo morocco, fitted with Clipper

Michelin tyres and Lanchester safety tiller steering. Its accessories include two side oil lamps, one tail lamp, one gong. Extras include Lucas or Bleriot search lights, foot horn, cyclometer, waterproof tonneau cover, waterproof front seat cover, Lanchester luggage rail fixture and luggage case".

Nine other tonneaus were displayed, with daring colours of olive green, yellow, lake, and khaki. Of these, two were clothed in the new 'special' wide-bodied tonneaus for three passengers. The star of the Lanchester stand was undoubtedly the 10 h.p. with "light touring equipment", priced at £535. The specification was the same as the standard tonneau but, as well as two acetylene search lights, mudwings were included which were described as 'streamlined', a word coined by George which has since passed into the English language. To allow access to the front chair, the pivoting leather-covered dashboard was raised, which simultaneously dropped a section of each wing and its small vertical panel serving as a 'door'. After using the now visible footstep to alight, the dashboard could be returned to its normal position, allowing the wings and draught-excluding vertical doors to regain their former position. Fred had patented his idea of the pivoting dashboard in 1901 as well as a section which was arranged to fit onto the top of the dashboard, "in which position it forms substantially a continuation of the lower portion". In this raised position "the upper dashboard is complete front protection for the driver, and it is provided with windows". At the whim of the driver it was also able to be angled inwards or outwards, secured by a telescopic strut.

A Lanchester's dashboard and multi-angled windscreen.

Possibly to compensate for the Company not attending and advertising at the Agricultural Hall Exhibition in March, two cars were sponsored to compete in the 'Glasgow to London Non-Stop Trial' in May 1903, each driven by its owner. The two-day event was divided by a compulsory stop in Leeds and, of the nineteen cars which started in Glasgow, only seven made trouble-free runs. One of these was the 10 h.p. Lanchester entry, number 5, which gained maximum marks. Lanchester number 4, however, had to withdraw just before Leeds on the first day, due to a broken reverse gear.

The following month saw Frank in a 10 h.p. competing in the Phoenix Park Trials knockout competition in Dublin, and although he was officially timed at an impressive 42 m.p.h., he was beaten in the third heat. He was nevertheless able to secure good advertising for the model in the quest for further sales.

Unfortunately, 1903 had started with more ill-feeling between Fred and the Directors. It was on February 16th that the board appointed a sub-committee consisting of Charles Pugh, Hamilton Barnsley and John Pugh, to consider the feasibility of producing a small car, similar to the American 'run-abouts' of the Stanley steam car and the Oldsmobile petrol car. Fred received instructions to commence design work and, in his ever-enthusiastic manner, worked continuously from early one morning, right through the night, until the following afternoon. In this time, Fred laid down the entire design, consisting of more than a dozen working drawings complete with dimensions. One would assume the Board were pleased with this massive effort as, one week later, the Chairman announced that the design submitted by Fred had met with their approval and authorisation was given for the further details of his design to be proceeded with. Much work was obviously put in by the Lanchester brothers and the whole design team and when this was well advanced, a Board meeting was called in May to discuss the 'voiturette' project. Fred had worked remorselessly on the venture, and naturally took the view that the Directors would have shown equal dedication; sadly this proved not to be the case. Fred later recalled, "I had assumed that the Board had considered the financial strain that such a programme might entail, pointing out that the resources of the Company were scarcely sufficient, even for its existing programme." It is not difficult to imagine his frustration with the Board of Directors, and he later wrote, "When it was mentioned that it would take at least £40,000-50,000 to finance the most modest programme, the interest in the proposal seemed to wane." There was then a long period with no decisions being made on production, but in August, long before October's official Board resolution for the manufacture of the 'voiturette' to be abandoned, the Board decided instead to accept the Midlands agency for Oldsmobile cars.

One of these was duly purchased, and "was put in charge of a member of our staff as a traveller, to obtain orders. For about two months, he 'stumped' the country - a forlorn looking object - and never succeeded in pulling off an order". How that must have amused Fred!

Another source of light-hearted relief was provided by the efforts to evade the long arm of the local constabulary, bent on enforcing laws which seemed archaic and irrelevant to many people, the Lanchester brothers among them. A memorable episode took place in 1903, at the occasion of a Midland Automobile Club rally. Fred and George both attended

A War Office 10 h.p., outside The Lanchester Engine Company's main door.

The off-side view (top) and the near-side view (bottom) of the twin-cylinder 10 h.p. engine in chassis no. 88. This car was originally owned by the Lanchester Engine Company director, James Whitfield. Years later, under different ownership, it provided the power to drive all the machines in an engineering workshop. (See appendix.)

A 10/12 h.p. Lanchester, chassis no. 91, with increased wheelbase in order to take the water-cooling system. (See appendix.)

The 12 h.p. car, chassis no.339, showing the movement of the streamlined mudguards when the dashboard is raised, in order to allow easy entry. Formerly owned by Lord Charnwood and Anthony Bird. (See appendix.)

A 12 h.p. tonneau, chassis no. 339, showing the useful luggage deck.

With the luggage deck removed, the engine of 12 h.p. chassis no. 241 becomes readily accessible.

Bird's-eye view of the 12 h.p. chassis no. 241 with tonneau coachwork by Lanchester. This car was used as a hire car until 1915 and from then on as a towing vehicle, regularly retrieving broken-down army vehicles during the First World War. From 1923, after having covered some 100,000 miles, it was abandoned. Francis Hutton-Stott bought it in 1932 and restored it; it remains with the same family. (See appendix.)

the rally in different cars, Fred in a car on final test before delivery to a customer, and George testing a new car which was the first 10 h.p. to be delivered to the War Department, complete with the King's Cypher in red emblazoned on the panels. The car being 'on test', George had no doubt exceeded the speed limit several times, but not being completely satisfied with the car he asked Fred to pass an opinion on it, and they exchanged cars accordingly. Just as Fred and his passenger, Sidney Pinsent, the Chief Draughtsman, were about to depart a policeman, who had been waiting nearby, demanded of Fred his name and address. After some dispute, Fred started to drive off. Not to be evaded, the determined policeman scrambled over the side of the tonneau into the car, badly bending one of the wings and scratching a panel with his boot as he did so.

Fuming at the expense and delay that repairing the damage would cause, Fred drove at well over the legal limit of 12 m.p.h. and took his unwilling passenger into the depths of the Cotswolds before stopping, whereupon he explained that in order that a charge could not be brought against him for his excessive speed, they must wait an hour or so before their return.

Whilst waiting, Fred and Pinsent indulged in considerable play-acting as they discussed what should be done with the policeman. One suggestion was to take him to Scotland Yard and report him for damaging the King's property, namely the Cypher, "which rather shook the fellow's nerve".

After enough time had elapsed, they embarked on the return journey via Stratford-upon-Avon, where they stopped for a meal. Fred reminded the policeman that, as he was still on duty, he could not enter licensed premises, and the unfortunate man had to wait outside whilst Fred and Pinsent managed to spin their meal out for nearly an hour, although they did relent sufficiently to provide sandwiches for their kidnapped passenger.

When they eventually arrived back at Solihull Police Station, Fred gave the policeman his name and address, and asked what offence he was charged with. "Exceeding the speed limit on the road between Olton and Solihull", came the reply.

"But," said Fred, the picture of injured innocence, "I did not drive this car on that road nor, as my friend here is witness, was I in the car. Some other driver was in charge of this car before I took it over." No charge against Fred could be made, and no subsequent police action was taken. (This famous adventure was duly taken up by Rudyard Kipling, by now a great friend of the Lanchester brothers, and his marvellously embellished story of the same event, entitled 'Steam Tactics', is surely a literary classic.)

There were numerous clashes with the police and magistrates over the years, as many police remained anti-motorist and eager to secure convictions. When George was wrongly convicted for a particular offence and fined £2 plus costs, he careered his Lanchester along the road bordering the magistrate's estate, bagged six pheasants and shared them out among his workers for a tasty revenge.

Chapter Seven

THE TWIN-CYLINDER 16 H.P.

In May 1903, a new company was set up called the Lanchester American Patents Syndicate Limited, which was registered to deal with the American patents for Fred's vehicle inventions. The first directors of this new company were Fred Lanchester himself, Allen Whitfield and John Pugh. They no doubt hoped that money could be earned in the United States of America to assist the programme in Britain, as the Lanchester Engine Company was, as ever, short of funds.

Work stops in the Service Department for a photograph.

Despite this, there were serious efforts to expand, with two new cars on the drawing board and the corresponding problem of limited available space within the factory at Montgomery Street to consider. The Company had recently taken over the adjoining galvanised iron sheds, known as Radix Works, and re-located the engine and road test shops as well as the service department there, allowing the main works to concentrate on the production of chassis. It became clear that further expansion was still necessary and additional premises were sought. One of the reasons for the anticipated increase in output was the new 16 h.p. car, about to make its public debut. The press had already been notified and in January 1903 reports of the imminent appearance of the new car were published.

In the spring of 1903, the carriage coachbuilding department and the repair and service department moved into the Alpha Works factory in Liverpool Street, Deritend, Birmingham, with Archie Millership appointed overall Manager at a salary of £500 per annum. As with most factories of the period, including Armourer Mills, the building backed onto the canal system. A contemporary Lanchester booklet explained, "This factory is very extensive and is replete with every modern appliance for the proper construction of the high-class motor

THE TWIN-CYLINDER 16 H.P.

The twin-cylinder engine and petrol tank on the test-bed.

The flapping overhead drive-belts of the machine shop.

body. When working at full pressure, this factory is capable of turning out six or seven motor bodies per week and from this may be judged its size and extent." This quantity may be laughable in today's world of high-tech, computerised production, but three or four cars a week constituted the whole of the Company's output, and it was to be another two years before the booklet's boast began to be seriously tested.

The 16 h.p. car was Lanchester's second production vehicle and, in George's words, "was essentially a modified 10 h.p.", produced mainly to answer the criticism from some quarters that the 10 h.p. was not as fast in a straight line as it might be. The 16 h.p. incorporated the same chassis as the 10 h.p., namely 7ft.9ins. wheel base by 4ft.10ins. track, and although the weight was marginally increased, the wheel size remained the same also. The body design was identical to the 10 h.p. light streamline tourer, but instead of rectangular air-scoop boxes mounted on either side of the engine compartment, the two air intakes were each via a long, horn-shaped air-scoop which also incorporated a sidelamp.

George wrote that the engine was similar in design to the 10 h.p. other than its increase in bore size from 5.25ins. to 5.5ins., the incorporation of Lanchester's cheese-roller bearings for the crankshaft and an increase in the speed of the air-suction fans. The 16 h.p. air-cooled engine proved to be a success, although a few owners later had water-cooling fitted.

The Lanchester low-tension magneto ignition system was used, and employed two massive bar magnets of tungsten steel secured within the engine's flywheel. What would normally have been a prohibitive static weight was therefore cleverly utilised to advantage, as it made possible an equal decrease in the weight of the cast-iron flywheel. (Although not of low-tension ignition, the system is still in use on lawnmowers and other small utility engines some hundred years later.)

The armature was fixed centrally and the magnetic flux circle was made by the flywheel rim. Current was induced in the two double armature coils by the revolving magnets and passed through insulated steel bars, thicker than wire, to the tweaker, which acted as a switch. The tweaker was secured to the camshaft and so was actuated every second revolution of the crankshaft.

When the flat igniter leaf-spring was deflected, an electrical contact was made within the combustion space. When the tweaker moved past and allowed the igniter spring to rebound suddenly, the whole current which formerly passed through both internal and external contacts was shunted through the internal contact as it opened. This produced a strong

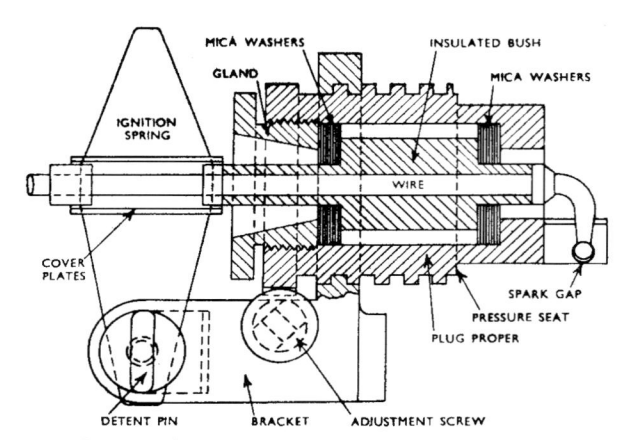

The Lanchester low-tension ignition assembly.

spark with great force and overcame any resistance due to corrosion of the internal contact points.

Other than a fracture of the igniter spring, there were no parts, wires or connections to give trouble. The whole device was easily removed, without tools, by releasing the insulated handles holding the igniter assembly in place. A quick clean or change of igniter would thus only take a couple of minutes.

Nominally rated at 16 h.p., this 4,500 c.c. engine developed 23 b.h.p. when run at its exceptionally high speed, by 1903 standards, of 1,750 r.p.m. Also using the high-ratio back axle, the difference in performance between this car and the 10 h.p. was significant, as a top speed of 48 m.p.h. was possible.

This was recorded at the first public showing of the new 16 h.p. model at the end of June 1903, at the Phoenix Park Trials, run in conjunction with the Gordon Bennett Races in

The ignition mechanism, complete with the insulated locking handle.

Two 16 h.p. Lanchesters at the 1903 Gordon Bennett races.

Dublin. Joined by Frank in a 10 h.p., both 16 h.p. cars performed well in Ireland. George was beaten in the quarter-finals of this knock-out competition and Archie Millership lost in the semi-finals. The idea of the exercise was to show the new cars to the press and public alike, and in this respect the event was a complete success.

Good spirits continued and many rallies and events were attended. On 1st August, six Lanchesters competed in the Midland Hill Climb event, the drivers being George in a 16 h.p., with Archie, Max Lawrence, the Pugh directors and an owner, H.Hemmings, all driving 10 h.p. cars. As it was noted that Fred and Frank were also seen amongst the spectators, no doubt a refreshing break from the heavy workload was had by all.

Although the Lanchester Engine Company still suffered from lack of working capital, these diversions must have been good for the morale of the Lanchester brothers. Indeed, Fred may well have thought that a new era of communication was approaching when, at the annual report of 1903, the Directors expressed their gratitude and appreciation of his services. Fred recalled, "It seemed to be, as the saying goes, 'Roses all the way'. I felt that it was well-deserved in view of the fact that I had put in something like ten or twelve hours per day, and often Sundays included." After making ample reserves for the Company, the Directors were able to report a net profit of £8,669 and the turnover for that year was nearly £80,000, which amounted to about one-tenth of the total British car trade.

Realising that these sales were encouraged by press articles on motoring events, another race meeting was attended in early October 1903. Two 16 h.p. cars were entered for the Southport Speed Trials, one an official works entry driven by George and the other driven by one of the first private owners, a Mr.Williamson. Surprisingly, George's car did not fare as well as that of the private entry, who achieved first prize in his class for the fastest time of the day, beating a 24 h.p. Georges Richard by a massive nine seconds. Good advertising was gained as a result of the event, and the Motor Car Journal reported, "Mr.Williamson came along on his new 16 h.p. Lanchester, a car with which he is exceedingly pleased. A short run on this vehicle around the town convinced us that it is worthy of the high

A 16 h.p. Lanchester about to race a Peugeot on the Southport Promenade, during the 1903 trials.

reputation of the firm, and we may say that its speed surprised us. As in all Lanchester cars, it is air-cooled, quiet in action and free from vibration."

First prize at the Lancashire Inter-Club Handicap competition in October 1903 was followed by more good publicity when an article appeared in the Autocar, detailing a reliability trial undertaken by Charles Dixon later in that month. After having driven his Lanchester 10 h.p. with great pleasure for some eighteen months, he had taken delivery of the new £650 16 h.p. model and had decided to test its performance, taking Archie Millership along as co-driver.

A brand-new Lanchester 16 h.p. prior to delivery. Photographed at Smallheath Park, a venue used by Lanchesters for publicity work for over 30 years.

They drove in six-hour, alternate shifts and covered five hundred miles in less than twenty-three hours' driving time. The 16 h.p. returned about 20 m.p.g., a creditable figure for a car with a weight of 1.25 tons. Charles Dixon was obviously impressed, as was the Autocar editor, who wrote, "It conclusively proves that an air-cooled engine of even so high a power as 16 on the brake, may be successfully run for very long distances without the aid of water-cooled jackets to the cylinders."

In fact, Charles Dixon was so delighted that ten years later he drove the same car over the same route, again taking Archie with him. The account for this well-used car described the run as 'plain sailing' and the Autocar summarised, "The liveliness, pace and road-holding of this old car speaks for itself. 24 hours non-stop at the commencement of its career and only trifling stops then for plugs etc., 10 years of active service, then to break its own record by covering its old route plus 10 miles, in less time, without a single attention whatsoever beyond replenishment, and with an extra occupant and luggage. Truly a remarkable performance upon which the Lanchester Company owe Mr. C.W.Dixon much for his sportsmanship." The magazine continued, "..... of course, the car does not do all that one is led to expect from an up-to-date car, but certainly it compares favourably in a good many respects with many present-day low-grade cars, a comparison no doubt aided by the commonly acknowledged fact that the early Lanchesters were years ahead of their time". The Autocar poured more praise on the attributes of the Lanchester 16 h.p., which was still in its original working condition, then concluded, "It is difficult to see why these 10-year-old Lanchesters are not among the popular cars on the second-hand market today, for their present market value is only £25. Surely a poor appreciation of the car which provided modern cars with many of their best talking points." This low resale value in 1913 highlighted the problem that then, as ten years earlier, Lanchesters were so unconventional.

Fred had always tried to 'educate' the motoring public to the science of correct automobile design and would not readily forgo these principles, even though it resulted in merely twenty 16 h.p. cars being sold. Some sympathy, therefore, must lie with his Board of Directors in the early years of the 1900s, as they appealed for a more 'ordinary' design.

Chapter Eight

THE TWIN-CYLINDER 12 H.P.

Even with Lanchester's air-cooled engines powering the 16 h.p. and 10 h.p. vehicles, as well as being used as stationary engines in Lanchester's own woodwork shop and elsewhere, much of the public and motoring press were still fiercely opposed to air-cooling. History has recorded that Fred Lanchester was against the implementation of water-cooling for his cars, but this was only true in part. Like any production engineer, he did not wish to introduce a change in the schedules for fear of it interfering with and upsetting the programme, or compromising reliability and quality.

However, it will be recalled that Fred had already successfully designed and built a water-cooled 8 h.p. twin-cylinder engine, which was installed in his second river launch. There were no major problems in water-cooling his 10 h.p. engines as patterns had already been made and some castings were already in stock. Indeed, Fred did not oppose the introduction of water-cooling for his engines at the Board Meeting of 9th July 1903. After explaining that they had carefully considered the design of Lanchester cars and those of its contemporaries, the Board themselves arrived at the decision, "air-cooling to be retained". As well as the advantages and disadvantages of this method, water-cooling was also discussed and Fred's acceptance of change was recorded in the minutes: "The General Manager informed the meeting that he had a number of sets of water-jacketed cylinders available, and undertook to have a car fitted up with water-cooling for the purpose of trial." This work was done shortly afterwards, although the first cars did not have full water-jacketing to the valve chest. The new water-cooled engine became available from August 1903, supplementing the air-cooled versions.

As the two large air-suction fans and 'windchest' of the 10 h.p. air-cooled engine were not necessary on the water-cooled version, the release of the horsepower which they had absorbed meant that the new engine was able to develop 12 b.h.p. at its nominal speed of about 1,000 r.p.m. The car therefore became known as the 12 h.p. Some owners, however, had their existing air-cooled 10 h.p. cars converted to water-cooled at Armourer Mills, and these cars were known as the 10/12 h.p. model.

All the new 12 h.p. cars with tonneau coachwork were produced with the wider body as a matter of routine, to take three passengers in the rear, the third passenger sitting on a finely-engineered folding seat on the inside of the central rear door. The two-seater tonneau coachwork could still be made to special order. It was probably at this time that the rear tonneau door was increased in height to become the same level as the main coachwork, whereas before a separate panel had been available for use above the lower door.

The overall width of the car was 5ft.7ins. and the overall length was 11ft. Mechanically, only slight alterations from the 10 h.p. were needed. In order to allow for the radiator and water-cooling system, the front cantilever springs were extended by 1in., so making the wheelbase 7ft.10ins. Also, the steering track rod had to be moved from behind to the front of the axle. In order to cater for some customers' unfounded doubts about the strength of Lanchester's wire wheels, Fred patented his own design of artillery wheels; these were advertised later as an option, at extra cost.

When the 12 h.p. engines were fitted into the Lanchester chassis, certain other changes

THE TWIN-CYLINDER 12 H.P.

The summit of Orme's Head, with George Lanchester at the tiller of the Works' 12 h.p. car.

were made. The 10 h.p. engine dust tray was replaced by an undertray to collect the oil which fell from the engine, where it was held until drained off.

In spite of the fact that this 21 cwt. Lanchester was not the fastest vehicle on the road, it nevertheless performed well. The engine had a similar bore and stroke to the 10 h.p., being 5.25ins. and 5.687ins. respectively, giving 4,029 c.c. and a compression ratio of 4.2:1. At its maximum engine speed of 1,700 r.p.m. it produced 18 b.h.p.

Roller bearings for the crankshaft main bearings were implemented from late 1904, otherwise the engine remained the same throughout its production life. There had been little reason for change in the car as it proved most efficient. Under varied road conditions it returned over 20 m.p.g. and, in one specific instance during 1905, a 12 h.p. achieved 38 m.p.g. under test. However, this was eclipsed the following year when another 12 h.p. was awarded a Gold Medal for its "wonderful performance of 42 miles to the gallon".

Some 52 years later, the Autocar used an electric speedometer to measure the performance of a 'well used' 1903 Lanchester 12 h.p., and the following timed figures were recorded:

From 0-20 mph = 10.8 seconds.
From 0-30 mph = 26.5 seconds.
From 10-30 mph, using top gear = 20.7 seconds.

The best level speed recorded was 36 m.p.h. and the maximum speed recorded down a gentle hill was 43 m.p.h., wholly acceptable figures for 1903.

It should be noted that while the Lanchester could hold its own with most of its contemporaries, although certainly not all, the figures for straight line acceleration speeds are perhaps misleading. To travel fast, other than in a straight line, requires more than just a powerful engine, and the comfortable, safe and reliable design of the Lanchester made

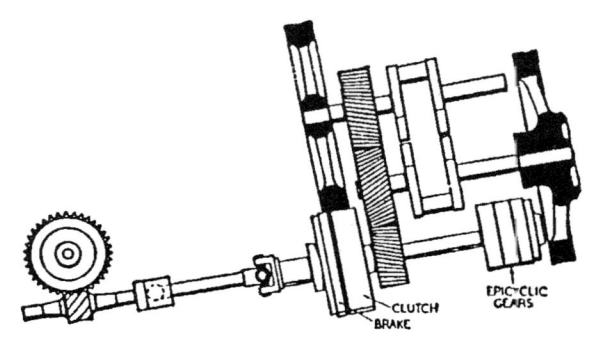

The twin-cylinder engine, showing the gear-train linking the upper and lower crankshafts, clutch, brake, gears and underhung worm-gear rear axle.

for very creditable average speeds. A surprised reporter summed up in the following sentence, "This fine old Lanchester has really endeared itself to me, and I was most agreeably surprised by its willing and easily manageable performance." One of the reasons for his observations was the ease of control of the clutch, gears and brake. The car was certainly capable of hill-climbing also, as a 1903 publicity stunt had seen two standard 12 h.p. cars ascend the infamous Orme's Head.

The engine was angled downwards towards the rear, where its power was transmitted by the lower of its two crankshafts, via helical gearing, through the countershaft to the drive shaft and rear axle.

The countershaft had five separate functions: the main brake, direct-drive top gear, intermediate gear, low gear and reversing gear. It employed three separate trains of epicyclic gear, all similar to each other and housed in brake drums of 6.125ins. diameter. The principle of the operation was that by braking the different gear trains, which were constantly in mesh, different rates of speed were obtained in perfect silence and ease of operation.

This gear system was a direct descendant of that used on the 1895 Lanchester, although the extra, intermediate, second gear made use of Fred's 1900 patent of compounding gears. Fred put to good use more of his patents, such as roller bearings for the planetary gears and splined ends on the shafts to secure the rotating elements. It was a very expensive

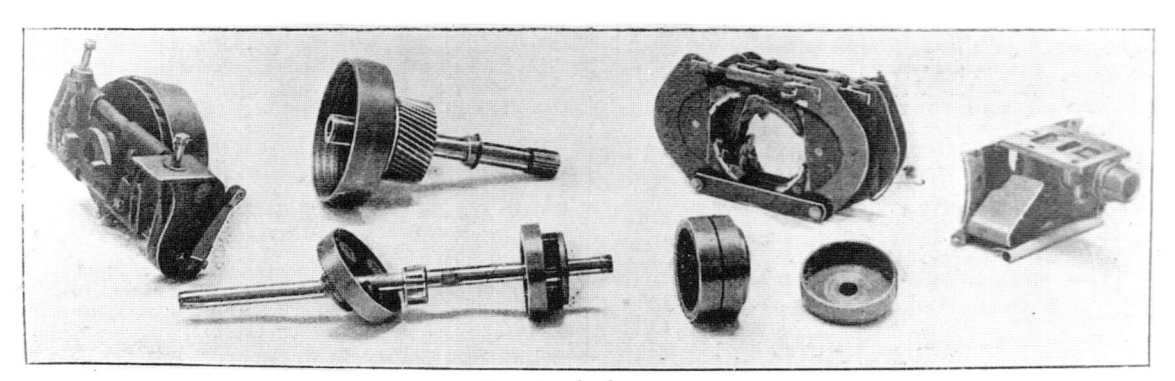

Countershaft parts.

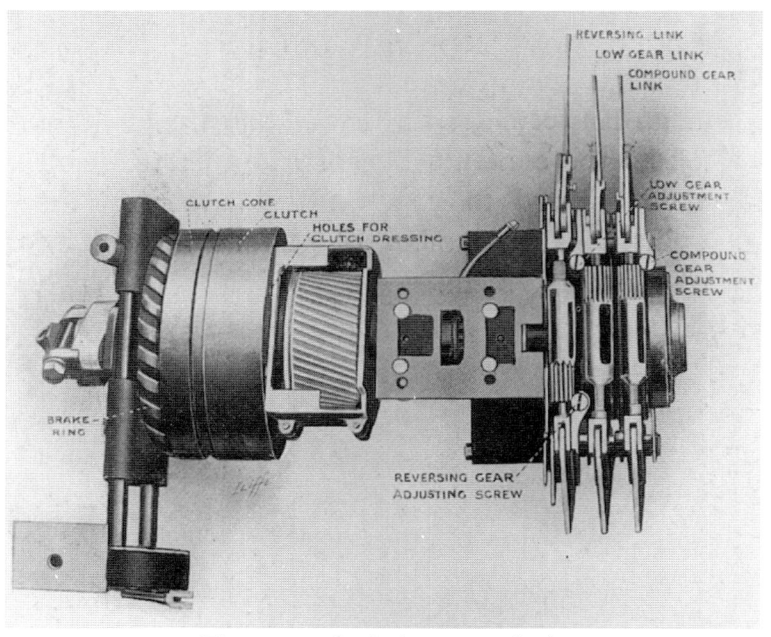

The twin-cylinder's countershaft.

system to produce and quite heavy compared to the alternative sliding-gear 'crash' type (but was successfully employed by the Lanchester Company until 1931, and is universally accepted as the direct predecessor of the modern automatic gearbox).

Major W. G. Wilson, a renowned engineer and good friend of the Lanchester brothers, will best be remembered for his early 'Wilson-Pilcher' version of Lanchester's compounding gears (and later his 'Wilson pre-selector' epicyclic gearbox). Referring generally to these, Major Wilson wrote, "The earliest effort was a most far-sighted and ingenious compound epicyclic gear train patented by F.W.Lanchester. It is of special interest as it was the first, and one of the only designs of true compound epicyclic gears, and consisted of using the annulus of the first speed as the planet carrier of the second and vice versa. It had a great many most important features, noticeably the low speeds of revolution of the planets, it demonstrated clearly that the running gear of the true epicyclic possessed such a degree of silence and length of life as to make it a practicable proposition."

Montague S.Pilcher himself wrote to the Company in praise of the 12 h.p. Lanchester he had bought: "Gentlemen, the car you have supplied me with has been quite a success. I have had no trouble as yet with it, and it is one of the best hill-climbers I know, though I have had several cars of different makes." Another letter of praise was from the Marquis of Anglesey, in 1903. "The car you supplied me with has given me entire satisfaction; in fact, I have had more use and pleasure from it than any of my other big ones."

After studying the 12 h.p. car in detail, the journal 'The Engineer' stated, ".... a motor car which, if it stood alone, would relieve English manufacturers of the reproach of not being able to do anything but copy continental designs". Although the staff at 'The Engineer' could no doubt comprehend the technical intricacies of Lanchester's unique car,

Archie Millership drives George Lanchester in a 12 h.p., on a winter's day.

Fred knew that the vast majority of his existing and potential customers would understand very little of this jargon.

To overcome this lack of knowledge, the methods of operating the gears, clutch, brake and other unique features were explained during the tuition offered by the Lanchester Motor Company. When giving the order for a new car, the customer would often request that it be accompanied by a fully-trained driver. With his family, the driver would then move to live at the home of the new Lanchester owner, to become his motor attendant and chauffeur. Before this, though, the driver had to prove himself suitable for the task, being tested first of all on his eyesight. In Fred's words, "Rejections other than for eyesight are almost entirely on account of such matters as general slovenliness or bad timekeeping, and the inference is that with good training a man of the most ordinary intelligence has it in him to make a fairly good driver."

After becoming thoroughly familiar with the car's general layout, the potential driver would have been shown the sequence for starting the vehicle. After inserting the starting handle into the shaft protruding from the coachwork near the driver's running board, an adjacent small lever was lifted to its horizontal position. In another example of Fred's ingenuity, this lever performed several functions: the starting handle shaft was brought into mesh with the matching bevel gear on the lower flywheel, the worm driven camshaft was re-positioned to bring into operation the half-compression cams, the ignition was retarded, and finally it shifted the flywheel magneto armature to enable the retarded spark to correspond with the maximum flux in the generator.

After giving the petrol lift pump a few operations and adjusting the rotating vapour regulator to the anticipated mixture strength, the two governor levers were lifted to achieve an increased tickover.

A few positive turns of the starting handle would have brought the engine to life, whereby the small starting lever would be returned to its vertical position to disengage the special

View of a new 12 h.p. showing the rear steps for entry into the tonneau. The third seat hinges down from the central door.

THE TWIN-CYLINDER 12 H.P.

starting functions, and lastly the governor levers would be depressed to give normal tickover. After this it was time for the open road.

The squad of drivers would be taken out in one or more cars where it was "customary to give them charge of the steering alone, until they had rendered themselves thoroughly 'automatic', the instructor taking charge of the brake and gears, and from then on gradually building up over three or four lessons to be a competent driver". In a relatively short time, therefore, the new driver could become proficient at controlling the car. At first, no doubt, he was confused, with an array of levers and only one pedal, the accelerator, other than the foot-operated bulb horn or that wonderful Lanchester item, the 'ting-tang' gong. He would have learnt the use of the vapour regulator control and the petrol hand-pump, which gave about two miles of travel for each stroke of the pump. The three remaining levers operated the three forward and one reverse gear, as well as the brake. Of the two main, adjacent levers on the central control panel, the longer right-hand lever operated the high-speed, direct-drive gear and the main brake. Its partner, the shorter, left-hand lever, was known as the low-speed lever and controlled reverse, first and the intermediate gear. Both levers could notch into the central position along their quadrants, so giving an idling or neutral position. The third lever, situated behind the left-hand, low-speed lever, was the compounding pre-selector gear lever known as the change-gear trigger. The car could be started from rest in any one of its gears, depending on the gradient of the road.

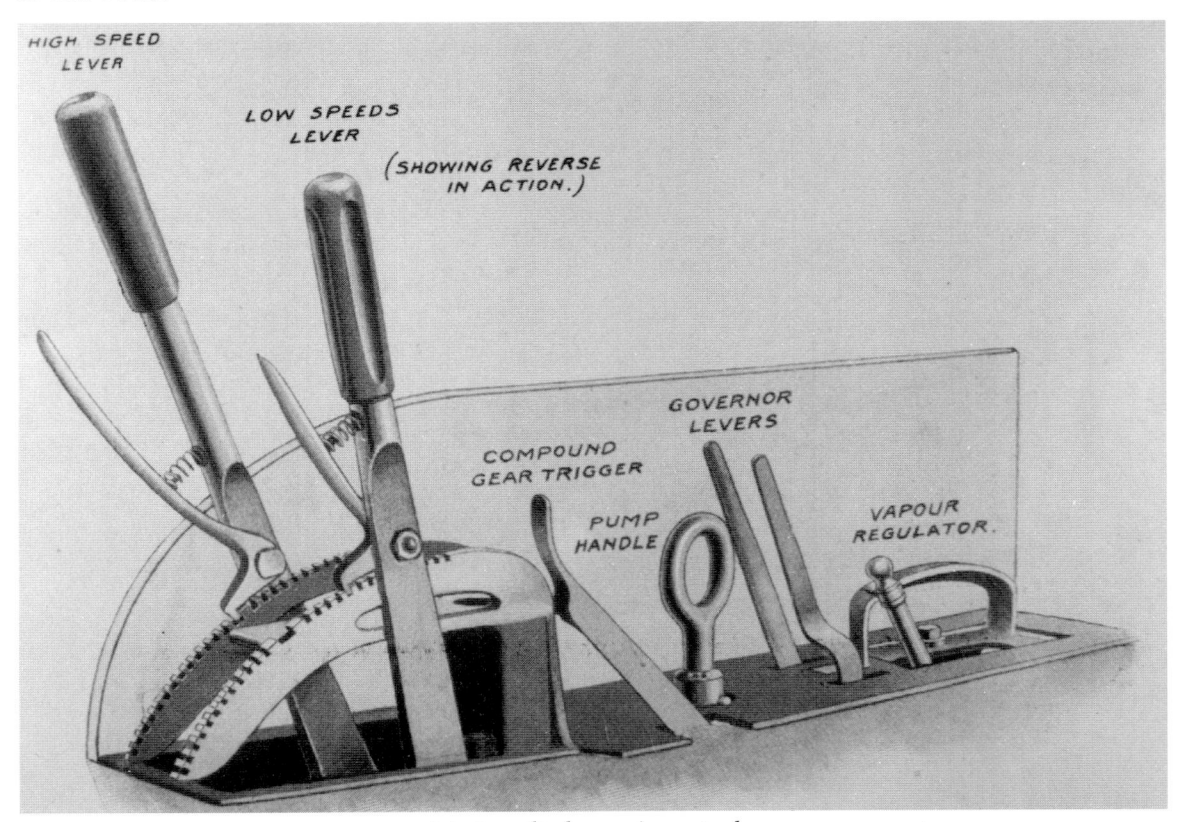

A twin-cylinder car's controls.

To pull away, the change-gear trigger was pushed into its forward position, and the high-speed lever to its central, neutral position. The low-speed lever was eased forward, simultaneously depressing the accelerator pedal to gain motion. In general terms, the change from low to intermediate gear would be required at about 7 m.p.h. Before this was reached, the gear-change trigger was pushed to its backward position to pre-select the intermediate, compound gear. To actually engage this gear the low-speed lever was pulled back to its central position and then eased forward again to its forward position. The accelerator would power the car up to around 16 m.p.h., when top gear would be required. The low-speed lever, controlling just the epicyclic gears, was then returned to its central idling position, and immediately the high-speed lever controlling the direct-drive clutch arrangement was pushed forward, so selecting high-speed gear.

When travelling on a downhill gradient or when anticipating the need to stop at road junctions or corners, both gear levers needed to be in the central position, allowing the engine to idle. The car could then free-wheel at will, with slowing down and stopping controlled by the high-speed lever being pulled backwards to engage the conical brake, which was sufficently powerful to slow or stop the car under any circumstances.

However, under some conditions, it may have been desirable to use the epicyclic gears to gently slow the car. Pulling the low-speed lever rearwards selected reverse, the only selection with no locking ratchet, and the car would steadily come to a halt before travelling backwards if reverse gear was held in position.

The Lanchester-trained chauffeurs would have been surprised by the ease of driving of this fine car, with its top speed well in excess of the new 1903 speed limit of 20 m.p.h.

However, for owners who took advantage of the Company's offer to deliver their new car by rail, a different method of tuition was obviously required. Fred therefore produced a trilogy of manuals: part 1, a Descriptive Manual; part 2, a Driving Manual; part 3, an Engineer's Manual. These were warmly welcomed by the press, who had sensibly been sent preview copies by the Lanchester Engine Company, presumably at the instigation of Frank, Chief of Sales. Their importance was stressed in descriptive terms by the Motor Car Journal at the end of January 1903. "Purchasers in many cases must learn for themselves the way to operate their cars, a task which may be simple enough when properly undertaken but, for a man who does not know the exact difference between a 4-cycle engine and an 8-day clock, it is no easily self-tutored task. It is not possible for the maker to send an expert with every vehicle dispatched from the works, hence the need for a clear, written manual which may act as a silent instructor. The production of a really sucessful manual is not an easy task, it is almost habitual with many technical writers to forget that the elementary reader may know little or nothing of the elementary facts concerning the subject at hand." Fred's work was certainly well-received, as the Motor Car Journal, referring to the Descriptive Manual, continued, "It gives a clear description in non-technical language of this novel vehicle; a good deal of time and trouble has been spent on the manual, which will be invaluable to Lanchester owners." The other two manuals were equally clear, and all three had revised editions published from time to time. Between them they covered every facet of owning, using, servicing and repairing a Lanchester motor vehicle, often in the most eloquent of terms.

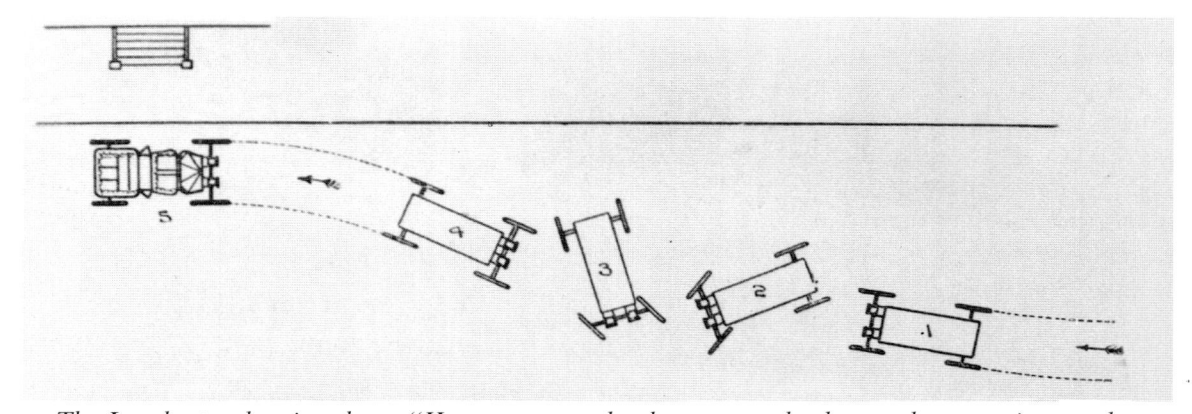

The Lanchester drawing shows "How a car may be thrown completely round at stopping, ready to start again on the return journey. It should on no account be attempted in the presence of other traffic even by a proficient, as it is a performance that is thoroughly disconcerting to other users of the road."

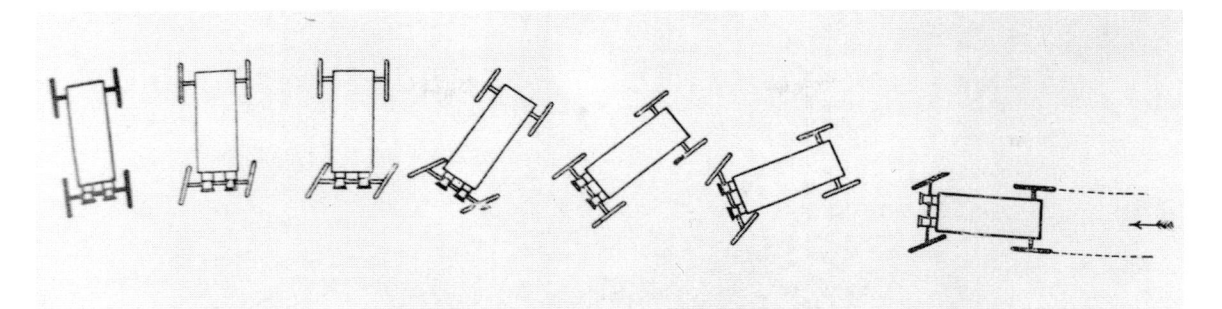

This Lanchester sketch shows that a sideways stop is "a manoeuvre which is very difficult to judge and execute with accuracy. The car may be brought up in this way in greasy weather in about half the distance that is possible when only the adhesion of the driving wheels is employed by using the brake in the normal way".

Fred Lanchester described the automobile world at that time as "in the age of novices". He dismissed the majority of tips and hints given in motoring papers as "frequently entirely inaccurate, both as to suggestions given and theories advanced". When mentioning himself as author, however, he reminded people, "The instructions have been compiled from the collective experience of men who have been motoring on the road ever since, and even before, the Act of 1896." The advice given was essential for pioneer motorists, many of whom had not even been in a car before, let alone driven one. Although it raises a smile today, Fred offered good advice at the time to the uninitiated: "..... experience in horse driving, bicycle riding etc. is all of the utmost value to the would-be automobilist ..... a bad side-slip is apparent to the merest novice by the uncanny sideways motion of the rear portion of the car .... he must humour his car by steering in the same direction as the side-slip, so as to keep his car in track ..... a skilled driver will never, under any circumstances, collide with a kerb broadside on. If necessary, he will exaggerate a side-slip sufficiently to negotiate the kerb backwards, rather than risk the consequences of a lesser side-slip which will result in broadside collision ..... remember that whatever the state of the road, it is bad driving to navigate the car sideways". Fred assured that, "the Lanchester car cannot be over-turned by violent steering, as the centre of gravity is so low". However, when instructing the new owner on driving technique in slippery conditions, he warned, "We do not take the responsibility of recommending drivers to practise these grease evolutions."

Equally entertaining is the section on how to cross expanses of water on the road. After

4,000 Miles--Repairs 3d.!

Burwash, Sussex, December 3rd, 1904.

Dear Sirs,—The 12 h.p. water-cooled Lanchester which I bought from you this year has run now for six months without flaw or failure. During this time she has covered over 4,000 miles in cramped and hilly country, often with a full load of luggage on the water deck.

She has never been out of order when I wanted her, she has never delayed me five minutes on the road, and no day has been too long nor any hill too steep for her powers. Her repairs since June have been confined to one or two igniter springs.

I am, after three years' experience, rather a believer in the Lanchester type. It seems to me that in ease of suspension, strength of brake, luggage room, and sobriety of appearance, the type embodies many of the things that will be considered essential in the car of the future.

Yours faithfully,

RUDYARD KIPLING.

The **LANCHESTER ENGINE CO.**, Ltd., SPARKBROOK, BIRMINGHAM

Selling Agents and Official Repairers—
LONDON—Lacre Motor Co., Poland Street, Oxford Street; Goy & Co., Praed Street, Paddington. MANCHESTER—R. Ramsbottom, Market Street. HANTS—Imperial Motor Co., Lyndhurst. GLASGOW—North British Motor Co.

Rudyard Kipling extols the virtues of a Lanchester.

A 12 h.p. car with a canopy top.

informing the novice driver that "it is an unpleasant experience to find yourself in the middle of a ford with your engine drowned out", instructions were given on how to achieve success in a brave and confident manner. "A Lanchester car may be easily driven through a ford or flood water which is not more than hub-deep. In crossing water not exceeding a foot or 18 inches deep, a car may be driven through in low gear without the occupants getting wet." He further explained that, "It is unwise to attempt the crossing of a ford over 2 feet 5 inches deep, but where the depth is as much as 18 inches the best plan is to rush it at 20 - 25 mph, at the expense of a considerable wetting!" With cheerful abandon, he continued, "..... in rushing a ford in this way, we have known the front road wheels lifted almost clear of the ground by the impact of water".

This written advice was also essential to customers from abroad, as they could hardly call at the Works for information. One Indian owner, His Highness Prince Ranjitsinhji the Maharajah of Nawanger, obviously felt completely at ease driving his 12 h.p. and performed exceptionally well during the Delhi to Bombay Trials in December 1904, covering the journey without a single problem at an average speed of 30 m.p.h. and losing only one-and-a-half points during the entire eight-day trial; and this was for pumping up his tyres! Indeed, the Maharajah was so impressed that he honoured the marque with his custom from then on, building up to a fleet of forty-two Lanchesters. He may also have enquired about a Lanchester 14 h.p. car, which was referred to in the 1904 booklet from 'The Motor News' although no other instance of the model's existence was recorded. (This is considered by the author to have been a printer's error.)

A December 1907 12 h.p. mail van providing reliable service in the Torquay district.

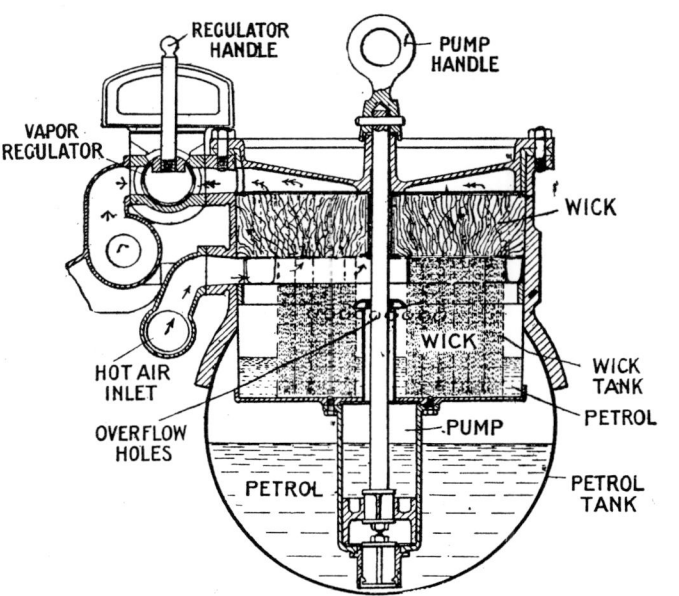

An advertisement following the Lanchester's Gold Medal Award in the 1905 Consumption Trials.

To encourage sales, from January 1904 the price of the 10 h.p. and 12 h.p. standard, four-seater tonneaus was dropped to £500. The following year, the price of the larger and more expensive brougham was decreased to the same figure.

The Company was still committed to production of the 'twins' and this was demonstrated at the Agricultural Hall Motor Show in March 1905, where working models of the engine and transmission were exhibited, along with three complete 12 h.p. twin-cylinder cars. One of these showed the updated tonneau coachwork, giving a raised section of bodywork at the rear to allow for a higher back to the three seats. Apart from increased comfort, slightly more protection from the road dust was gained. Another of these cars was a wide tonneau body clad in basket-weave decoration, voted "one of the smartest things in the show" by The Autocar, and which boasted the new body design of an open-sided canopy top above the rear curved glass windows, as well as a vertical windscreen over the car's radiator. The Motor Car Journal, reporting on the event, expressed the opinion that the Lanchester exhibits were, ".....full of interest and would appeal strongly to mechanically-minded visitors to the show".

More praise came with the Midland Automobile Club's awards of Gold Medals for both the May 1905 and 1906 Fuel Consumption Trials. However, full production of the 12 h.p. ceased within that year, due to falling sales, although the car still had its supporters even after this date. The Autocar of May 1909 confirmed this by including one writer's comments: "The 12 h.p. Lanchester with horizontal engine is today still far ahead of many new cars of the same class and power, for example, in sweetness of running, petrol consumption, hill climbing (their forte), and ease of control, and the material and work put into those cars were so good that, provided proper attention be given to them, they will last almost indefinitely."

Chapter Nine

THE TWIN-CYLINDER 18 H.P.

The 12 h.p. car developed into the twin-cylinder 18 h.p., just as the 10 h.p. had earlier evolved into the 16 h.p., although in both cases the smaller models remained in production. With the former, however, it was not only a bigger engine but a completely re-designed car, with an 11ins. increase in the wheelbase to 8ft.9ins.; the track remained the same at 4ft.10ins.

The water-cooled 18 h.p. was the ultimate development of the twin-cylinder engine; with its bore size of 5.75ins. and stroke of 5.687ins. it developed 18 h.p. at 780 r.p.m. and nearly 30 h.p. at maximum revs. Due to the greater body size and weight, it only achieved a similar maximum speed to the 16 h.p. air-cooled car, of just under 50 m.p.h. For the first time, Fred inventively installed a fixed starting handle for the engine which was able to be rotated by the driver whilst sitting in his seat. It seems most probable that its ease of operation would have been as effective as every other asset on the car.

Fred himself obviously thought highly of this apparatus as, since 1902, there had been another option available to him for starting an engine. His patented invention detailed an 'improved construction of turbine suitable for employment on the exhaust of an internal combustion engine for utilising the residual pressure'. It was very advanced, and incorporated machined pockets (the Daimler fluid flywheel used a similar design some twenty-five years later) able to withstand very high temperatures as they were made in high tensile steel. This gave a sturdy construction and allowed a car's exhaust gases to be used, therefore, to rotate the turbine.

Fred proposed to use this "coupled to high-speed ventilating fans or pumps, or to a dynamo for lighting or ignition purposes". He wrote that it could be "employed to pump up pressure in a reservoir to be used for starting ....." It is not recorded which motor

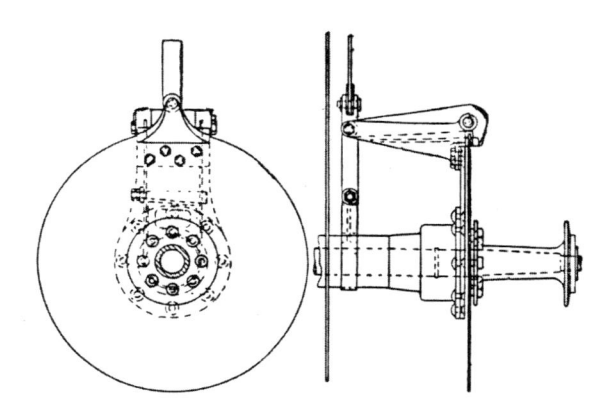

In 1902 Fred Lanchester applied for a patent for his invention of a mechanically-operated disc brake, with mechanism "which somewhat resembles a pair of dentist's pliers".

manufacturers used this facility, or indeed the reservoir's other important use of ".....assisting the main engine when required". (The latter incredibly appears to be the world's first instance of a form of turbo-charging.)

The radiator was fitted in front of the car and accordingly the steering track-rod was positioned forward of the front axle. The rear axle was of the high-ratio type used on the 16 h.p. car, with Lanchester artillery wooden wheels again provided as an option. The unladen weight was 19.75 cwt.

Without doubt, the most remarkable feature on the 18 h.p. was the rear braking system. From the list of fifteen various patent applications of 1902, one of them specified 'improvements in the brake mechanism of power-propelled road vehicles'. This invention was in fact the disc brake (it was next used experimentally on the AC car in 1919, and on the production 1949 Chrysler). Fred wrote, "The object of the present invention is to construct a brake which is applied direct to the road wheels and is unaffected by oil or mud, and which has very little elastic yield. I arrange a sheet metal disc securely riveted to the wheel hub and I operate on the edge of this disc by a pair of gripping jaws actuated by a direct bell-crank lever. These gripping jaws and bell-crank lever form a piece of mechanism which somewhat resembles a pair of dentist's pliers, the jaws being at right-angles to the direction of the two handle levers. Mounted on a plate securely riveted to the axle or axle frame, one of the handle levers is tied by a stay on to the said axle frame, while the other is actuated by a tension link from the car body. The stay by which the one handle of the pliers is secured to the axle frame is approximately lineable with the direction of the tension link, so that when tension is exerted on the link by suitable operating mechanism there is no tendency to distort the attachment plate or the brake disc in a lateral direction."

The disc brake was duly fitted to the 18 h.p. model and later to the other twin-cylinder cars as an optional extra, classified as a "metal to metal reserve brake, fitted direct on road wheels". The farsighted design was not completely successful in use, however, as the copper disc pads wore quickly on the roads which "in those days were either thick with dust or deep in mud", related George. (No Lanchesters remain today incorporating the disc brake,

George Lanchester at the tiller of an 18 h.p. fitted with the side starting handle and disc brakes on the rear wheels.

THE TWIN-CYLINDER 18 H.P.

A side-entrance 18 h.p. tourer, without disc brakes, ready to leave the works.

which is a great pity as they would doubtless have performed far better on today's road surfaces.)

The purpose of the increased chassis length, incorporating a change in design of the chassis-frame side members, was to give more scope for the latest body style of side-door entry. Ladies and gentlemen in their finery disapproved of walking across the mud-covered road to get to the traditional, central rear door of their luxurious limousine. However, the rear entrance tonneau was still available, and the increased length of chassis allowed for two extra seats in the front corners of the tonneau compartment, giving seating for five in the rear. Any other style of coachwork could be fitted to customers' requirements, but all included the folding dash and sidelamp-cum-airscoops.

An 18 h.p. with 'D-fronted' brougham coachwork.

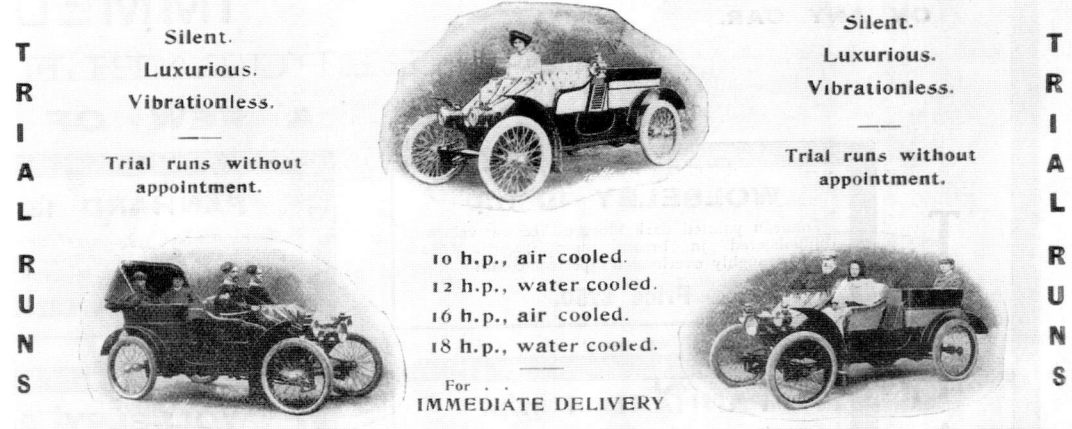

The Lacre Company's advertisement of May 1904.

An invention of Fred's from 1902 might well have been considered for use on the 18 h.p., the largest and heaviest twin-cylinder model. Submitted for patent approval in February 1903, for the first time in motoring history it provided a system of power-steering. It utilised a cylinder, piston and valves, no doubt with water as the medium, to "cause the steering mechanism to be actuated by hydraulic power". The system also cleverly incorporated a "balance weight, by which a sense of direction is imparted to the steering operating lever or gear to enable the driver to feel the centrifugal pull of his car". Thinking of every eventuality, Fred also made the provision that "the car may be steered by direct action without the use of hydraulic pressure if desired". (Unfortunately, no cars remain which include this advanced design.)

After the initial trials by the Lanchester Engine Company, an 18 h.p. was delivered to Rudyard Kipling for further inspection. He usually received new models for road testing, as his constructive layman's comments were much valued by the Lanchester brothers. This first car, however, had some teething problems whilst in his care, and George laughingly recalled, "One day we got a telegram from Kipling, 'Jane disembowelled on village green Ditchling. Pray remove your disorderly experiment'." In a letter to Frank in 1921, Kipling recalled the saga of the 18 h.p., which ".... was taken to pieces once on a vacant plot at Worthing which she completely filled with her disassociated entrails".

Once the problems of the "disorderly experiment" had been resolved, others found the car to be in the best of Lanchester traditions. Indeed at Crystal Palace Automobile Show in February 1904, the Company proudly displayed an 18 h.p. seven-seater touring car, and an 18 h.p. seven-seater, side-entrance landau tonneau; as well as one 10 h.p., two 12 h.p. and one 16 h.p. car. The Motor Car Journal impressed on the public, "Those motorists not conversant with the Lanchester system should make a point of visiting this stand, as it is one of the most distinctive in the show."

Another source of praise came from Filson-Young's 1904 book 'A Complete Motorist', which stated, "Plenty of luggage accommodation is more than a convenience in touring cars; it is a necessity. With such a car as the 18 h.p. Lanchester, one could go on a tour for months and carry with one practically all necessaries and not a few luxuries." This statement surely supported Fred's own theories on how a car should be designed, as opposed to the ideas of other contemporary companies, whereby the engine and transmission took up a large part of the usable space, so leaving less for passengers. The big 18 h.p., selling for £720, had arrived too late in the day to achieve good sales, however, and only six were built. Production of the 18 h.p. and the three smaller twin-cylinder models ceased around early 1906, although they were all available by special order.

Chapter Ten

THE EDWARDIAN YEARS

Alongside the continuing workload of car design and production, the personal and social lives of the Lanchester brothers had been equally full. A case in point was the Midland Automobile Club, which they had founded with friends and colleagues in 1901, with the aim of promoting both serious and social motoring events. Frank, the main driving-force behind the formation of the club, became the first Honorary Secretary; in fact, the minutes of the inaugural meeting of the Club in 1901 had been written by Frank in his elaborate copper-plate handwriting. Also elected to the organising committee had been, amongst others, Archie Millership, Charles Pugh, Herbert Austin and John Siddeley, and all had enjoyed participating in many of the Club's events, including its annual Hill Climb competitions. In 1902 two 10 h.p. Lanchesters won gold and silver medals for their ascent of Gorcutt Hill.

As with all motoring occasions in those times, however, there was a constant battle of wits with the local constabulary. The confrontations reached a climax in 1904, when police traps were set near Sunrising Hill, where a hill climb was being held. As George delightfully explained, "In consequence, many competitors and spectators received urgent invitations to attend unpleasant appointments before unsympathetic magistrates a fortnight later." This, coupled with the fact that the local anti-motoring farmer "turned out his carters and drovers to trundle hay waggons, carts and cattle up and down the hill during the competition", encouraged the decision to hold the future hill-climb events on private land.

George drove the Treasurer of the M.A.C. in a 10 h.p. Lanchester on test runs up and down a potential route. They judged it to be highly suitable and, after certain improvements, the famous Shelsley Walsh Hill Climb was born. George competed in the opening event on the 12th August 1905, no doubt sporting his new design of motorist's goggles, which he subsequently patented. For the following few years, he was accompanied by the Company's Managing Director, Hamilton Barnsley, in another Lanchester. Although he never competed at Shelsley Walsh, Frank Lanchester remained the Honorary Secretary of the M.A.C. for many years.

Frank had also been engaged in other interests during the early months of 1902 when, together with various respected members of the automobile industry, he formed the Society of Motor Manufacturers and Traders. In July of that year, at the first general meeting of the S.M.M.T., Frank was appointed a member of their council. He was also one of the first group of A.C.G.B.I. members and was subsequently involved throughout his working life with both organisations. Obviously much travelling was undertaken by Frank during these years as his home remained in Birmingham; he had moved to 48 Sandon Road, Edgbaston in November 1903, following his marriage to Miss Minnie Thomas.

George, who had also moved from Lincoln's Inn at this time to take bachelor 'diggings' near Armourer Mills, had become a regular visitor to Frank's home, where his brother used to entertain both George and Minnie's sister, Rose. The subsequent engagement of these two led to many happy occasions, frequently with the M.A.C.; indeed, under George's instruction, Rose became one of Britain's pioneer women drivers. They were married in the April of 1907 and George left his bachelor diggings in number 6 St. Alban's Road, Moseley, for the house 'Staverton', in nearby Stoney Lane (now number 144 Yardley Wood Road).

These personal and social occasions for George and Frank, resulting as they did in time away from work, did not gain full approval from their elder brother. Fred continued to labour relentlessly on many varied and different projects, spending his limited time for relaxing in a different way. He had patented ten inventions in the year 1900 and another seventeen in 1901 on motor vehicles and ancillary design, as well as one thrown in for good measure on 'roofing for marquees'. 1902 and 1903 had continued in the same vein, with another forty-two patented inventions of various descriptions, including one consisting of 'a self contained motor and screw propeller to an oar or scull of special construction'; seemingly the world's first outboard boat engine.

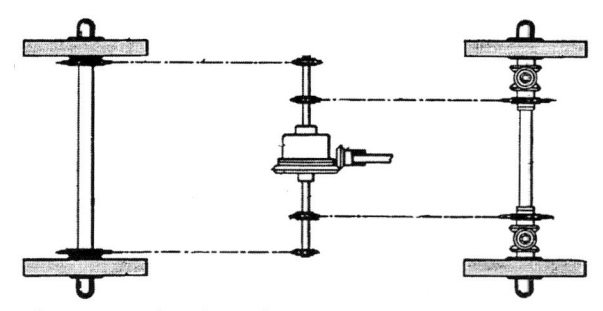

One example of Fred's 1904 patent for four-wheel drive. He wrote, "My present invention has for its object to permit the distribution of power to each and all of the road wheels, thereby securing greater adhesion under all conditions of running and less liability to skid or skip."

His inventive mind spawned sixteen patents in 1904, one of which was the first four-wheel drive system. (The next such design came from Jenson Motors in 1966.) Another six patents came in the year of 1905, with work on subjects ranging from sound-reproducing equipment to aerodromes, as well as, of course, motor-vehicle design. All this was an incredible achievement considering his Lanchester Engine Company workload, as well as the continual work on his pet subject of aerodynamics.

Alongside these, he had formulated a study of the relationship between the bore and stroke of an engine and its subsequent horsepower. This profound piece of work was read the following year at a meeting of the Institution of Automobile Engineers, and was later to lead Fred into conflict with many eminent persons and eventually the R.A.C. itself.

His relatively small total of thirteen patents in 1906, 1907 and 1908 reflected his lack of spare time, although he remained dedicated throughout to his experiments and theories on flight. Somehow, he also found time to submit his back axle worm gear to the National Physical Laboratory, but arguments about the efficiency of worm-gearing rumbled on until Lanchester's paper was set before the Institute of Automobile Engineers in 1913.

A project which took up much of Fred's time in 1907 was that of his 50 h.p. engine and chassis. The Lanchester Directors had always longed for a conventional design, a concept which Fred had never given them. Against his better judgement, Fred bowed to their appeals and designed an engine of massive proportions, with six separate cylinders of 4.75ins. bore by 4.75ins. stroke. He had been ordered to produce this with a side valve 'T-head' combustion space, where the inlet and exhaust valves for each cylinder are set diametrically opposite each other on horizontally-protruding sections from the top of each cylinder. This was despised by Fred as being both inefficient and a retrograde step. It is reasonable to conjecture that Fred seized this occasion to experiment, as he located the camshaft beneath the crankshaft in the sump. George recalled. "The motive was that, the cams and rockers being immersed in oil, it would make for silence. He (Fred) completely overlooked the effects of expansion in that very high engine. The crank case and cylinders expanded more than the push rods so that the valve clearances became greater, and the mechanism got noisier than would have been thought possible."

However, no doubt delighted by their 'fashionable' engine, the Directors proudly insisted on its display at the November 1907 Olympia Motor Show. The motoring press caught their mood of excitement when they reported, "The new 50 h.p. six-cylinder engine will assuredly attract the notice of all who search for something in the shape of the departure from the absolutely normal ..... This engine does not follow upon either the Lanchester horizontal or vertical engine, but more nearly approaches the general type, though differing therefrom in detail likewise." Differ it certainly did, and not for the better. George termed the engine, "An aberration in the brain of a genius."

Certainly the engine was in the true 'Lanchester position', with the dashboard running across its centre. George related, "As far as I remember, there were three experimental chassis put in hand, only one of which was fitted with a body and went on the road. But I do recall a terrifying crankshaft vibration on the occasion of its first emergence from the Works' entrance." This experimental failure would surely have done nothing to improve the relationship between Fred and his Directors.

Away from the Company, Fred's other major project of 1907 was his work on flight, leading up to the publication of his two books in that and the following year. Much of their content evolved from the paper he had written in 1894, entitled: 'The Soaring of Birds and the Possibilities of Mechanical Flight', which he had revised and submitted to the Physical Society in 1897. They had been unable to fully understand his work, and it had therefore been rejected. This had meant that his inventive work had not previously been brought to the attention of other pioneers of flight, but the publication of 'Aerodynamics' and 'Aerodonetics' went some way toward redressing the balance. Although he had lost the credit and recognition he deserved during this time, his books now conveyed to the world his theories of air flow, stability and control, streamlining and skin friction, drag and resistance, trailing vortices, as well as designs of elevator fins, rudders and aerofoils. His vision for the future even included such ideas as the deployment of liquid hydrogen as a fuel for powered flight.

1909 saw the dawn of a new era for Fred Lanchester. Apart from his regular points of interest, such as other important papers to the Institute of Automobile Engineers, he patented fifteen new inventions, one of which was for the manufacture of piston rings, which was most successful and remunerative. His life took a fresh turn when he was appointed Consultant and Technical Adviser to the Daimler Motor Company of Coventry. The February 1909 issue of the Daimler Bulletin, the works' magazine, displayed the following notice, "Mr. F.W.Lanchester has been appointed Consulting Engineer to the Daimler Motor Company and during the course of the next few months he will commence to actively interest himself with the many engineering matters which the Company has under consideration."

Compared to the Lanchester Motor Company, where he continued to work as before, the Daimler Motor Company was massive. As well as motor cars, the firm constructed commercial vehicles, agricultural motors, omnibuses, and many more engineering products. The scope for Fred's range of abilities was therefore much larger, and Fred wryly admitted that his appointment was "a far more important position than would have ever been open to me at 'The Lanchester', and a trifle more remunerative".

The Napier motor company had earlier been urging Fred to work for them as a consultant, in order to find a cure for the uncontrollable vibrations which occurred within their six-cylinder engines. They had been struggling to combat the scourge of the inherent problems associated with six-cylinder engines without success for some years; Napier's S.F.Edge had managed to successfully market their engine despite its vibration, by his brilliant ploy of terming the resulting noise 'the power rattle'. Due to heavy demands on Fred's time, however, he had been unable to assist them in what was a major concern not only for Napier but for many other companies also. Even though, in 1904, Napier had been the most famous firm to produce six-cylinder engines, it was not the first. Spyker, Brooke, Sunbeam, Maudsley, Ariel, Wilson-Pilcher, and Durkopp had produced them earlier, with varying degrees of success. Daimler's first attempt at producing a six-cylinder engine, the 56 h.p., came in 1909 and resulted in a torsional vibration so great that it frequently gave rise to crankshaft fractures.

These were caused because the impulses of the firing strokes tended to make the crankshaft oscillate about its axis during rotation. When these impulses coincided with its natural frequency of vibration, the result was serious and "quite audible harmonics" occurred. These Fred poetically described as "changes of voice at certain speeds". Fred's experiments showed that a crankshaft could actually fracture after only two hours' work when subjected to its critical speed of torsional vibration. After much experimentation he overcame this by his invention of a torsional vibration damper, whose objective was "to destroy or diminish the synchronous vibrations in which the elastic portion of the crankshaft plays an important part, and so to render an engine of light weight and comparatively non-rigid construction free from vibration periods".

The housing for the torsional damper, which was also utilised to drive the pulley for the cooling fan, consisted of a small flywheel which rotated freely, driven via the engine crankshaft through a multi-disc clutch arrangement composed of thin steel plates, all lubricated by oil. Although it offered no resistance to crankshaft rotation, the number of plates and their corresponding oil-surface area formed the resistance necessary to eliminate any irregular oscillations of the crankshaft.

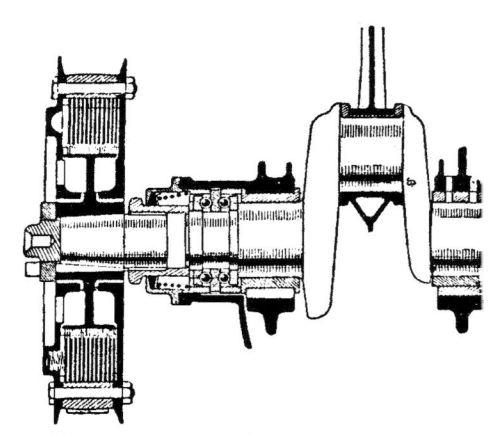

The Lanchester Crankshaft Vibration Damper.

The Daimler Motor Company must have been overjoyed by Fred's speedy response to their problem, as not only was their six-cylinder engine already designed, but it had rashly been put into production and installed in their cars. These were, however, totally unsaleable as, besides the major risk of crankshaft failure, the vibrations felt were "totally intolerable to the occupants of the car". The Company eagerly incorporated Lanchester's damper into these cars and Fred wrote that, "The application of this invention immediately released something like £20,000 worth of frozen assets."

He patented his invention, along with nine others, in 1910, and it proved an immediate and unbounded success. According to the acclaimed engineer Harry R. Ricardo, "Within a very few years every engine with six cylinders or more, and many with less than that, were fitted with Lanchester Dampers as a matter of course." In fact, Fred's device was used on engines of four, six, eight and also twelve cylinders.

Receipts from patent royalties were "very substantial" in Great Britain and many countries throughout the world. The American rights to Fred's patent were sold, via a patent agent, to the Warner Company during the First World War for the sum of £4,000, although this again demonstrated a lamentable lack of financial acumen on Fred's part, as he later learned that the purchaser would gladly have paid £20,000 for this re-named Warner Crankshaft Damper.

However, the device greatly furthered Fred's reputation, as scores of companies as diverse as Humber, Bentley, Gardner, Willys Knight and Packard were to prosper from its use over the years. Even the British Royal Navy used it on their 750 b.h.p. diesel engines, although on a somewhat enlarged scale!

Yet another company, Rolls-Royce, had been plagued by torsional vibration from January 1906, when their new 30 h.p. six-cylinder engine had the habit of breaking crankshafts. To preserve the firm's reputation, an immediate lightening of the engine's internal components was undertaken to avoid the damaging synchronising of the engine's critical speed with its usual revolution range, and this measure proved effective.

Rolls-Royce's next model, the acclaimed 40/50, saw its debut in November 1906. Over 600 were sold prior to the decision to install a torsional vibration damper to the forward end of the engine's crankshaft in 1911.

One historian, C.W.Morton, wrote in the 1960s that it was Henry Royce who had actually made "the first-ever successful slipper-flywheel vibration damper", not Fred Lanchester. To reach this conclusion, he referred to a mock-up wooden distance piece which was added to a 30 h.p. crankshaft's front flywheel to try to eliminate timing-gear rattle. It was successful until produced in metal, when subsequent investigations of the original "revealed some evidence of charring at the interface between wood and metal, and realising what had been happening, he (Royce) hit upon the principle of the 'slipper-flywheel' damper". This statement of how Henry Royce beat Fred Lanchester to the concept of torsional vibration damping "has been verified", Morton wrote.

However, he also advanced a second explanation, which told of one engine ".....being noticeably quieter than the others of its batch and stripping and inspection revealing an apparently badly fitted key and signs of movement of the flywheel on the shaft, the securing nut being partially tight only." He conflictingly added, "There is documentary evidence of this."

Both these theories have subsequently been echoed many times by various authors, but were not supported by Rolls-Royce expert, Anthony Bird, who commented, "Without further evidence, it is impossible to reconcile the details ...."

Jonathan Harley, a world authority on Rolls-Royce cars of this era, wrote, "It seems unlikely to me that Royce would have produced such a poor idea, as described. It is just not compatible with the way he worked and thought." He also discounted another commonly held theory that a torsional damper was installed by Henry Royce on the magneto drive for the early 40/50, which again would have proved that he had beaten Fred to the idea. He wrote: "The device fitted was designed to reduce 'chattering' of the gears in the drive train caused by the intermittent pulsing of the magneto through its cycle, not torsional vibration. It would be absurd to suggest that Royce anticipated Lanchester's torsional vibration damper by the use of this device."

If indeed Royce had invented a crankshaft vibration damper for the 30 h.p. or the early 40/50, it would surely have been a monumental error of judgement not to have patented the device and accepted royalties from the host of motor companies, including Daimler, struggling to find a cure to this problem. The benefits from marketing the damper would have been enormous and it seems incredible that this golden opportunity would not have been exploited. Although all historical research is open to conjecture, analysis supports Jonathan Harley's statement that "of course Royce did not invent the slipper damper before Lanchester". The first damper fitted to a Rolls-Royce was on 27th February 1911 and the very next day Henry Royce and Rolls-Royce Limited applied for their own patent for this 'Means of Damping the Vibrations of the Crankshafts of Fluid Pressure Engines'. This encompassed all the benefits of Fred's patent but was different enough to qualify for patent acceptance in its own right.

The production of Rolls-Royce 40/50 cars with torsional vibration dampers brought a legal challenge from Fred. An internal Rolls-Royce memo of July 1913 was headed: "Re Lanchester. Infringement of Slipper Drive Patents." No facts remain to show the end result of this dispute, but it is worth considering that Fred Lanchester's patent was marketed throughout the world and was never challenged by Rolls-Royce.

Over the decades, the Lanchester Crankshaft Vibration Damper evolved to take many forms, the most common being thin discs of hard-friction material, called Halo, forming the linings of the 'clutch', some lubricated with oil and some remaining dry. More modern dampers incorporated the use of rubber, but George pointed out that, as this is technically a viscous fluid, it was still covered by Lanchester's original patent.

Daimler's next challenge for Fred was the sleeve-valve engine. An American, Charles Knight, had patented his sleeve-valve type of engine in 1905. Although still working on the four-cycle principle, the normal poppet-valves were dispensed with and concentric vertical piston sleeves, which reciprocated independently, had ports machined into the upper ends of the cylinders to allow for the flow of inlet and exhaust gases. The conventional piston secured to a connecting rod provided the compression, working inside the inner sleeve.

The Daimler Company had completely abandoned its own poppet-valve engine in favour of this Knight sleeve-valve engine in 1909. As Harry R. Ricardo wrote, "The Daimler Company had pinned their faith, their fortunes, and a vast amount of propaganda and publicity to the Knight double sleeve-valve engine and Lanchester had the unenviable task of trying to make this rather unsatisfactory engine do what the Company and its publicity agents claimed for it."

Fred certainly did not approve of the basic concepts of the Knight engine, but in line with his position as Consultant Engineer, he immediately set to work. Fred himself commented, "So far as the Knight engine was concerned, I had a perpetual task in endeavouring to solve problems of a technical character, in order to maintain some semblance of advantage over the poppet-valve type." He succeeded very well, as the Daimler sleeve-valve engine became a great commercial triumph. Although often associated with smoking exhausts, its refinement and silence proved extremely popular and firmly installed Daimlers as the acknowledged Royal Car. (The sleeve-valve engine was used on Daimler and some B.S.A. cars for the next twenty-three years.)

Over his next twenty years of employment with The Daimler Company, Fred was to work on many and varied projects. In February 1909, the Daimler Company designed a new Omnibus, with the accepted standard form of thirty-four seats. Although its weight had already been reduced from 5 tons to 4.5 tons, the Metropolitan Police, who controlled the construction details of London omnibuses, unexpectedly announced that the maximum unladen weight from then on would be restricted to 3.5 tons, thus making the Daimler Omnibus immediately obsolete for use in London. Obviously anxious to obtain the lucrative omnibus contract, the Daimler Company asked Fred to draw up a design for them. Fred revelled in the challenge, and became so interested in the project that within a few days he had sketched a completely new layout, and within the month the construction and machine drawings were in the workshops. To design a standard omnibus in so short a time was in itself an extraordinary achievement, let alone one so radical in its approach. Its unladen weight was within the new requirements, made possible by employing an "all-metal body with separate units attached to it".

This new public service vehicle was called the K.P.L. Omnibus; the initials standing for Knight (as two sleeve-valve, four-cylinder engines were deployed), Pieper (whose patented electric-auxiliary power unit was adopted), and Lanchester. It is odd that the Daimler Company was not highlighted in the name of the new omnibus, and that the 'L' came last, for the project was very clearly another innovative Lanchester design. Years in advance of his time, Fred constructed the chassis and body as one unit (now termed monocoque construction). All the body sections were in steel and were clothed in aluminium panelling. For the first time in omnibus construction, all the seats faced forward.

The K.P.L. petrol-electric omnibus.

Each of the two sleeve-valve engines was directly coupled to a dynamo-motor with magnetic clutch and brake mechanism. Of great benefit was the fact that no gearbox was necessary. Depending on circumstances, these 'dynamotors' acted either as dynamos to recharge the batteries, or as motors to assist the engines in moving the vehicle.

The whole power assembly was secured centrally along the wheel base below the chassis achieving a low centre of gravity. One unit was positioned on the outer edge of each side of the omnibus and in each case transmitted the power through its own propeller shaft via a Lanchester underhung worm drive. Differential gears were not needed, as no transverse drive shafts linked the two systems. The worm drive unit and twin rear wheels were suspended on semi-elliptic springs. For good measure, Lanchester designed and employed a four-wheel braking system, a feature not even in production on private cars (Austin being the first to do this, in 1913).

Twelve of these K.P.L. petrol-electric omnibuses were produced and sold and the vehicles seemed set to become a good commercial proposition. However, the project was abruptly cancelled in order to concentrate on a more lucrative project, supplying thousands of engines for a conventional omnibus design of the London General Omnibus Company. Fred's frustration at this belated decision must have been immense after all the effort he had put into the project.

The merits of his design must have been recognised, however, for in the following year Fred was asked to lay out a mechanical design for the construction of a 48ft.-long railcar unit. As before, two separate power trains were employed, each using direct-coupled dynamo-motors and worm gearing, this time connected to four-wheel bogies. Although the design work would have been done in a reasonably short time, it seems there was little urgency by Daimler to complete the project, as the first trials did not take place until 1914. Immediately curtailed by the First World War, the trials were re-opened in 1919. The railcar, with its two 105 h.p. Daimler sleeve-valve engines, caused a stir by recording a faster return time from Coventry to London than the London North West Region express service. Although encouraging the scheme initially, L.N.W.R. began to lose interest in the project, and it was abandoned in 1921. The mechanical features initially demanded of the project had been met, and Fred was again left to ponder on much wasted time and effort. He insisted that all that was needed to make a commercial success of the railcar was the substitution of diesel engines in place of the Daimler sleeve-valve petrol ones, but accepted the situation with the wry comment, "a successful but commercially unpromising mechanical frolic".

Many other grand schemes were put to him for development work, such as the infamous Renard Roadtrain, which he described as ".....invaluable in one or two places on God's earth, but that is all". Similarly, the non-use of his designs for seven-speed epicyclic gearboxes and other such endeavours, must have been a great drain on Fred's enthusiasm. As Harry R.Ricardo wrote, "These must have been years of frustration for Lanchester, for his time and his great talents were devoted either to flogging what he regarded as a dead horse, such as the Knight engine, or to the development of schemes for which he foresaw but little future."

Although Fred regretted Daimler's lack of vision with regard to his development projects, which he felt was due to apathy on the part of those commercially responsible, he nevertheless enjoyed his time with the firm. There would have been no way that the Lanchester Motor Company could have entered into such grandiose schemes as those instigated by the Daimler Company, and Fred's agile mind obviously revelled in his association with them.

This was to remain the case when, in October 1910, the Birmingham Small Arms Company took control of the Daimler Motor Company. Fred was appointed Consulting Engineer and Technical Adviser to the whole group of B.S.A.-owned companies, although his office and most of his work involvement was still at the Daimler headquarters in Coventry.

Since the publication of Fred's two books on Aerodynamics and Aerodonetics, his reputation had spread; he followed the developments in flight throughout the world, and gave lectures on his aeronautical theories. Indeed, he had already received a Bronze Medal award for his paper to the Aeronautical Society in 1908, and in April 1909 he became a founder-member of the British Government's Advisory Committee for Aeronautics. Despite his inventive achievements, Fred was not well-known to all, as he recalled: "One of the daily papers published my brother Frank's portrait instead of mine and did not seem to be inclined to insert a correction. We talked the matter over and came to the conclusion that neither of us had cause for complaint; my brother got the credit for my knowledge of aeronautics and I gained the reputation of being a much better looking fellow than I am!"

Fred's aeronautical work was not limited to theory. From 1909 through to 1911, he held the post of Aeronautical Consultant to the small pioneer firm of White and Thompson. Fred's initial design for the company was a single-seater aircraft. This had to be altered at the last minute to conform to a directive from the War Office, stating that it was interested in two-seater planes only. Although strongly against the change in design, Fred did make alterations to accommodate the extra seat and equipment. Due to the additional weight involved, the design had to be modified to take two engines, which he inter-connected by a crossed belt in case of failure of one. Much of this work was done at the Daimler Company's works in Coventry, with Fred overseeing the engines' construction and conducting the acceptance test there himself.

In typical Lanchester style, the bi-plane's wings were clad in aluminium rather than fabric and were connected to each other in triangular form by tubular steel struts. Many other novel features were also employed. Before the flight, a storm ruined their chosen stretch of beach, leaving boulders and debris strewn on the sand. Tragically, the plane crashed on its take-off run, with the firm's pilot "running it onto one of the patches of exposed rock". Due to the subsequent undercarriage failure the plane turned upside down and was badly, but not irretrievably, damaged. However, White's financial backing was exhausted and Thompson had other distractions. As John Milligan, Fred's right-hand man at Daimler, later explained: "Thompson became far more intent on the wife of a famous actor and after the smash, which could have been rectified, he abandoned the whole project."

In 1910, Fred was elected a Member of the Institution of Civil Engineers, and also President of The Institution of Automobile Engineers, both honours being richly deserved. Each year he had given the I.A.E. a comprehensive lecture packed with deep knowledge and controversial statements, all laced with his own brand of dry humour. The resultant discussions on his papers showed the high regard in which he was held by his contemporaries, although the debates were often heated. This year of his Presidency was dominated by his forcible arguments against the new R.A.C. horsepower formula for cars. It was Fred's great friend, Sir Dugald Clerk, who was the main instigator of the new formula, but Fred felt that it was not based on a scientific foundation and should be rejected.

Fred had first presented a paper to the I.A.E. on the relationship between the bore, stroke and weight of an engine compared to its horsepower, in 1906. He had analysed the effects of the bore/stroke ratio when comparing engine performances, and had produced his own horsepower formula. F.R.B.King, who made a study of engine output, wrote in the 1980s, "The objective of Lanchesters work was rather limited in its scope but the work itself was extremely wide in its implications." These implications were obviously not taken into account by others, as F.R.B.King added, "..... but what Lanchester failed to appreciate, was the inability or unwillingness of his listeners to follow his reasoning."

Basically, Lanchester wrote that engine power was proportional to "the square of the bore multiplied by the square root of the stroke/bore ratio". This was contrary to the formula applied by the R.A.C., which concluded that power was related solely to the square of the cylinder bore area, taking no account of the length of the stroke of the piston.

The dispute intensified when the Government used the R.A.C. formula as the basis for its new taxation levied on the purchase of motor cars. British car manufacturers therefore became inclined towards engines of long stroke proportions rather than large bore size, in order to help their customers escape from increased taxation on purchase.

All this had a direct bearing on Fred Lanchester, who deplored the fact that car designs would be effectively bound by conditions which he felt were inherently incorrect. His letter to the R.A.C. protested against their lack of foresight (which was to determine engine design for over three decades). Referring to their formula, Fred stated: "..... it would serve if all engines were of the same D/S (piston diameter/stroke) proportion, but is manifestly absurd as an attempt to properly represent the horsepower. This fact is clearly recognised in a circular issued recently to the trade, suggesting that manufacturers should adopt the R.A.C. rating for purposes of catalogue description. It is ingeniously stated in this circular that the formula is not intended to be a scientific statement of horsepower. Having gone wide of the mark, it is most convenient to explain that the shot was not aimed at anything in particular."

At the 1910 I.A.E. meeting, Fred gave the 'Presiding Chair' to a deputy, so that he would not be restricted whilst venting his disbelief and anger. In response to the I.A.E. testing of a hundred various engines, Fred mocked, "I have repeatedly expressed the opinion that little is to be gained from the statistical examination of quantities of motors. The present report is a monument to the futility of any such course, though the fact remains, not without its humorous side, that the committee themselves appear to be blandly unconscious of that fact."

The 1911 I.A.E. meeting saw the continuing debate coming to a head, when the famed Laurence Pomeroy read his paper entitled 'Engine Design For Taking Advantage Of The H.P. Rating Rules'. In it he declared that the permissible speed for an engine of large bore was the same as for one with a small bore. It seems that his conclusion was reached, not through scientific study, but by referring only to car engines seen at a recent speed event which he had attended. Unable to bear such an imprecise statement, Fred contradicted this, ending with, "..... the fallacy of the author's assumption is therefore manifest."

Fully aware that his paper would have provoked Fred, Laurence Pomeroy answered, "I

A 20 h.p. tourer with canopy top, on chassis no. 578, with Veteran Car Club registration of 1904. (See appendix.)

A 1907 20 h.p. Lanchester landaulette with double passenger comfort of front windscreen and folding dashboard, set behind the updated sloping radiator. (See appendix.)

Sunshine but no draughts on this 1908 20 h.p. landaulette, chassis no. 771, used regularly by the same lady owner and her chauffeur until 1939, when it was involved in an accident. The lady bought a new Austin and transferred her beloved Lanchester coachwork to it. Francis Hutton-Stott purchased the chassis and mechanics. Four years later, he was contacted by the chauffeur after the lady's death. He duly collected the car body and the two halves were re-united. (See appendix.)

Lanchester landaulette coachwork on a 1910 28 h.p. chassis, no. 1013. (See appendix.)

1910 28 h.p. chassis no. 1047, with Lanchester limousine coachwork, originally owned by the Lanchester Motor Company director, Joseph Taylor. (See appendix.)

The Maharajah of Rewa's 1912 38 h.p. detachable-roof limousine, on chassis no. 1154. (See appendix.)

The Maharajah of Rewa's 1912 38 h.p. car, unveiled as a luxurious touring car, for which a folding hood was provided. (See appendix.)

A 1912 38 h.p. limousine, chassis no. 1367. This car has been in the safe hands of the National Museum of Science and Industry for some 60 years. (See appendix.)

Fred Lanchester at work.

am sorry Mr. Lanchester is such a busy man, for I would presume a certain part of the paper missed him. I will just read a portion of it again, because I knew Mr. Lanchester would come along with exactly the words he has done, so I am not disappointed ..... The fallacy underlying Mr. Lanchester's theory of inertia limitation of piston speed is that it does not apply to engines at the sizes used in motor cars. When the speeds of such engines are two or three times as great as they are now, or when the cylinders are two or three times as large as the largest now used, it will be time to apologise to the Institution for my impertinence in demonstrating the fallacy of his doctrine I have referred to."

Pomeroy's clever debating reply won the day in the eyes of most people. However, the researcher, F.R.B.King, summed up the facts thus: "Pomeroy's conclusions, of course, follow logically from his argument, but it can be shown that this was based on a number of fallacies ..... The case of the 1914 Grand Prix Vauxhalls (which Pomeroy designed) is no exception. Following catastrophic engine failures in the T.T., it was found necessary, amongst other things, to re-design and strengthen the connecting rods on a rush basis. Pomeroy was thus hoist with his own petard, so demonstrating fairly convincingly who really won the battle which started with the skirmish not long previously at the I.A.E. meeting." F.R.B.King also added that, "Lanchester's theories, far from being worn out and only fit for the history books, are in fact in advance of most modern engineers' and scientists' ideas, still."

Although Fred was greatly irritated by Pomeroy's arguments, he did not bear a grudge, and indeed he later recommended Pomeroy for a high position within the Daimler Motor Company. The implications regarding the R.A.C. horsepower formula and its corresponding basis for taxation were ironically felt more by the Lanchester Company than their competitors, as they were to have a great effect on their current models, which had superseded their twin-cylinder cars.

Chapter Eleven

THE FOUR-CYLINDER 20 H.P.

Even though the Lanchester Engine Company had increased its paid-up capital by £12,070, in July 1903, to £57,595, the finance required to design and develop the proposed new 20 h.p. model was a major factor to be taken into account by the cash-starved Company. The original Board of Directors, consisting of the Chairman, Charles Pugh, with Hamilton Barnsley, John Pugh, Joseph Taylor and James Whitfield, showed insufficient confidence in their Company and decided to bring in outside finance rather than invest further themselves. The authorised capital was increased to £100,000, and in December of that year a prospectus was issued inviting subscriptions for the further 42,405 £1 shares.

However, this was a period when many motor companies were in financial trouble. There was not much confidence in the motor industry amongst investors and no support for The Lanchester Engine Company was forthcoming. The issue failed to materialise and in late March 1904, the Company, having passed the point of no return, had no alternative but to declare bankruptcy. It was a tragic disappointment to Fred and his brothers, and was yet another cause of dispute between them and the Board. The Receiver appointed, Arthur Gibson, was in full charge of the Company, but declined to make any sweeping changes as, much to the embarrassment of the Directors, the business continued to show a profit.

During the Receivership, a further confrontation occurred when Fred was not asked to help conduct the annual stock-take as he had always done in the past, the Directors and the Works Manager undertaking the task themselves. Fred was later asked to sign for the valuation as if he had attended, which he politely refused to do. The Chairman's consequent rage prompted Fred to sadly remark, "From that time onward, I realised that I should not be able to continue to serve the Company on the old footing." The fact that the Receiver agreed with Fred, merely inflamed the situation, and another nail in the coffin for compromise was hit home.

No doubt Arthur Gibson was pleasantly surprised by the fair condition of the Company, which had good products, loyal customers and great potential. New and second-hand Lanchesters were being sold by the London agent, Lacre Motor Car Company Limited, from their offices in Oxford Street, and also by agents in Manchester and Glasgow. Sales were good, with exports to India, South Africa and many other countries around the world.

Indeed, a review of the British automobile industry in 1904 listed about twenty companies producing cars, and made comparisons of the top six:

Estimated value of cars produced per week

	1900-1	1903–4
Napier	£600	£6,500
Daimler	£1,400	£4,500
Lanchester	none	£3,000
Wolseley	£500	£2,500
M.M.C.	£1,000	£2,000
Ariel	£150	£1,000

The overall 400% increase in total British car sales related strongly to the position of the Lanchester Company, who were also mentioned in the list of firms who "have done considerable work in the manufacture of engines for petrol launches".

This independent report surely made a mockery of the lack of confidence shown by the Board of Directors in their own Company. On December 19th 1904, the firm was reconstructed under the name of The Lanchester Motor Company Limited. The former Lanchester Engine Company had been in debt to the tune of £20,000 and, in view of the bankruptcy, the creditors accepted preference shares instead of cash. The former Company's Directors now did what Fred felt they should have done a year earlier, and took up additional shares themselves.

Fred had invested everything in the old Company and, sadly, was now unable to subscribe further in the new. The old shares were cut down to half their value, two old £1 shares being replaced by a £1 share in the new Company. Some felt the Chairman's subsequent action was harsh, as he demanded that Fred surrender two thousand of his shares, to be distributed amongst those who had found the extra money. Furthermore, it was recorded that Fred, with typical generosity, gave one-tenth of his shares to George and Frank. Evidently, Fred was reduced to being a relatively poor man.

In a seeming act of spite by the Board, he was informed that he was to be relieved of his position of General Manager of the new Company. It appeared that Fred was to be made the scapegoat, although it will be recalled that on the formation of the Lanchester Engine Company, he had deliberately passed over the offer of Managing Director so that the financial aspects of the Company could be looked after by the experts in that field. Fred's response to the Chairman was simple, and showed a greater loyalty and conviction than that of his Directors. "I said that whatever conditions were necessary to the continuation of the business, I would accept." In his personal notes he wrote, "I was not prepared to haggle with a man whom I had begun to hold in contempt." It was the Chairman who had informed Fred that his position was terminated and that Fred would serve merely as Designer and Technical Consultant under the Managing Director, Hamilton Barnsley, with a corresponding reduction in pay from £350 to £250 per annum. In Fred's acceptance of this sad situation, he made the stipulation that, "..... it is understood that if in the future the Company is so successful as to demonstrate that the sacrifice on my part was not justified, the shares surrendered or their equivalent should be returned." (Eight years later, with the success of the Company assured, Fred was to remind the Chairman of this and requested that the stipulation be honoured. In Fred's own words, "....with something of a sneer, he replied 'Well, you didn't expect it did you?' I replied 'I cannot say that I did, but I thought it fair to give you the opportunity'; perhaps one of the bitterest things I have ever said.")

Arthur Gibson was appointed to the Board of Directors and it was at his instigation that certain concessions to Fred were made. These were that "...... the clause in the agreement making over any further inventions or improvements to the Company should be abrogated, and that nothing in the new agreement should be deemed to affect my right to associate myself as consultant or otherwise with other firms."

Arthur Gibson admitted to Fred that, "The Receivership had been a great blunder and the Chairman and Directors had to save face, otherwise they would have been the laughing stock of Birmingham for not supporting a prosperous business." By retiring Fred from his position, they were able to keep up appearances.

Once Fred had got over the shock, the change in his management position actually turned out to be advantageous for him in many ways. Whereas he had earlier worked an average of seventy hours per week, he now found that he was called upon to undertake only one quarter of that. His reduced salary, therefore, actually turned out by the hour to be a wage increase. Although bitterly disappointed at his treatment by the Board, Fred Lanchester found it had actually been a blessing in disguise. His home for the past five years, 53 Hagley Road, saw more of him, and he was able to devote more time to his work on aerodynamics and aviation, which he had been forced to neglect during this time. One of his 1905 patents related to this, entitled 'Aerodromes', and his agile mind still left time for five other patented inventions on different topics, to add to the previous year's tally of sixteen.

To Frank Lanchester, also, the loss of the Company must have been a great blow. Apart from seeing the end of an era, he also lost his natural position as Company Secretary, the responsibilities being taken over by Noel Thornley. Frank did, however, continue in his previous post as Chief of Sales.

George retained his position, and was essential in implementing Fred's design concepts at the Works when Fred was absent.

Max Lawrence lost his position as Works Manager to W.F.Rainforth, and sadly Archie Millership was also forced to leave, taking employment at the Wolseley Tool and Motor Company.

The senior staff rest-room with, from right to left, Fred Lanchester, seated, Sidney Pinsent (Chief Draughtsman), W.F.Rainforth (new Works Manager) and George Lanchester.

Despite the traumas of the Lanchester Engine Company under Receivership, Fred had continued with the design of his new car, the 20 h.p. Although the model was, in effect, in direct competition with his current 18 h.p. twin-cylinder car, it heralded the new era of multi-cylinder engines for the Company and boasted many outstanding features.

The Lanchester Motor Company's first advertisement for the 20 h.p. was published in January 1905, and the car was shown to the public at the Olympia Motor Exhibition the following month. Strangely, the car was a pre-production model produced in 1904, of which George later wrote, "There were very few of these experimental types made." It was displayed on the Lacre Motor Company's stand, in their capacity as Lanchester's London agents. Presumably due to lack of time at the Lanchester works, Lacre themselves had been entrusted with building the coachwork, probably to the Lanchester Company's general design. The result was a refined town car, featuring a collapsible forward section of the passenger compartment and a landaulette roof at the rear, thus allowing for conversion to a completely open car. The front compartment was not fitted with side doors and this, together with a lack of lamps and horn, gave the car a stark appearance for its publicity photographs.

By the following month's Agricultural Hall Show in London, however, Lanchesters had produced their own landaulette coachwork on the production 20 h.p. chassis and the whole

THE FOUR-CYLINDER 20 H.P.

The 20 h.p. Lanchester in February 1905, with Lacre coachwork and pre-production front axle.

A pre-production 20 h.p. showing the four-cylinder engine located between the seats and the two-lever gear selection as used on the twin-cylinder models.

was displayed on their own stand. It created an immense amount of interest, with The Autocar stating, "As great an attraction as it was at Olympia." The body was similar to the earlier Lacre construction, but Lanchester's aluminium panels were shaped to produce a more balanced appearance, with side panels to the seats and front doors. (This coachwork design was so popular that it was retained for many years.)

The mechanical detail on this production car incorporated a variety of novel features, which resulted from a combination of past experience and forward thinking. For a number of years, designers had known that the engine power produced could be increased by a corresponding rise in the engine's piston speed, such that twice the normal piston speed would result in the engine developing twice the horsepower. It also became clear that a weight saving could be gained by decreasing the size of each piston cylinder and increasing their number accordingly. These two principles, coupled with the fact that the greater the number of cylinders the greater the advantage gained by the corresponding reduction in the engine's flywheel weight, meant that the multi-cylinder engine design was undergoing continual development by many motor companies.

Fred would have inspected the 8 h.p. four-cylinder in-line engine in the 1896 Panhard et Levassor, the four-cylinder 'V' engine by Mors of 1897, and experiments from other companies, but it remained a fact that practically every engine up to the turn of the century, whether British or foreign, was of the one- or two-cylinder design. From then on, the feasibility of producing multi-cylinder engines became more realistic, and engines of three and four cylinders gradually began to supersede single- and twin-cylinder engines of a greater cylinder size. One major problem for all companies to overcome was the manufacture of crankshafts of such an increased length, and their not uncommon fractures frustrated the ambitions of designers for many years. Only when the technical developments had caught up with the theoretical aspirations did Fred Lanchester consider employing the multi-cylinder engine design himself.

An off-side view of the pre-production 20 h.p. engine, no. 501, showing the box-shaped inlet manifold and long brass slide-valve throttle control immediately below. (See appendix.)

Near-side view of pre-production engine no. 501, set in the later chassis, no. 727. (See appendix.)

His 20 h.p. vertical in-line engine was of the 'over-square' design, with a bore size of 4ins. diameter and a stroke of 3ins., each cylinder being independent and separate. George himself wrote of this high-revving engine, "The capacity of the four-cylinder engine was 2.6 litres and taking the normal factors of that period - 6 b.h.p. per litre per 1,000 r.p.m. - gave 15.6 b.h.p. at 1,000 r.p.m. and had a straight line increase to 20 b.h.p. at 1,500 r.p.m. Allowing for a reduction due to breathing efficiency, this resulted in 25 b.h.p. at 2,000 r.p.m. and at 2,200 r.p.m. the output was 28 b.h.p.; one could safely exceed 2,000 r.p.m.

THE FOUR-CYLINDER 20 H.P.

as piston speed then was only 1,000 feet per minute." The engine's performance was outstanding, giving an output which other makes would only have expected from engines of over 4 litres capacity.

The special light-weight pistons which Fred employed in his new engine were made from high-grade steel stampings, rather than the normally-used cast iron, in order to lessen the weight of reciprocating parts, to reduce vibration, and to eliminate the risk of the pistons breaking up. They had four compression rings at their top and one oil-scraper ring at their base.

The piston to crankshaft connecting rods were precision-made from stampings in high-tensile nickel-steel. Each was drilled inside its length to provide a route for the pressure-fed oil supply to the little end as well as, importantly, to save weight. The connecting rods were accurately machined externally to provide perfect balance and again to reduce weight.

To avoid any dangers of irregular heat treatment processes, the crankshaft was machined from a solid billet of high quality, 3% carbon-steel. Very strong in its proportions, it was machined to the highest of standards in order to be free from balancing problems. The relatively small 'throw' of the crankshaft, plus the employment of five white-metal main bearings, was part of Fred's overall concept to reduce torsional vibration, as was his innovative idea of positioning the heavy flywheel at the forward end of the crankshaft while retaining the countershaft components at the rear.

Off-side view of an early production engine and countershaft, showing the side-lever gear change, linked throttle levers, and front flywheel housing the low-tension ignition gear.

As on the earlier twin-cylinder cars, the 20 h.p. flywheel incorporated the low-tension ignition system, where the magnets were mounted in the flywheel and rotated with it, round a fixed armature, although on the 20 h.p. the system was re-designed and housed inside a much smaller-diameter flywheel. The armature differed from the twin-cylinder type in that it had a special winding to give a strong spark for starting at very low rotational speeds, and was called the starting coil. After the engine was running, the coil was switched out of action by the ignition lever, in favour of the running coil, although the former could be used at 'dead slow' speeds. The sparking plugs, formerly known as igniters, were identical to those used on the twin-cylinder engines but, owing to the faster-running speed of the 20 h.p. engine, the insulation required cleaning every few hundred miles. This apart, the ignition system proved itself extremely reliable and free from maintenance, save for an occasional 're-flashing' of the magnets to ensure their effectiveness. This was achieved by applying a D.C. current of between 20 and 110 volts, using the armature windings to energise the magnets for fifteen or so contacts, each of one-second duration. However, from March 1905, the improving quality of high-tension magnetos allowed customers the option of either system.

Two camshafts were employed, one on either side of the engine, and from these, via vertical rocking levers, the overhead valves were actuated. These were arranged in oil reservoirs, which Fred called cam-boxes, in order that the cams' and the levers' points of contact remained immersed in oil. In order to keep the width of the engine to a minimum and also to reduce the weight of moving parts, flat leaf-valve springs were employed instead of coil springs, and they proved to be very quiet in operation, strong, reliable and not so readily affected by heat. To ensure the long life of Fred's separate inlet and exhaust horizontally opposed valves, they opened within an ante-chamber above the combustion chamber, an idea used much earlier by Fred on his gas engines. The valve layout proved highly efficient and, besides helping to achieve a good fuel economy, increased the car's performance. (The idea was used from 1911 on the Delage racing cars.) Fred provided a ready means of adjustment for the valves and, indeed, no special tools were needed for their removal.

The valve areas were adequately cooled by the water-cooling system, which did not incorporate a water pump but instead used the thermo-syphonic system. This worked on the principle of hot water rising, so ensuring continuous circulation of the cooling water. Big-bore pipes were used in order to offer as little resistance as possible.

An early 20 h.p. with high radiator and pre-production axle.

The radiator was constructed of brass parallel circulation chambers of more than ample cross-section. It was termed 'the Pugh-Lanchester', and was constructed according to their earlier patent of 1903. Situated behind the radiator, which was set high up at the front of the car, was a single, eight-bladed fan, driven by a belt.

Setting a precedent for modern automotive practice, Fred designed a pressure oil-lubrication system operating at a pressure of 40-45 lbs. p.s.i. This advanced feature contrasted with other companies' designs, which employed oil splash-feeds or gravity drip-feeds from an overhead reservoir, supplied by either a small hand pump, a mechanical pump, or by exhaust pressure. The closest competitor to the Lanchester direct pressure-fed system did not register even a quarter of the pressure which Fred obtained. The oil pump was driven by the crankshaft and fed oil to each of the main bearings through copper pipes. The oil was then forced, via oilways in the crankshaft, to feed the big-end journals before being forced up the tubular connecting rods to the little-ends. From there the discharge lubricated the cylinder walls, in which small holes transferred oil to the camshafts. Completing this highly-efficient arrangement were oil supplies for the ignition shafts as well as for the gears. From 1906, for the first time ever in motoring history, pressure-fed lubricating oil was applied directly to the gears, clutch and brake parts. The whole surplus of oil was allowed to drain back into the crank case for further use.

In order to continue his design principles of giving maximum room for the passengers without lengthening the wheelbase, Fred had carefully considered the engine's position within the chassis. The engine, therefore, was of exceptionally narrow design, with the inlet and exhaust manifolds and water circulation pipes arranged above it rather than at its sides. The engine's central location allowed the driver's and front passenger's legs plenty of room on either side of it, and they were protected from the engine heat by a one-piece removable bonnet. (For the production cars this was altered to a three-piece bonnet.) Fred's design was a technical success and provided much better passenger space than other makes of car.

To gain acceptance and recognition for this unique engine, transmission and chassis design, Fred offered an analogy: "It has frequently been stated that the proper position of the engine has been settled once and for all by racing experience. Accepting as a fact that the position of the engine in a modern racing car is satisfactory, it is quite certain that it is the endeavour in every motor-racing design to get the engine position as well back in the car as possible. With the introduction of the side entrance bodywork, a fundamental difficulty has arisen in design. If the racing position of the engine is adopted, and a side entrance rear body is provided, a wheel base of about 12ft. is necessary; this is always inconvenient and sometimes out of the question. If the motor is pushed out forward, it is no longer in its best position, the weight distribution is bad, being insufficient on the driving wheels to give proper adhesion ..... in the new Lanchester engine position, the difficulty is disposed of; .... the engine position shows it to be identically that of the typical racer, it is also seen that full and ample side entrances are provided, and the wheel base, in spite of it being a four-cylinder car, does not exceed 9 feet centres."

The chassis was completely new and rendered non-flexible due to its box tube members

The production 20 h.p. chassis and mechanical layout.

measuring 6ins. by 2ins., pressed from 0.062ins. nickel-steel plate. (In the 1950s, George made the comparison that these "closely resembled some chassis frame members that came into vogue about 1937 and are used in some present day cars".) The longitudinal members of the chassis did not extend over the axles, as was normal practice, but instead connected to the rear axle by means of the cantilever and parallel link suspension. A strong bulk-head spanned the front of the chassis frame, but the main transverse member was in fact the petrol tank, made of heavy-gauge sheet steel and giving great torsional rigidity.

Mounted centrally inside the top of the petrol tank was the surface carburettor arrangement, known as the wick-vaporiser, similar to that used on all previous Lanchester cars. It employed a series of cotton wicks which drew up petrol at their base and gave off petrol vapour at their top, thus avoiding the inefficiency of conventional carburettors which instead permitted droplets of petrol in the induction manifold and cylinder combustion space. Another major benefit of the Lanchester vaporiser was that the wicks would act as intended, regardless of the density of the fuel; a definite advantage in the early days of the century when the quality of fuel was more than questionable. The fact that quantities of dirt, dust and water were likely to be taken into a petrol tank, due to the constant decanting of fuel, was also a problem with contemporary makes. On the Lanchester, only petrol vapour emanated from the wicks, located in their separate tank of about one-gallon capacity. Prior to 1906, this tank, which was situated immediately above the main tank, was replenished by the use of a hand pump, a few strokes being necessary every 10 or 15 miles. Over-pumping could easily occur, but the simple design took any surplus back to the main 17-gallon petrol tank.

The Lanchester machine shop, with 20 h.p. rear axle casings and engine crankcases being attended to.

The mixture-regulator control moved a rotating valve, numbered from 0 to 6, at its base, which simply permitted less or more air to mix with the petrol vapour. When the mixture-regulator pointer was set at 0, the vapour was off, and the engine, when rotated, sucked pure air. When set at 6, the engine sucked saturated air. The richness of the petrol vapour was less necessary as the engine became hotter and the heated air passing through the wicks caused the petrol to vaporise more readily. Under all normal circumstances, the vapour regulator attended to every condition. However, in really cold weather or if the car had not been used for some time, the top of the wicks could be doused with a small quantity of fresh petrol to assist vaporisation, as separation of the petrol's constituent parts might have taken place. (Some eighty years after the last Lanchester wick vaporiser was produced, informed opinion suggests that its development, by using a porous diffusion barrier to

THE FOUR-CYLINDER 20 H.P.

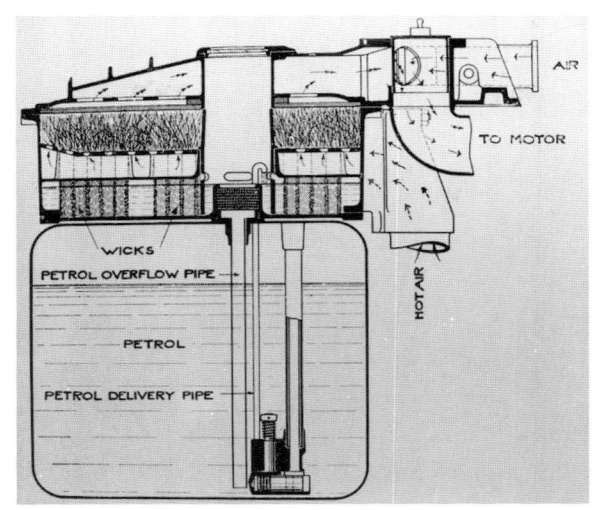

The wick vaporiser and petrol tank of the 20 h.p.

vaporise precise quantities of petrol, is the way forward to reduce exhaust gas pollutants in both petrol and diesel engines.)

One of the very first examples of 'unit construction' was employed, as the counter-shaft was bolted directly to the rear of the engine (a system which did not come into general use for another twenty years or more). The unit was, in effect, a single rigid casing in which the whole of the mechanism was enclosed and, importantly, shielded from dust.

Greatly influenced by the current fashion, the initial design parameters for the 20 h.p., set by the Directors, had been, "It should be a light, speedy, open, touring car, able to do all its normal running on top gear, with a single, emergency, low-speed for quite exceptional hills." Accordingly, the gearbox, which was the first portion of the countershaft assembly, employed just one forward and one reverse epicyclic gear. These were basically the same as those of the 12 h.p. twin-cylinder but of greater size, the 20 h.p. gear drums being 8.25ins. diameter as opposed to 6.125ins.

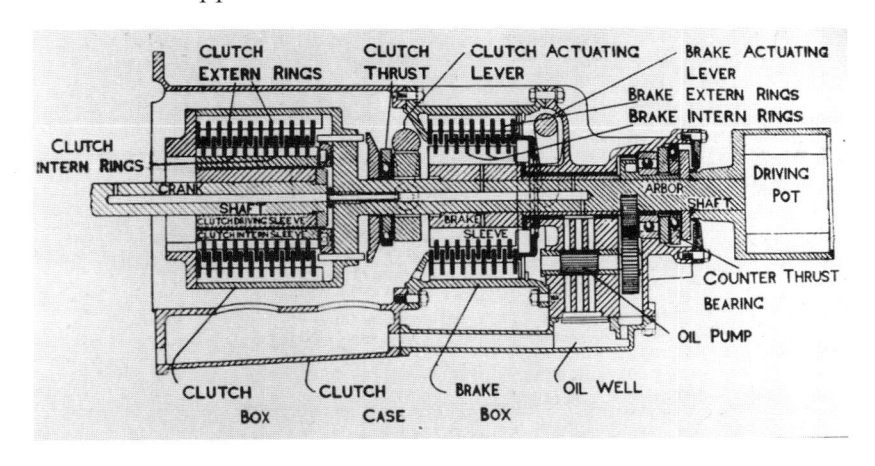

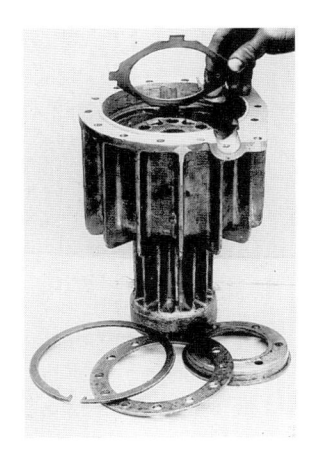

The clutch and brake boxes of the 20 h.p.

An extern disc being inserted in the 20 h.p. brake-box.

The second portion of the countershaft casing was termed the 'clutch-box' and contained the direct-drive clutch for the top speed. Instead of the wedge-shaped, single-cone design of the earlier twin-cylinder direct-drive clutch, Fred designed a multi-disc type, composed of several steel discs called interns and externs, mounted at the end of the crankshaft. As the clutch in its enclosed casing was partly immersed in oil, any which had penetrated between these steel discs had to be squeezed out to allow complete surface contact, thus giving a progressive and smooth clutch action.

The oil bath extended to the third, rear section of the counter-shaft casing, which was termed the brake-box, and was again of the multi-disc type. It acted on the arbor shaft, which transmitted the drive from the crankshaft extension to the propeller shaft. The arrangement of outer discs, called the brake extern rings, were mounted in grooves in the brake-box casting and were unable to revolve, whereas the inner set of discs, called the intern rings, were carried on the revolving arbor shaft. The system proved very effective and reliable.

There had been problems with the front axles of the pre-production cars, as George described: "The original front axle broke down in one of the cars on an early test run with Archie Millership testing, and F.W.L. immediately designed a stiffer one with a 3ins. tube similar to those used on the 10 h.p. and 12 h.p. cars." The car shown in February 1905 at the Olympia Show was thus equipped.

A 20 h.p. with traditional Lanchester tonneau body and a canopy top.

The turning circle of the 20 h.p. was an impressive 33ft., steered by the side-lever which "has thoroughly established itself on our own type of car and is universally popular with our customers". However, as previously, wheel-steering was fitted when specified by the purchaser, at no extra charge. The steering track-rod was positioned forward of the front axle, which used ball bearings to locate the stub axles and the wheel hubs. The design of the wheels, whose tyres measured 875mm. by 105mm., was another area where many people followed the fashion trends and the preference was for those constructed of wood. To meet this demand, wooden wheels were supplied to insistent customers and again no extra charge was made. However, they would have been provided in the most reluctant fashion, as the Lanchester tangentially-spoked wire wheels, "always advocated by Mr. Lanchester", had proved themselves to be of outstanding strength as well as very light. Indeed, the Lanchester brochure simply stated, "wire wheels built on our system never fail".

Confirming this, the Motor Car Journal reported that the owner of a new Lanchester had a disquieting adventure when driving along a country road. He unexpectedly encountered a herd of gypsy ponies and, taking avoiding action, ".... the car took a steep bank which bordered the road, but as the Lanchester is provided with an unusual amount of lateral movement, and is very low and broad as well, it went a long way up the bank before threatening to overtop its centre of gravity .... the car seemed poised in mid-air and just upon the point of toppling over, when the wire wheels sprung back, the body righted itself, and the driver found himself back on the road again, not underneath the car as he expected, but driving normally instead".

Alas, the disc brake design of the current 18 h.p. was proving incompatible with the poor road surfaces, and the 20 h.p. was instead equipped with internally expanding cast-iron brake shoes acting on the rear axle brake drums and operated by a foot pedal.

Although one later report suggested that the pre-production 20 h.p. cars employed the twin-cylinder model back axle, certainly by March 1905 a new one had been designed and was in use. In order to save weight, the rear axle casing was machined from aluminium castings, the joining faces of the two halves being vertical. Plain bearings were tried for both the front and rear axles on pre-production models but, from the beginning of full production, ball and roller bearings were employed.

Tom Birkett Barker's wife and family in their 20 h.p. tourer.

A number of other points were also altered as full production commenced in, it is believed, March 1905. The Pugh name on the radiator was dropped in favour of the sole name of Lanchester. The rubber mountings for the engine and countershaft supports were found to be unnecessary as the engine did not vibrate. Their removal allowed the components to be bolted directly to the chassis frame, where they acted as additional cross-bracing members. However, by far the most important change was the alteration to the chassis length. The reason for this was that the original design concept of "a light 20 h.p. to appeal to the sporty, two-seater customers", as George described it, had subsequently been amended. George continued, "The salesmen very soon reported that there was very little demand, but if it could be made a four-seater they thought it would sell. Of course, when it was made a four-seater, they wanted three or four gears and more power." From the time of full production, therefore, all the 20 h.p. cars included three forward speeds and an increased chassis length of 9ft.5ins., although one note mentioned an early chassis length of 9ft.3ins. Fred also incorporated his highly successful compounding, intermediate gear to the assembly. Even though the new chassis with its heavy formal coachwork had raised the total weight from 16cwt. to 23cwt., the new gear arrangement allowed a good, flexible

20 h.p. engines on test.

performance from the car, and speeds of 10 m.p.h. in first gear, 20 m.p.h. in second and 50 m.p.h. in top were attainable.

To complement this, Fred designed a new gear-change system, which he named the 'single lever control', to replace the pre-production type which had been copied from the twin-cylinder cars. The long lever, whose base was connected to the countershaft, had a sliding and rotating handgrip at its top. This was turned to engage with four notches, to pre-select the three forward gears and one reverse. The actual forwards and backwards movement of this lever engaged the already selected gears, a clever design which ensured that two gears could not be selected at the same time. A Lanchester manual described the typically unorthodox system of operation: "Changing 'up' - The low gear is in. The handle will be found to have a certain amount of rotational backlash, turn as far as it will go, counter clockwise. Try to slide handle upwards, and while thus trying, unnotch gear and take out and put in again quickly; it will be found that the gear has changed. Don't continue to try to turn handle, it will not slide freely while you are twisting it. Don't dally about the neutral position feeling for the gear to change, take it for granted, just out and in again as quick as you like. Changing 'down' - Turn handle as far clockwise as it will go. Press handle downwards, unnotch lever, take out and replace quickly; the gear will be found to have changed."

Once the knack of using this unique system had been gained, gear changing was simplicity itself, and very refined when compared to the 'crash' gearboxes of rival makes. Another advantage was that the car could still be braked either by selecting reverse, or by using the transmission main brake, operated by pulling back on the gear-change lever, or by depressing the foot brake to operate the rear brake drums; the car therefore had three separate braking systems.

One customer later testified: "My experience of the Lanchester dates with The Exhibition of 1905, after carefully looking over most of the makes of that period. The satisfaction and pleasure derived from your cars are greater than many car owners are aware of; one of the numerous advantages is the silence in changing speed. In a Lanchester, one is not aware of the change, except by noticing the varied speed at which the engine is running. Then, again, they are better hung and better sprung, adding to the pleasure of the occupants and, I fancy, to the durability of the tyres. The absence of the bonnet gives a more graceful look to the carriage, while the low, easy step into the car and wide doors are charming."

The November 1905 Olympia Motor Show featured no twin-cylinder cars but instead a complete exhibition of the new 20 h.p., a side-entrance tonneau-bodied touring car, a landau, and a complete chassis showing the arrangement and detail of the mechanics. All of these generated great interest from the press and public alike, proving that events such

THE FOUR-CYLINDER 20 H.P.

A 20 h.p. on meat deliveries.

as this were essential to the Lanchester Motor Company, whose publicity and advertising fund was small.

With ten or more different body designs as standard, the 20 h.p. car sold well and proved most popular, which was especially welcome to the Company after their painful experience of Receivership. The steady profits accrued earlier by the Receiver had continued through 1905, to show a net profit for the period from 19th December 1904 to 30th September 1905 of £2,204.19s.2d. Although a small amount, this must have been a figure to inspire confidence, as the whole of the development and building costs of the new 20 h.p. car would have been taken from this. Lanchester's terms of business for cars sold required payment of a 10% deposit on order, and this amount from each new customer was most sorely needed.

A guarantee against any item, excepting the tyres, found to be defective within three

All-weather protection on the 20 h.p.

months of the date of delivery was offered to the customers, although this did not include delivery of the car to the Works, or the charge for labour. Most customers would either repair or replace the damaged item themselves, or utilise the skills of their chauffeur or the local garage. Lanchester's parts service was most efficient, as one customer affirmed: "As I only had small breakages I attended to them myself. I always found, no matter what part of England I might be in, or what part required renewing, on wiring to the Firm the spare part would be sent first post. On two occasions I wired in the afternoon and received the goods next morning."

Although "hiring out purposes" rendered the guarantee invalid, some firms still bought Lanchesters "to meet the wishes of customers who desire something considerably superior to the ordinary hiring car". The Company may have been protecting itself against long-distance motorists by giving such a short guarantee period, even though the cars were renowned for their ruggedness and reliability; a richly deserved accolade achieved through strict and involved testing procedures.

Before being installed into a chassis, each engine was mounted on a test jig and subjected to several hours' running, using its own electrics and transmission gear but without any load, and then for further hours with loading being applied by dynamo meters. When the required b.h.p. was achieved, the engine was classified 'off test' to allow its gears, clutch and brake to be given progressive loading tests to ensure that they also met the exacting standards.

When passed, the engine and transmission were installed in the chassis. With a test body in place, a drive averaging some 80 miles over rough and hilly ground enabled the tester to record any deficiencies. After correction the car would be put through the same procedure to ascertain that all was well. The normal practice was for a team of testers to go out in convoy, supervised by a foreman. Few faults were discovered, although George did become suspicious about one particular foreman who kept sending back 'clear' reports. George recounted, "One day Fred and I followed him out and found all the testers in the pub playing darts!"

When the testers signed the cars 'off test' they were driven to the coachbuilding works in

A chauffeur-driven 20 h.p. landaulette.

Liverpool Street for the customer's chosen body to be fitted, painted and trimmed. After many weeks, the finished product would then be tested again. "By this method, it is obvious that the utmost precautions are taken to prevent any car leaving our works if not in the most perfect condition", stated the Company's brochure.

On occasion the Company used freelance agents to sell their cars for them, but it seems that not all these agents were reliable. George related: "One buyer sent a man to chauffeur the car back to his home. When the car didn't arrive after three days the police were called. They found the agent and the chauffeur in a public house, both too drunk to drive, tossing the chauffeur's glass eye into the air to see whose turn it was to buy the next round!"

In 1906, the reliability of the high-tension magnetos had continued their improvement to such an extent that Fred standardised them on his cars, abandoning the Lanchester low-tension ignition system which had served them so well. High-tension magnetos which were guaranteed by their makers were subjected to stringent tests, and those supplied by the Bosch Magneto Company Limited were adopted, driven from the off-side camshaft. A small flywheel was fitted to the shaft, all to Lanchester's own design, in order to balance the magnetic pull on the magneto armature. Conventional high-tension spark plugs were used to complement the new system. It seems probable that the 20 h.p. cars fitted with the earlier ignition system were able to be returned to the Works, or recognised Agents, for the high-tension magneto system to be fitted.

It was also decided to make another improvement. The petrol feed to the wick vaporiser was altered to become automatic, in preference to hand-pumping, although this method was still used for initial priming. The car's own exhaust system maintained the tank pressure to force petrol to the wicks.

The experimental 20 h.p. cars had been fitted with slide-valve throttles but, due to the suction effect created by the intake, these had tended to stick. They had been replaced, probably in 1905, by a butterfly-valve throttle in each cylinder's inlet pipe, all linked together. George wrote that these throttles were "operated simultaneously by means of a centrifugal governor, actuated by a lever at the driver's side, and by a foot pedal which can be regulated from zero to full open". He wrote later that "the hydraulic governor was always put out of action by drivers who used the accelerator as a positive control". The governor was termed hydraulic as servo-assistance was provided by the oil lubrication system. The historian, Anthony Bird, wrote that: "As the concept of car control in the

Hamilton Barnsley (Managing Director) driving the leading 20 h.p. car, taking the politician, Joseph Chamberlain, on the procession through Birmingham in July 1906.

modern manner by a throttle pedal gradually gained ground, motorists came to accept the idea of an ungoverned engine. In 1904, when the car was designed, most motorists were still in favour of governors, at least for large engines, as the accelerator or foot throttle was still quite a rare feature and it was feared that if a driver had to de-clutch suddenly in some emergency, an ungoverned engine might race to its destruction before he could spare a hand from brake, horn, gear-lever or steering wheel to fumble with a hand throttle. By 1905-6 accelerator control was becoming much more acceptable and the governor, in consequence, less and less necessary. Therefore, all the early 20-hp Lanchesters were recalled to the factory and had their governors removed. As the Lanchester theory of driving always had been that the governor, or hand-throttle control, should maintain a slow tick-over and that the engine should be 'opened out' by the pedal, this was a logical step."

In July 1906, the Lanchester 20 h.p. received more acclaim and good advertising when seven landaulettes headed a convoy of some ninety vehicles to parade the politician, Joseph Chamberlain, and his party around Birmingham over a two-day period. The cars ran well, with the Autocar reporting: "The Lanchester cars surpassed themselves, as they were used by Mr. Chamberlain and his friends to accompany the torch-light procession, and for one hour and twenty minutes they never exceeded two and a half miles per hour, and yet never stopped for a single second. In fact they ran like clockwork throughout the entire proceedings." The reporter cleared up an obvious controversy with, ".... and more than justified the exclusion of the horse".

Fred Lanchester driving the second 20 h.p. car in Mr. Chamberlain's procession, in July 1906.

Probably the low speeds during the tour of Birmingham would have been welcomed by the passengers, as an earlier Autocar had reported, "The Birmingham streets are very bad (most of them) and we know just how much they hurt on an average car....." Their road test of a 20 h.p. then commended the Lanchester suspension: "We can only say that the Lanchester more nearly obliterates their painfully prominent inequalities than any other car in which we have ever driven over them." Similar tributes for other Lanchester features led to the glowing summary: "We have always had the greatest admiration for Mr. F.W.Lanchester as a designer and an originator but we must say that we think he has surpassed himself in the 20 h.p. which, by the way, was put through about as exhaustive a series of trials as any car which has ever been made, before it was actually supplied as a standard product to the automobile world."

For all the praise given to the Lanchester's unique design and novel fittings, there were a substantial number of potential customers who were worried by these unorthodox features. The earlier twin-cylinder models had acquired the reputation of being intricate pieces of precision machinery which were difficult and expensive to work on, and this doubt had carried forward to the new era of vertical-engined cars as well. Some wrote to the motoring magazines for advice and guidance and were answered in reassuring terms by existing

Fred Lanchester, in the leading car, waits for eight other Lanchesters to prepare for the annual foremen's outing from Armourer Mills, in July 1906.

owners, one of whom wrote: "I have driven my 20 h.p. over 20,000 miles. Never stopped on the road. Let me emphatically state that the engine is an exceedingly simple and non-complicated example of the ordinary vertical motor engine. All parts are easily get-at-able, and for any adjustments needed on the road the bonnet can be lifted and engine examined without moving from the driving seat..... On the hottest day the position of the engine between the occupants of the front seats is not noticeable at all."

Another satisfied customer remarked, "I drove my 1906 20 h.p. seven-seater touring car for about 9,000 miles in all parts of England, mainly the Peak District, and consider it light on tyres, the average being about 4,000 miles before retreading. It is not the complicated piece of work many people believe. The engine can be removed entirely from the frame in a very short time if required. The clutch and main brake can be taken off and new discs put in by any novice and put back in a few hours, and with ordinary care this would not be required for some thousands of miles. The gears require no attention, but there is an inspection plate to them. The carburettor gives absolutely no trouble whatever. The valves can be got at and taken out more easily than any I know, and a new valve spring can be put on in a few minutes. Now that high-tension magnetos are fitted one of the troubles the novice had to contend with is done away with, as the old igniter gave a little trouble until it was properly understood. The magneto itself gave no trouble. There is no oil feed to worry about, as provided plenty of oil is put in and the gauge is working, you can be assured every part is getting the right amount. The carburettor is now pressure fed, and so does away with the old way of pumping ..... My petrol consumption averaged about eighteen and a half miles to the gallon, but on good roads would do more. No inconvenience is felt by the front seat passenger or driver on the hottest day. A great deal of prejudice exists against the car by people who have never driven one, and who have no knowledge except what someone else has told them. Anyone has only to drive one for a short time to appreciate it. I forgot to mention the back axle; it is very seldom that one goes wrong. I consider the Lanchester car one of the soundest and best made on the market, and as reliable as any, and provided it is not too much overloaded with a body will climb any hill easily. The car I drove was running as well as ever and as quiet as ever after 9,000 miles."

As with the earlier Lanchester models, there had been no pretence of record-breaking pace, but the 20 h.p.'s steering, suspension, gears and other features combined to give a more than acceptable mean speed. One customer, after singing the praises of his car, candidly remarked, "We have scarcely ever been passed on the road by any other car no matter how big the power. I do not mention this as evidence of great speed, so much as of reserve of power, which enables us to keep up over long distances a good rate of travel, no matter how hilly the country may be."

The Olympia Motor Show in November 1906 offered the 20 h.p. car for display with five-seater landaulette coachwork, "finished in most luxurious style", for £150 above the chassis price of £450, whereas the five-seater touring car, also built on a short chassis, was

A 20 h.p. with commercial van body.

an extra £100. These were probably offered, as were all subsequent Lanchesters, with the radiator angled slightly backwards in order to give a more streamlined appearance.

A contemporary Lanchester brochure recorded that a 1ft. longer chassis version measuring 10ft.5ins. had been introduced, still retaining the front track of 4ft.10ins. This was only £25 more expensive to purchase than the shorter chassis and was ideal for the six-seater landau or the six-seater brougham. The coachwork of each cost an additional £200, with various screens, hoods and canopies available at extra cost. These prices remained the same from the first Lanchester 20 h.p. offered for sale until the last.

The Lanchester coachwork was to be the cause of a fierce argument with the Chairman,

A production run of 20 h.p. landaulettes inside Armourer Mills.

when Fred realised that there was no need to continue producing doors to meet the criteria set by horse-drawn vehicles, namely the standard width of 19ins. He designed a body with doors 28ins. wide and work was started on the first six. When the Chairman saw one of these, in Fred's own words he, "went up in smoke". He ordered that only the three bodies that were practically finished should reluctantly be allowed for sale, and the remaining three were to be altered back to traditional standards. When the public saw the three cars which had escaped his censure, however, they demanded more and, much to Charles Pugh's chagrin, the wide door became a popular Lanchester feature which other companies then tried to emulate.

For all Fred's earlier comparisons of the 20 h.p. car with the specialised design of the period racing cars, he never felt inclined to race his cars on the track. He was happy, however, for them to be entered for general performance trials, and in this they achieved much success. A first prize at the 1906 Somerset Automobile Club's hill-climb was equalled later by a first prize awarded for the best overall performance at a Town Carriage Competition. The Vapour Emission, Incline and Road Test events, also in 1906, were organised by the Automobile Club of Great Britain and Ireland and two 20 h.p. Lanchesters were entered. In two Emission Tests, samples of the exhaust gases collected in a copper receiver, specially made for the purpose, were analysed. The Incline Test consisted of driving up and down a 1 in 7 hill for ten minutes, whereas the Road Test consisted of a run of 150 miles. As a tribute to Fred's design of the Lanchester 20 h.p., which was recorded as weighing 1.25 tons and which consumed petrol at the rate of 20 m.p.g., one received the Gold Medal Award while another was given First Place; the cars were declared as being "absolutely smokeless". In March 1907, the A.C.G.B.I. changed its name to the Royal Automobile Club, by permission of King Edward V11, and in that year also a Lanchester 20 h.p. won the Gold Medal Award.

A foremen's Works outing leaving Armourer Mills, in tiller-steered and wheel-steered cars.

Other aspects of this popular car were described in the period terminology of 1907 by the magazine entitled The World And His Wife: "Ease of control in dense traffic is essential. Hence the success of the Lanchester tiller steering, which is admittedly the best for this kind of work. It is hardly necessary to comment on the silent running of the Lanchester, the phrase 'silent Lanchesters' is known all over the world. Hitherto, one of the greatest objections of the motor car as a shopping vehicle was the smell and smoke emitted from the exhaust. Many cars, however, showed great improvement in this respect, and this is in a great measure due to the vapour emission trials held by the R.A.C. .... The wide side entrances and low step level of the Lanchester cars have always met with unqualified approval from the opposite sex, owing to the ease of entrance and exit; a very great feature indeed when one considers the number of times during a morning's round of calls, or a shopping expedition in Bond Street, one has to step out of, or into, the car. In the ordinary type of car, with narrow side entrances and a high step and floor level, the effect upon temper and toilette is quickly apparent. There is little of the machine about the Lanchester, for just as the highest art is to conceal art, so the highest engineering science aims to show only the results without obtruding the means whereby the results are obtained. Beautiful coachwork, luxurious appointment, and an engine which can neither be seen nor heard by the merely casual observer, are the obvious points in the Lanchester design. Once seated in the Lanchester car, one is immediately struck with the amount of room available; the dressing case, library books, or similar luggage, may be put on the floor, and still there is all the space comfort could require ...."

The Lanchester exhibition stand itself received extra special attention at Olympia in November 1908, when Henry Vaughan Lanchester, Fred's elder brother and an acclaimed architect, designed a new frontage for it. This displayed to advantage the 20 h.p. five-seater landaulette exhibited, a similar model to the one shown the previous year, dark green with leather-upholstered front seats, the interior compartment being finished in a "soft, comfortable green cloth of a shade well chosen to harmonise with the general tone of the garage".

The unique Lanchester suspension, allowing full movement with no chassis twist and no wheels off the ground.

A 1908 report describing the 20 h.p. stated, "Detailed improvements have been introduced." These items included the optional fitting of a coil and battery ignition system to accompany the Bosch magneto ignition. There was also a change in the method of petrol feed, as carbon deposits affected the seal of the valves within the exhaust pressure system, causing it to become occasionally unreliable. It was therefore altered for a camshaft-driven, mechanical lift pump.

Other improvements were inspired by George Lanchester, who, when he was ordered to bed for a long and complete rest to recover from illness, defied his doctor and spent the time designing a completely new system of gear selection. He realised that the Company needed to change from the single-lever control to something more acceptable to the general motoring public and, on his return to work in the autumn of 1908, put forward his ideas and suggestions to Fred who, after improving a few items, willingly accepted them.

The subsequent control selection mechanism cleverly imitated a conventional, right-hand 'crash' gearbox lever and 'gate', and worked in conjunction with the clutch pedal. Along

The detachable hard-top, raised from the car by pulley-block, to give the owner an open touring car.

with the brake and accelerator pedals, it was made to be adjustable for height to suit different drivers. Many thought it unfortunate that Lanchester's innovative mechanics had to be disguised as a conventional gearbox but, Lanchester purists apart, the decision was commercially sound. Beneath the parts on show, Lanchester's epicyclic gear engagement was still silky-smooth and silent, as opposed to the grating noise of most other gearboxes when under the control of the less proficient.

Coinciding with the new method of gear change was the introduction of a wheel-steering system, which was described as "very light and positive in action". The Lanchester Company put the reason for this change partly down to fashion. They explained: "Owing to the popularity of wheel steering amongst those accustomed to the ordinary type of car, we have standardised our special system of wheel steering and foot control." George later admitted that, "Due to the growing popularity of the closed body, tiller steering became impossible." The completely enclosed bodywork was indeed making life difficult for right-hand turns with the tiller and this, coupled with the increased weight of the large coachwork styles, made the change inevitable. However, the two new systems were not generally available on the 20 h.p. model until the autumn of 1909.

A certain ambivalence was shown to the complete abandonment of the Lanchester tiller and single-lever control. The November 1908 Motor Exhibition report stated, "Hitherto, wheel steering had been fitted to the cars to individual order only, but this season they have standardised a foot-control and wheel-steered car which in no way supersedes their popular tiller system, but has been designed to meet the requirements of customers who own cars of the 'orthodox' pattern, or who do not wish to learn a system of driving different from that to which they are accustomed. The new system is claimed to be quite the most silent and simple form of ordinary control as yet placed on the market, and in action it is exceptionally easy to manipulate."

However, alongside a wheel-steered five-seater phaeton at the November 1909 Motor Exhibition was a tiller-steered and single-hand-lever-controlled Lanchester landaulette. At this time the Company made the comment that "once a tiller steerer, always a tiller steerer", and remained "convinced that the tiller steering is preferable" for this small, light 20 h.p. car. Despite this, the end was in sight and at the following year's Motor Exhibition no reference at all was made to the tiller steering or single-lever-control gear change.

Further improvements had been made to the Lanchester models by November 1909. Fitted to each of their sloping radiators was a sight-glass, which allowed for a quick and simple inspection of the cooling water level. (This became a feature on all Lanchesters until 1931.) Other changes included ball bearings, dirt-excluding seals for the stub axles, a non-detachable engine starting handle, and separate oiling arrangements for the engine and gearbox, which allowed lubrication of the gearbox even when the engine was idling. The magneto was re-positioned across the rear of the engine and ball bearings were employed to locate its worm drive.

The interior was further refined by the introduction of a spring mattress base frame underneath the rear seat squab. A novel foot-warmer was secured to the rear floor, its heat being taken from an exhaust pipe by-pass, controllable by the occupants.

As these were the only alterations necessary, the Autocar was prompted to comment: "Indeed, this is not remarkable, for as a whole no car upon the market to-day has given more satisfaction than the Lanchesters of 1909."

The last public showing of the 20 h.p. model was in 1910. Fred Lanchester's car of unique design, born in the days of Receivership, was destined to be phased out in late 1911, after helping to provide the backbone for the ever-more-profitable Lanchester Motor Company.

Chassis no. 979 with wheel-steering, delivered new in 1910 to a German. He was later imprisoned as an alien for the duration of the First World War, and his Lanchester was commandeered as a staff car. It was later used as a hire-vehicle by two brothers of the Shrapnell family. When it was worn out, they stripped it to the last nut and bolt and put it in storage. Decades later, George Lanchester identified all the parts for Alen Warner, to enable the Lanchester enthusiast to rebuild it. (See appendix.)

One grateful customer writing to the Company in 1910 said, "I have now run my 20 h.p. touring car 40,035 miles and it is exactly 23 months since I bought her from you second-hand. I often do as much as 800 miles in a week and on several days have done over 300 miles in a day, and seldom have any trouble except an occasional spark-plug failure or tyre puncture, and it is only a month ago I travelled from Bournemouth to North Wales with only one compulsory stop to change a sparking plug. (The writer used Benzol instead of petrol.) This run, after what the car has already done, speaks volumes for its sound construction and durability. My car is now running quite well, and as sweetly as the day I had her .... your marvellous carburettor, which has never given me the slightest trouble on a single occasion, my average consumption working out at about 25 to a gallon. The car is quite easy on tyres, averaging about 4,500 miles on the back wheels before needing re-treading and I have often got about 4,000 miles out of a re-treaded cover. The front wheel tyres easily do from 8,000 to 10,000 miles .... I have been using your cars now for the past seven and a half years, and I have done over 150,000 miles and I feel sure no other make of car would have given the same reliability which yours has done for me."

Edgar N. Duffield, the motoring correspondent, wrote of similar remarks from a more famous admirer of Lanchester cars. "Henry Ford, when he visited Armourer Mills and saw how their cars were built, said that he had never previously, either in Europe or America, seen such fine work, such sanely scientific exactitude, as characterised every process in the production of what continued to be to all intents and purposes a hand-made car." Henry Ford, who was a good friend of Fred, George and Frank, bought and shipped a 20 h.p. to the United States of America for inspection, and subsequently utilised a foot-pedal-operated epicyclic gearbox with forward mounted flywheel on his own range of cars. He also, of course, took great notice of Lanchester's interchangeability of both mechanical and body parts, and utilised the system for mass production of his own cars.

Chapter Twelve

THE SIX-CYLINDER 28 H.P.

The plaudits for the Lanchester were not limited to the 20 h.p., however. Its sister car, the 28 h.p., utilised many of the same fittings and also shared the praise. Work had begun on the design of this six-cylinder car in 1905, but a delay had been caused by Fred's early efforts to produce an engine which maintained the high standards of its predecessors. Fred wrote, "The first six-cylinder engine of my design never got beyond the experimental stage and was abandoned."

The off-side view of the 28 h.p. engine and countershaft.

With Fred's impressive track record, it is strange that he was not among the first of the six-cylinder engine producers. However, when he was completely satisfied with his design, his resulting six-cylinder 28 h.p. engine was almost certainly the first to be completely free of the scourge of torsional vibration. The 28 h.p. engine's bore size and stroke of 4ins. and 3ins. respectively, as well as its general design, were the same as on the four-cylinder 20 h.p. engine. The troublesome 'power rattle' was effectively avoided by a combination of the cylinder dimensions, the revolving masses within the epicyclic gearbox, clutch box and brake box, the forward-mounted balancing flywheel and the large bearings for the rigid crankshaft, all housed within strong castings.

George explained this Lanchester's technical aspects thus: "The six-cylinder, 28 h.p. engine had a capacity of 3.8 litres, but by reason that the cylinders were served by one common induction pipe and one exhaust manifold, there was an overlap between cylinder aspirations which reduced efficiency; hence the six-cylinder engine gave only 28.5 b.h.p. at 1,500 r.p.m., instead of 30, and approximately 38 b.h.p. at 2,000 r.p.m. with a maximum of 42 b.h.p. at 2,200 r.p.m."

THE SIX-CYLINDER 28 H.P.

George Lanchester at the tiller of the pre-production 28 h.p. chassis.

In early April 1906, the Autocar was given a preview of the 28 h.p. before production began. There seems little doubt that the reporter was impressed by the car, which was driven by Fred. "Described quite roughly, the car was a six-cylinder version of the four-cylinder type which is now so well known. The car we tried was merely a bare chassis without upholstery of any sort, but we were simply delighted with the power, pace and smoothness of running of the car. The engine ran like a dynamo, and this with the unique Lanchester suspension gave one the impression of floating at high speed with some invisible power towing the car. The sensation was about as nearly akin to flying as we imagine it is possible to attain with a motor car. Whether our impression is correct or not time alone will prove, but till we have actually tried flying and found it altogether different from what we expect it to be, we shall hold to our opinion that the sensation of being driven in a six-cylinder Lanchester on a long, straight, undulating road before a setting sun is very much akin to it." This choice of analogy must have been music to Fred's ears!

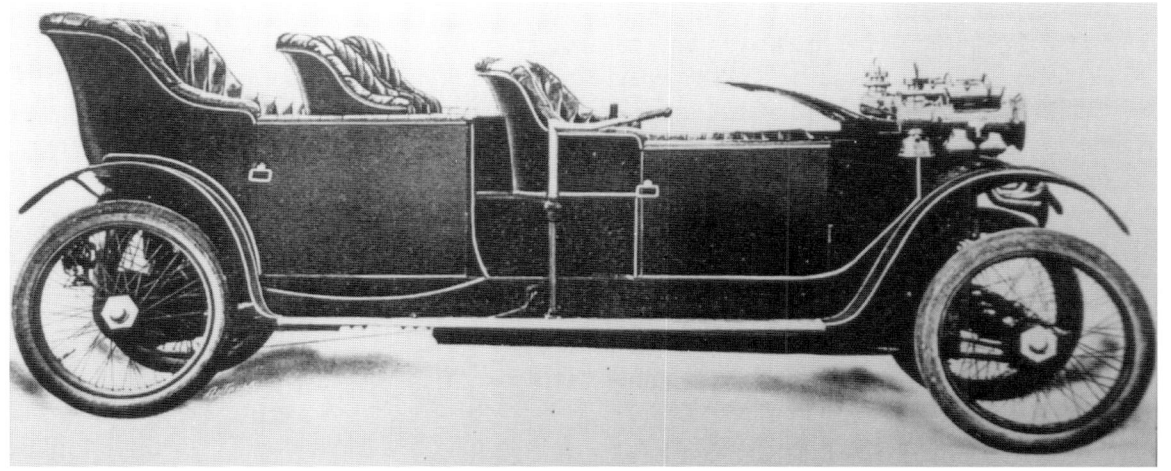

A seven-seater 28 h.p. touring car, coachwork by Lanchester, shown at the November 1906 Olympia Motor Exhibition.

The first public showing of the 28 h.p., at the November 1906 Olympia Motor Exhibition, featured a seven-seater touring car as well as a demonstration six-cylinder engine, complete with Bosch high-tension magneto, countershaft and change-speed gears. To allay any potential customer's anxiety about six-cylinder cars, the Company provided explanatory advice: "The flexibility of an engine is a term popularly used to denote the capacity of an engine to run satisfactorily over a greater or lesser range of speed. Flexibility depends upon two things. Firstly, the capacity of the flywheel to absorb the impulses of the explosion, without undue speed variation. Secondly, the capacity of the carburettor, ignition, etc. to function properly over the range of speed desired. If the weight of flywheel were no consideration, a one-cylinder engine might be made as flexible as we please. In a motor car, however, weight is all important, and the flywheel weight required for a given size of cylinder diminishes when the number of cylinders is increased." The Company then extolled the virtues of the Lanchester in particular, with their claim that, "The existence of an engine would never be suspected by an unsophisticated passenger."

28 h.p. phaeton coachwork.

Apart from the 28 h.p.'s excellent all-round performance and top speed of 54 m.p.h., the car would return 18 or 19 m.p.g. Indeed, one customer wrote to the Company, "The hill-climbing power of the 28 h.p. is wonderful, and it is very seldom, though my car weighs 31 cwt. unladen, that I have to use the second gear. I never recollect having used the first, the lowest of the three, since I have had the car, and we have some terribly steep hills about here." The weight mentioned related to the long-chassis model, with an 11ft.5ins. wheelbase, whereas the shorter wheelbase version, measuring 10ft.5ins., weighed 28 cwt. The chassis prices were £625 and £600 respectively and, with such a small difference, most customers purchased the longer. There was a corresponding difference in price for the coachwork as, for instance, a five-seater touring body on the short chassis would be £100, whereas a seven-seater touring body on the long chassis would be £125. The six-seater landaulette, brougham or limousine bodies made for the long chassis were £200 above chassis price, and the six-seater limousine with roof canopy was even more expensive at £235 above chassis price. The Lanchester Company made it clear that they would quote for any design of body at all, made to customers' own ideas and requirements, although no records show anything outrageous being requested.

This willingness to fulfil each customer's whim no doubt encouraged The Maharajah of Nawanager, Prince Ranjitsihnji, to order another two Lanchesters in July 1907. The first was a short wheelbase touring car for private use, which was provided with an elaborate Cape-cart hood with folding screens. The car was fitted with an additional screen which could entirely seal off the passenger compartment if royal privacy was desired. The second

The Maharajah of Nawanagar's 1907 28 h.p. State-carriage, in front of his palace which Fred Lanchester's elder brother, Henry, designed.

was a magnificent state carriage. The Autocar stated, "These two cars are splendid specimens of the Lanchester Company's work, and their elaborate fittings and upholstery will undoubtedly appeal to the love of luxury and magnificence which characterises the majority of Oriental potentates, and to which Prince Ranjitsihnji must bow."

Yet another Maharajah who succumbed to the lure of this "unbridled luxury" was His Highness the Thakore Sabib of Rajkot, and a 28 h.p. landaulette became the first of his fleet of Lanchesters.

The car's popularity was such that again their exhibition stand at Olympia was viewed with great interest. As one official report began, "The Lanchester Motor Company's exhibits are always looked for with keen anticipation on account of their excellent workmanship and originality of design."

One car exhibited was a six-seater limousine with dark green coachwork, relieved with black lines on the panels. The front seats were upholstered in green leather and the rear interior "is finished in an attractive shade of pearl-grey cord cloth, finished with silk lacings in two shades of grey, picked out with a minute green design". The 28 h.p. tiller-steered five-seater open phaeton touring car displayed a novel Lanchester design suitably described in period language. "A chauffeur's seat is provided on this car, a great convenience impossible on cars of any other type. The location of a 'correct place' for the chauffeur when the owner drives himself has always been a vexed question. The Lanchester Motor Company have evolved the only satisfactory solution of it, in placing a seat, facing sideways, low down in the foot space of the front passenger."

The description of the six-seater landaulette sounds equally impressive, ".....having two, folding, swivel seats in the interior, in addition to the luxurious seats within the hood,

A 28 h.p. touring car, incorporating a sideways-facing seat in the passenger's foot-well for the chauffeur's use when the owner was driving.

A 28 h.p. landaulette with prominent coach-lining.

designed to seat two, but wide enough to seat three comfortably. The car is fitted with glass screen and canopy of similar design to that of the limousine, the upper half of which is arranged to open outwards or inwards at the will of the driver. All the seats are upholstered in green leather and the interior of the hood and roof are finished in green cloth, the whole relieved with lacings of green silk in two shades, the upholstery being entirely in harmony with the paintwork. Both these cars are fitted with electric lights and indicator to direct the driver from the interior of the carriage, together with many other appointments likely to fulfil the requirements of the motorist to whom comfort is of first consideration."

The three 28 h.p. cars each had a different construction of road wheel: the Rudge-Whitworth wheel, the Lanchester wire wheel, and the wooden artillery wheel, although the customers would have been unaware of the evolution of events leading to this choice.

On all production cars from the turn of the century, Fred had realised the importance of having each wheel with the same sized rim, thus avoiding the need to carry different sized spare tyres and inner tubes. Any journey undertaken would have been plagued by constant punctures; indeed, George wrote that on one journey he "finished with all four tyres stuffed with straw and bound to the rims with cord". It would therefore have been insufficient to carry only one spare tyre on a wheel, ready for use, so instead a supply of tyres and tubes would always be carried. In view of this, there had been no need to make the wheel centres interchangeable between front and rear, as the wheels remained on the car and merely the tyres were changed.

Wheel-steered and tiller-steered Lanchesters at Armourer Mills, circa 1908.

In 1906, however, road improvements and tyre developments had made punctures less likely and both the Napier and Rudge-Whitworth firms produced a new type of wheel, which was made easily detachable and fully interchangeable between the front and rear hubs. The Rudge-Whitworth wheels incorporated eight 0.5ins. wide splines, to locate with similar splines on any of the vehicle's hubs. John Pugh, a Director of both this and the Lanchester Company, knew all about the use of splines through their many applications on Lanchester cars, even though, it seems, he had been one of the Board members who initially argued against their use. He had also revised his opinion of Fred's specially

Lanchester bodywork under construction.

configurated, tangent wire wheel, which again he had initially disliked. It was these two features, however, which formed the basis of the Rudge-Whitworth wheel, which was patented in early 1907. It is not known whether Fred was upset by this use of his ideas or by the Director's preference for this new wheel over his own, but it is a fact that after their debut, amidst growing popularity from motor manufacturers and the public alike, Rudge-Whitworth wheels were fitted as standard to all Lanchester cars.

Also fitted as standard to the 28 h.p. model in the autumn of 1908, were the wheel steering and foot control gear change systems. Their use was strongly discouraged on the 20 h.p. model in favour of retaining the tiller steering and hand lever control, but it was reported that "after exhaustive comparative trials of both systems the Lanchester Co. are of the opinion that for the heavier and faster car the wheel steering and new control are upon the whole the better". It seems that most customers accepted this change, although the option of the alternative system remained. When issuing a statement in January 1909 confirming that this would remain the case, the Lanchester Motor Company explained that, "As manufacturers we cannot afford to lose orders by neglecting the demands of a large section of the motoring public. Thus in adopting the wheel steering and foot control as a supplementary design, our car is now able to enter a market that was hitherto closed to it."

The Company took this opportunity to neatly secure their electric horn button to one of the four spokes of the new steering wheel. To avoid the unsightly appearance of the connecting loop of wire necessary to allow for steering wheel movement, they manufactured a contact ring to fit the steering column head, which engaged with a short wire from the horn button on one side and the horn on the other.

Until late 1908, customers would have given their various orders to Lanchester's London office at 311 Oxford Street, in Mayfair, but from this time the showroom was moved to nearby 95 New Bond Street. This corner building (which was retained until 1931) provided an impressive setting for customers to decide on their requirements. In this choice they were expertly guided by Frank Lanchester who, in order to be more readily available, had moved from Birmingham to Surbiton in 1906, and later to Newstead, The Drive, Belmont, in Surrey.

Customers were also offered the option of a visit to Armourer Mills in Birmingham. There, a complete tour of the Works would also have included an inspection of the separate coachbuilding operation at Alpha Works, Liverpool Street, which was divided into various departments and had a workforce of fifty men, working from 7.30 a.m. until 6.00 p.m.

A customer's order for a particular body style was usually held over until a batch of a dozen could be produced together. The drawing office would then issue explicit details of the required bodies to the shop-floor. Various pre-cut sections of bodywork, which had been built on separate jigs to ensure uniformity before being primed in the paint shop, were

Lanchester's torpedo tourer coachwork on a late 1912 25 h.p. car, chassis no. 1434, formerly owned by Professor A.L.Bird and subsequently by Lord Charnwood. (See appendix.)

1913 38 h.p. chassis no. 1460, with updated bodystyle classified as a touring car de-luxe.

Foot pedals and hand levers on the 38 h.p. and 25 h.p. models.

Torpedo tourer coachwork on the 1912 25 h.p. chassis no. 1434, showing a shorter bonnet than on the similar 38 h.p. model.

The 25 h.p. and 38 h.p. drivers' compartment, showing the narrow bonnet and the catch on the folding steering-wheel. The control panel shows, from left to right: the rotating ignition selector, the indicator plungers for the countershaft and engine oil pressures, the petrol vapour regulator, the hand-throttle lever, the ignition-adjustment lever, and the engine-starting lever.

Rear view of the 38 h.p., showing the folded steering wheel.

1921 40 h.p. chassis no. 1762, with Lanchester tourer coachwork. (See appendix.)

1921 40 h.p. chassis no. 1762, which was used regularly in its earlier years as a taxi and for motor tours in the Lake District.

The Lanchester tourer coachwork is now restored to its original colours. (See appendix.)

Converted to a limousine by Gill Coachbuilders, from the original two-door tourer bodywork on 40 h.p. chassis no. 1774. (See appendix.)

Chassis no. 1843, with right-hand-drive steering, was exported to the United States of America for the Brewster coachwork stand at the 1921 New York Auto Show. Comedian Buster Keaton purchased the car and commissioned Brewster to build a sedanca body for it. Years later, the chassis was found in a junk-yard with only the Brewster fenders remaining. A Lanchester tourer body was later removed from a Rolls-Royce to complete the car.

The State-limousine, complete with external servant's seat, on chassis no. 1915, ordered in 1924 for the Maharajah of Rewa. (See appendix.)

THE SIX-CYLINDER 28 H.P.

The sight-glass showing the water level in the 28 h.p. radiator.

The London showroom at 95 New Bond Street, Mayfair, in use from 1908.

already available in indexed bins. These were then mounted permanently on a light-steel jig which was "an exact facsimile of the chassis frame upon which the body will eventually be fitted". As the woodwork progressed, the aluminium body panels were secured to them. These had also been pre-prepared, so that no time was lost in construction. The process of painting the vehicle bodies was meticulous, to ensure durability. Removed from their jigs, the completed bodies received their priming paints. Each subsequent top coat of enamel paint was brushed on and took some twelve hours to dry. As ten or twelve coats were applied, this accounted for much of the total cost.

28 h.p. landaulettes ready for despatch.

The trimmers would then receive the bodies for upholstering. Again, the various colours and grain of leather would be available from the stores, pre-cut. When finished, each body, complete with its integral steel foundation base, was placed on the car chassis, which had been driven over from Armourer Mills. After the electrics had been connected, the final top coats of paint and varnish would be applied before the car was ready for its thorough road tests.

Tom Grimmett, who was employed in the Despatch Stores at Alpha Works, remembers these times of supreme craftsmanship and the pride felt by the workforce when a fully-tested car was pushed into the Display Room at the entrance of the Works, to await its purchaser. Incredibly high standards were expected, as the new owners would have compared the Lanchester against its contemporary rivals prior to purchase. The car's size was obviously an important feature to one customer, who wrote in praise, "I have now owned and driven a 28 h.p. Lanchester car for nearly eighteen months. During this time we have covered about 15,000 miles, over all sorts of roads, generally with eight people and a dog in the car."

Elaborate trimming in the Lanchester rear compartment.

Another customer also sang the praises of his 28 h.p. car: "My wife and her sister travelled all the way from Edinburgh to Birmingham to test the Lanchester car in the matter of comfort. After a trial run they had no doubt about the matter, and they have no doubts now. Friends remark that they are never thrown about in my car when going at a good pace on bad roads. I remember once driving into Edinburgh from Talla Reservoir, 40 miles, fast; when I pulled up at our door, I noticed the tumbler standing on the table we have at the back of the front seats. I thought it a dangerous place for a tumbler and made some remark. My wife told me, without any appearance of surprise, that it had been forgotten when packing up the luncheon basket at Talla, so she had left it on the table, where it had remained. To anybody who knows the nature of a great part of that road, it is a remarkable testimony to the Lanchester's suspension ..... I drive myself, and I have been driving or riding somewhere or other for 50 years (and more), and I think I know something about that indefinable thing called quality, and the Lanchester car has it in a pre-eminent degree. It is a luxury to drive a Lanchester."

The Autocar added to the praise by commenting, ".... it holds the road like a racing car." One of the main reasons for this was indeed that system of suspension. The springs which were made for Lanchesters by the best of specialist firms were produced not in identical batches, as with earlier models, but in individual sets of different strength to attain the perfect combination. Different spring sets were therefore labelled for use, such as: 'driver's side-lever steered', 'passenger's side', 'short chassis', 'long chassis' and 'extra heavy for rough districts and colonial cars'.

By 1910, the system was further enhanced by rubber cushioning at the spring ends to keep the dust out and the lubricant in. The quality of the suspension had already enticed the New Engine Company to sail close to the wind, by copying much of Lanchester's patented system. The resultant court case was monitored by Rolls-Royce's Chief Executive, Claude Johnson, who was always interested in the mechanical attributes of Lanchester cars. He enquired whether the N.E.C. had been ordered to pay royalties to the Lanchester Motor Company. The answer he received to the contrary was, no doubt, relayed to Henry Royce, who already had a Lanchester 28 h.p. at his home in the south of France, undergoing extensive road testing. They decided that the adoption of cantilever suspension would be beneficial, and the Rolls Royce 40/50 cars were duly modified.

A 28 h.p. colonial tourer model of 1910, with larger tyres to give a greater ground clearance.

By November of the same year, the Motor Show saw the Lanchester Company bowing to popular demand when none of the four cars on display was constructed with the famous, side-lever steering nor indeed the single-lever hand-controlled gears. The limousine shown had been re-classified as a seven-seater, and the price was correspondingly increased by £40 to £900. The landaulette was also classified as a seven-seater, no doubt counting the rear seat as three places rather than two. The price of this car was raised by £25 to £850.

George's growing influence on design had allowed him to modify the coachwork in late 1910, and he cleverly meshed together the two ideals of Fred's Lanchester engine position and the general public's demand for a bonnetted car. Whereas the driver's compartment had previously started behind the radiator, George's compromise brought the windscreen back close to the steering wheel, thus covering half of the engine by external coachwork to give a sloping bonnet effect. A practical disadvantage to this style was that any engine repair was made slightly more awkward, and also that the legs of the driver and front

Exquisite side-lamps on a 28 h.p. car.

passenger could have become uncomfortably warm in very hot weather. George alleviated this latter possibility by incorporating in the scuttle dash a controllable opening panel, called a ventilator. The hand brake and change-speed levers were placed inside the body, and allowed easy entry to the driving seat. For the same reason, the steering wheel was engineered to fold flat against the column, operated by a catch on one of its four spokes.

Despite these modifications, there were customers who still held the opinion that it was foolish to have so much of the car outside the passenger compartment and this resistance to change meant that many continued to order cars with the traditional coachwork.

Sales of both styles of car were encouraged by continuing good publicity, such as the report of the 28 h.p. claiming first prize in Australia's New South Wales Hill Climb competition in 1911, and the model continued in production until 1912.

Chapter Thirteen

THE GEORGIAN YEARS

The period around 1910 was extremely busy for Fred, and finding time for so many varied and demanding tasks was difficult. Partly for this reason, he gave up the position of Chief Designer and Consulting Engineer of the Lanchester Motor Company, although the various disagreements with the Directors were the major cause of his decision. His contribution over the years had brought the Company increasing benefits, as the annual financial reports showed. 1906 had seen a £9,793 increase in profit to £11,997, followed by net profits of £3,254, £6,308 and £14,261 respectively for the three years up to 1909. However, due to the 1904 bankruptcy debacle resulting in his loss of shares, little of this profit actually went to Fred. His over-riding concern had always been the benefit of the Company, but now he felt that the time was right for a change and therefore accepted the part-time role of Consultant and Technical Adviser. George again stepped into his brother's shoes and became Chief Designer of the firm.

1910, the year that George V came to the throne, saw the brothers working in full and amicable collusion in all aspects of design work. Their plans for expansion prompted the Lanchester Motor Company to purchase the strip of land in Montgomery Street adjacent to Armourer Mills, and build a new wing to their factory. This additional space was used to house the whole of the coachbuilding operation, or carriage-building as it was then termed, and included much larger paint and varnish rooms. This allowed both the body and chassis construction to be concentrated in adjoining works, and was considered to make the managing of operations both easier and more efficient. This move enabled the Alpha Works to be emptied of most of its equipment, although the site was retained and used as a storage area for finished cars prior to their delivery to customers.

One part of the new works in Montgomery Street was taken up by the oil department, which was engaged in blending imported oils to achieve a good constant quality. The Lanchester brochures at this time continued to stress the importance of using the correct lubricants for their cars, and naturally advertised their own blended oil as giving the best results. This process had been considered necessary by the Lanchester Engine Company at the turn of the century, when the many and varied makes of oils were equalled by the many and varied grades of quality. It became a condition of a Lanchester's guarantee that only their oils were to be used, but difficulties were encountered in supplying that oil to customers. It had therefore been agreed, since January 1903, that the Rudge-Whitworth group should henceforth stock and advertise Lanchester lubricants at their depots throughout the United Kingdom and the world.

The range of Lanchester oils available was extensive and included Lanchester greases and rubber-mat dressing. The applications of machinery oil and gas-engine oil were self-explanatory, but not so the pale cylinder oil, which was made for the two-cylinder, water-cooled range of cars. The air-cooled models used medium cylinder oils, probably to take account of the extra heat created. Later to be re-named as Grade A, this oil was issued for all subsequent Lanchester models. Dark cylinder oils, later re-named as Grade B, were otherwise described as heavy oils, and were blended especially for rear axles and worm gearing. By 1910, the Lanchester oil division at Armourer Mills was also producing

Lanchester oils through the years; the last was offered in the middle 1920s.

suspension-spring oils and greases to complete the full range of products, all of which were available from their showrooms and depots. The Company was not averse to selling oil for other makes of car engines, or for boats.

The Lanchester headquarters for the north of England was at 38 Kings Street West, Manchester. It had been opened in 1908, in response to the restrictions of having Birmingham as the sole Company-base north of London, and from then on it had acted as a focal point for the important Manchester Motor Shows in order to attract further sales.

Archie Millership took control of both the Manchester and Birmingham works as Sales Manager on his return to the Lanchester Motor Company in 1912, under the overall responsibility of Frank Lanchester. Since the bankruptcy, Archie had found success at the Wolseley Tool and Motor Company, where Herbert Austin was in charge. In 1905, he left with Herbert Austin when the latter started his own firm. It was Archie who suggested the Longbridge Works as a suitable factory for Herbert Austin, and due to his subsequent endeavours he was made Sales Manager there. However, he was more than pleased to return to the fold at the Lanchester Motor Company, although disappointed about Fred's partial release from the firm.

From Fred's point of view, however, the arrangement worked very well. His valuable and varied scientific research was recognised by the Institution of Naval Architects, who elected him as an Associate in 1911, and the seventeen patents credited to him in this year showed his enduring energy and enthusiasm. One of these patents was classified as 'window cleaning apparatus', seemingly a strange subject, but his patents covered many diverse topics. Of his twelve patents registered in 1912, no fewer than five related to pianos, and over the following three years another seven patents related to musical instruments. There were also such varied inventions as boat propellers, fuel storage, telephones, guns and signalling, as well as those for his profound work on aircraft and motor vehicle design. Not that Fred's agile brain was restricted to patent applications; his advisory appointments, lectures and meetings also consumed much of his valuable time. He was still contributing to the annual lectures of the Institution of Automobile Engineers, and one of these, delivered in 1913, was his important paper entitled 'Worm Gear'.

It will be recalled that Lanchester's first car of 1895 was fitted with a driving chain to transmit the power to the back axle. Fred had found this unacceptable in view of the noise, the wear due to dirt and dust, and the lack of proper lubrication. He therefore abandoned it when the car was re-modelled. An alternative proposal had been the bevel gear drive, but this was considered too noisy at that time. Fred had thought carefully about a more efficient way to transmit the drive from the engine to the rear wheels, and his ideas had led him to consider a little-used mechanism for clocks and musical boxes devised by Henry Hindley in the 16th century. After much calculation, Fred decided to adapt the gear for his rear axle drive and worked out a complete method of manufacture; the famous Lanchester worm gear was born.

He patented a system of cutting gears, and at great expense produced a matched pair of master cutters, from which further cutters were made to actually produce the gears. The master cutters, normally kept in a fire-proof safe, were of such high quality that they were used for all Lanchester cars produced up to 1931, as well as for many Daimlers, Sheffield Simplex, Siddley Deasy and other cars, buses and lorries. In fact, it was reported that the Lanchester Company made the manufacture of worm gears a profitable side-line. The Warner Gear Company of Indiana produced the Lanchester worm drive for the American market under licence.

THE SILENT CAR
FITTED WITH
THE SILENT WORM DRIVE.

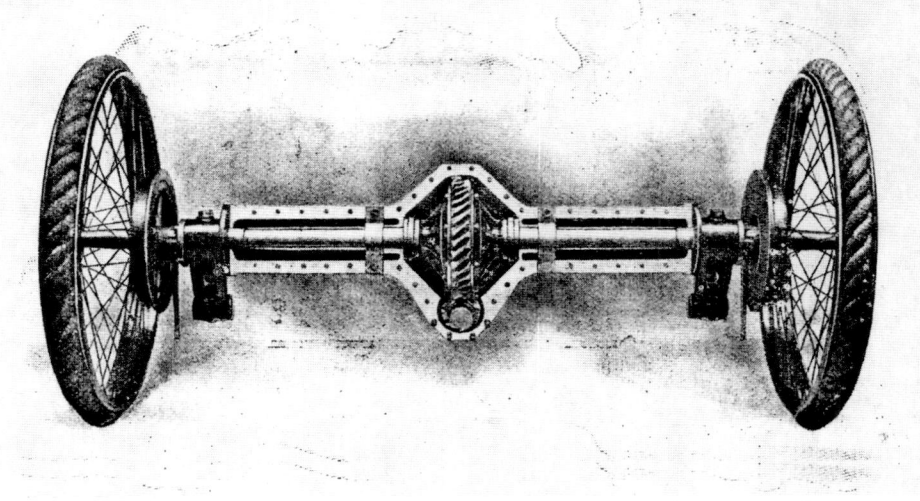

Lanchester worm drive advertising with the 20 h.p./28 h.p. rear axle.

The small worm gear, which closely resembled a screw thread, was made of steel. As it operated near the bottom of the rear axle differential casing below the matching worm wheel, it was termed 'underhung' (and later 'underslung'). The large-diameter wheel above it was made of phosphor bronze and rotated in axial alignment with the road wheels. Fred no doubt had considered adopting the alternative worm gear system, whereby the worm was located above the wheel but, probably due to the worry of insufficient lubrication at the sliding contact surfaces, this was never adopted.

Lanchester's worm and wheel profiles were curved as in an hour-glass, whereby both the worm and wheel were of a slightly greater diameter at their ends than in their centres. A greater surface contact was obtained, which in turn resulted in a lower pressure between the teeth, so reducing friction. As the teeth were in continuous engagement, the gear meshing was exceptionally quiet. The Company often used this fact to advantage, by referring to the drive as "the silent gear". Coupled with Lanchester's very fine working tolerances, the gears were designed to last for countless years (as the remaining Lanchesters of today testify). This was also true of the important bearings taking the end thrust of the worm and the side thrust of the wheel. Fred produced special ball thrust bearings to cope with this, where each ball was set to run in a different plane from the others.

Many competitors saw the benefits of the Lanchester worm gear but, to avoid paying royalties, they used instead a gear with parallel sides. Its easier and cheaper manufacture encouraged many firms, such as Rover and Standard, to adopt it. Harry R.Ricardo once termed the parallel worm gear "notoriously inefficient and generally irreversible". The two firms who manufactured these parallel gears were the David Brown Company and the Wrigley Transmission Company. When Fred was asked what he thought of the latter's worm gears he replied with his customary wit that, "The only use I would have for a Wrigley worm would be to go fishing!"

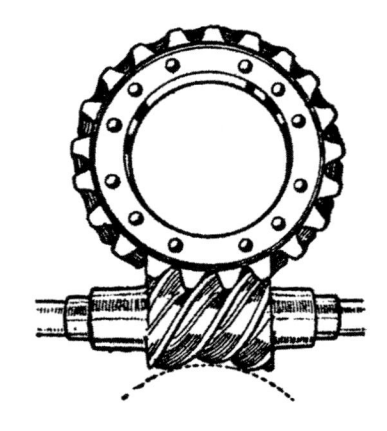

The Lanchester worm and wheel in their relative positions, showing full gear engagement.

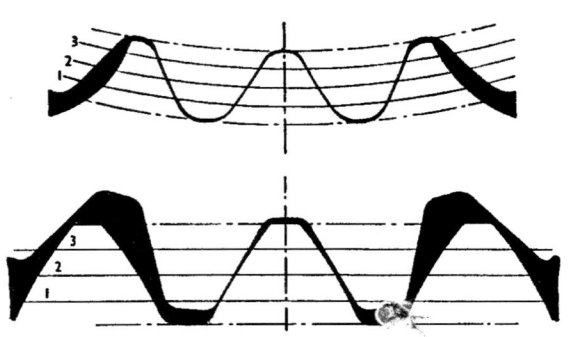

Showing the reason for the difference in efficiency and quietness between the two worm-drive systems, the dark section details the area of no surface contact between the worm and wheel. Top: the Lanchester hour-glass type. Bottom: the parallel type.

The debate over the respective merits of the parallel and hour-glass types of worm gear should have ended in 1907, when Fred had submitted his worm gear to the National Physical Laboratory for testing. Lanchester's hour-glass worm gear recorded a 97.6% efficiency; suspiciously, the parallel worm gear was not submitted and therefore not tested. The Autocar stated that: "The manufacturer who seriously attempts to meet the present demand for silence without adopting worm gearing, imposes on himself a very great and unnecessary handicap. Good results can be obtained with bevel gearing, but better can be obtained and maintained with worm gearing." Nevertheless, despite Fred's two patents of 1911 on worm gears, and his paper read to the I.A.E. in 1913, the controversy continued to simmer.

None of this was lost on the Lanchester Directors, one of whom, John Pugh, called in Fred for an interview. He told Fred that the Board had decided to give up the worm drive gear in favour of the more popular bevel gear for all future designs. Fred refrained from arguing against the decision, but later asked innocently, "...... whether the floor level was to be raised a matter of six inches, or whether the rear body was to have a tunnel in the floor for the propshaft, or whether a compromise was to be effected: the rear suspension being made stiffer or limited by buffers and the floor raised to a lesser extent?" Obviously none of these options had been anticipated by the Board and the directive was quietly dropped.

As if this slight was not enough for Fred to endure, the Directors also wanted to abandon the epicyclic gearbox, a feature of Lanchester cars for so many years. Fred's forceful character and strength of argument, coupled with the fact that he was proved correct more often than not, prompted them to cancel that directive also.

Although he won these battles, Fred felt increasingly distanced from the Board, some of whom he had come to despise. He felt that the Directors should have been experts at their jobs, as he strived to be at his; anything less was anathema to him. With so many unpleasant disagreements and misunderstandings between them, he felt that their doubt over his special worm drive and epicyclic gearbox was the last straw, and in the Autumn of 1914 he resigned.

He had received much praise during this time from many quarters, but sadly not from the Directors of the Lanchester Motor Company. In that year the Board consisted of Charles Pugh (Chairman), Arthur Gibson, John Pugh, James Whitfield, Hamilton Barnsley, Allen Whitfield and Joseph Taylor. It was indeed sad that their business and financial aspirations had never been fully in harmony with the scientific and inventive ideals of Fred, but a proportion of the blame for this must be attributed to him. Fred was able to grasp detail of the most intricate nature so quickly that he often became frustrated by the relatively slow reactions of others, and although he had the ability to patiently explain something, he would only do it once. The workforce respected Fred immensely and were in awe of his capabilities, as also were his Directors, albeit grudgingly, for they had allowed him to influence the Company's direction considerably.

The Company accounts at the end of the financial year of 1911 had shown a creditable profit of £30,755, an increase of £2,185 over the previous year. As well as loans having been paid off, the Directors and shareholders would have appreciated the £32,167 bonus in preference shares which had also been amassed. Although the overall profit for 1912 had fallen by £3,698 and then by £4,141 in 1913, dividends had still been paid, varying from 10% to 25%. The small profit of £1,032 for 1914 related to the worry of the impending War and a consequently shorter financial year.

It must have been fully realised by all involved that these profits were due to Fred's creative talent, for although George was Chief Designer, he was in effect just fine-tuning the legacy of Fred's earlier work. Accepting all this, the Directors' striving for a less radical design of car cannot, from a business sense, be seen as unreasonable. In their view, Fred's aim of educating the motoring public to adopt his "mechanically and scientifically correct design" would never be realised.

Fred's departure from the Company which bore his name was the end of a long and gradual process of declining trust and friendship, which had begun so brightly with the Company's formation just before the turn of the century. From this time on, although he was occasionally asked back to the Lanchester Works to give them the benefit of his advice, Fred was to inspire others with his original theories and inventiveness.

Chapter Fourteen

THE SIX-CYLINDER 38 H.P.

The new 38 h.p. Lanchester was, in effect, the product of co-operation between the two brothers, Fred and George, and the latter must have been very proud to have progressed to the position of "collaborating with Fred on equal terms".

The 1911 38 h.p. car was not brought out to replace the 28 h.p. six-cylinder model, but was produced alongside both it and the 20 h.p. model. The 38 h.p. evolved directly from Fred's earlier designs and so continued his influence in the development process. However, George's input as Chief Designer was far from insignificant, as he studied the whole programme and ordered changes which made the production methods easier.

A 1911 38 h.p. Lanchester-bodied landaulette, towing Britain's first streamlined caravan.

One alteration high on his list was the construction of the chassis. Although of the same general shape and pattern as the 20 h.p. and 28 h.p. models, George changed its construction from the earlier steel box-section to the latest technique of pressed steel 'C'-shaped girders. The standard wheel base of the 38 h.p. was slightly longer at 11ft.7ins. although the Company would build its 'special short chassis' of 10ft.7ins. at no extra charge. Torsional cross-bracing of the chassis was employed, with the sixteen-gallon petrol tank assembly forming the central main member and a sheet steel brace crossing the front, behind the radiator and the fans.

The 38 h.p. suspension also remained of the same general design, although the rear cantilever springs were relocated on the outside of the chassis. The parallel link system was revised to employ rubber-cushioned joints for the increased-length links, resulting in an even smoother ride than previously. The front axle was tubular in section and allowed for

THE SIX-CYLINDER 38 H.P.

The 38 h.p. chassis, with workshop rope-tyres.

a track of 4ft.10ins. The 42ft. turning circle was effected by a well-proportioned, brass, fold-flat steering wheel.

The rear axle construction was altered so that it could be produced in four separate parts, instead of the pair of aluminium castings as before. Two pressed-steel axle casings were secured to the two sections of the underhung worm gear housing. Ball bearing mounted half-shafts transmitted the drive to the standard Rudge-Whitworth wheels, which held tyres of 895mm. by 135mm. size.

The reserve brake system acted on the two rear-wheel drum brakes, and was actuated by the vertical side-lever operated by the driver's right hand; this brake was also used for parking. The cast-iron brake shoes working against the steel brake drum worked equally well forwards or backwards. The reverse brake was part of the epicyclic gearbox construction and was brought into operation by the engagement of reverse gear. As with previous designs, this brake could be applied when travelling forwards at any speed and would give the reversing gear only if its application was continued after the car had come to a halt.

The car's main brake was operated by the foot pedal, positioned to the right of the clutch and the central accelerator pedal. Housed in the rear of the countershaft unit, it consisted of a series of metal discs whose size was increased by half in the autumn of 1912. This main brake was both efficient and durable, and was constantly fed by lubricating oil from the gearbox oil pump. Late in 1912, the arrangement was further improved by re-designing the oil pump to enable its withdrawal from its casing without the escape of any sump oil.

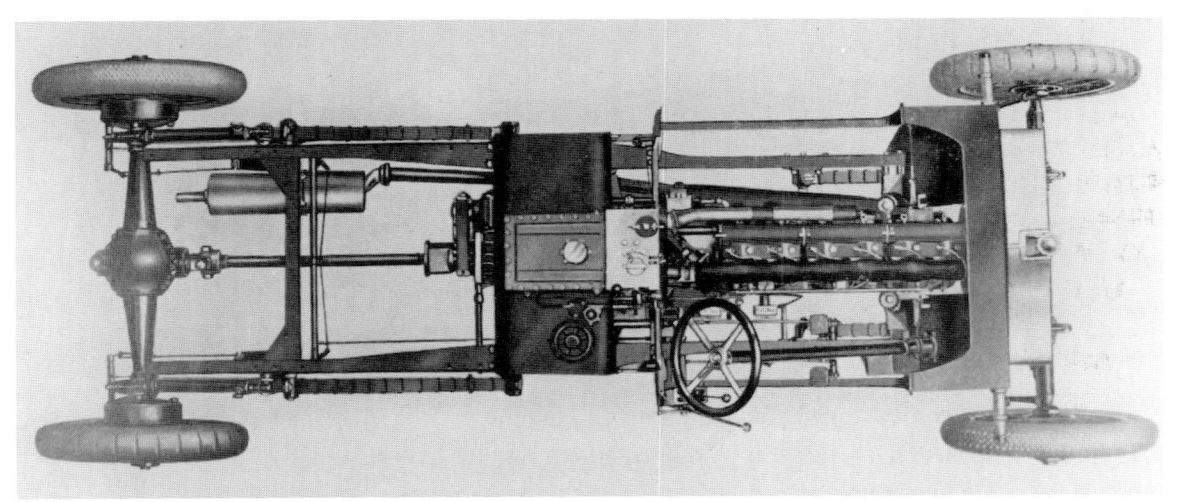

The 38 h.p. chassis, with pneumatic tyres of different tread patterns.

A 1911 38 h.p. with torpedo phaeton coachwork by Lanchester.

The throttle actuation was effected by a valve of the ordinary butterfly type and was operated by the foot accelerator for variable driving conditions, while the hand throttle was used for tick-over control. Fuel was fed to the wick carburettor by a pump located in the base of the tank, driven at its head by a spiral spring belt from the camshaft.

The tiller-steering and single lever gear-change systems had by this time been classed as obsolete, and the Lanchester wheel-steering and patented foot-control methods were the only systems the Company offered. The gear lever, termed the gear selector hand lever, was moved through conventional gate change slots, with three forward gears and reverse. Top gear operated through the direct-drive clutch consisting, as before, of a series of steel discs lubricated by the same system as the main brake. The other gears were of the familiar Lanchester epicyclic layout with compounding second gear. The combination of left-foot pedal and change-gear lever made the Lanchester system the most simple to use of any make.

The engine and countershaft were retained as one complete unit and were secured in the chassis in the accepted 'Lanchester engine position'. To avoid any tendency for the adjacent seats to become hot, the engine-covering panels were fitted with a double skin. The gap between allowed a flow of air to act as an insulator from engine heat and noise.

The engine itself was for the most part similar to those previously designed. In order to take advantage of Fred's new crankshaft vibration damper, employed at the front of the engine, the stroke was able to be increased by 1in. to 4ins. thus equalling the bore of the engine. The six cylinders were cast in pairs and the exhausts were taken out at each base through the water jackets; this had the effect of further quietening the engine. The cooling water pipes were cast within the tops of the cylinder blocks. The connections between the three blocks were equally simple, as an upper and lower rubber ring formed the seal between each; so successful was this method that Fred persuaded the Daimler Company to adopt it also.

The pistons were extremely light, made of high-grade steel in order to ensure a minimum of vibration and to "make a fracture impossible". Each piston had four compression rings at its top and an oil-excluding scraper ring near its base.

The connecting rods were made of high-tensile nickel steel, machined all over and carefully

THE SIX-CYLINDER 38 H.P.

The six-cylinder 38 h.p. engine and countershaft.

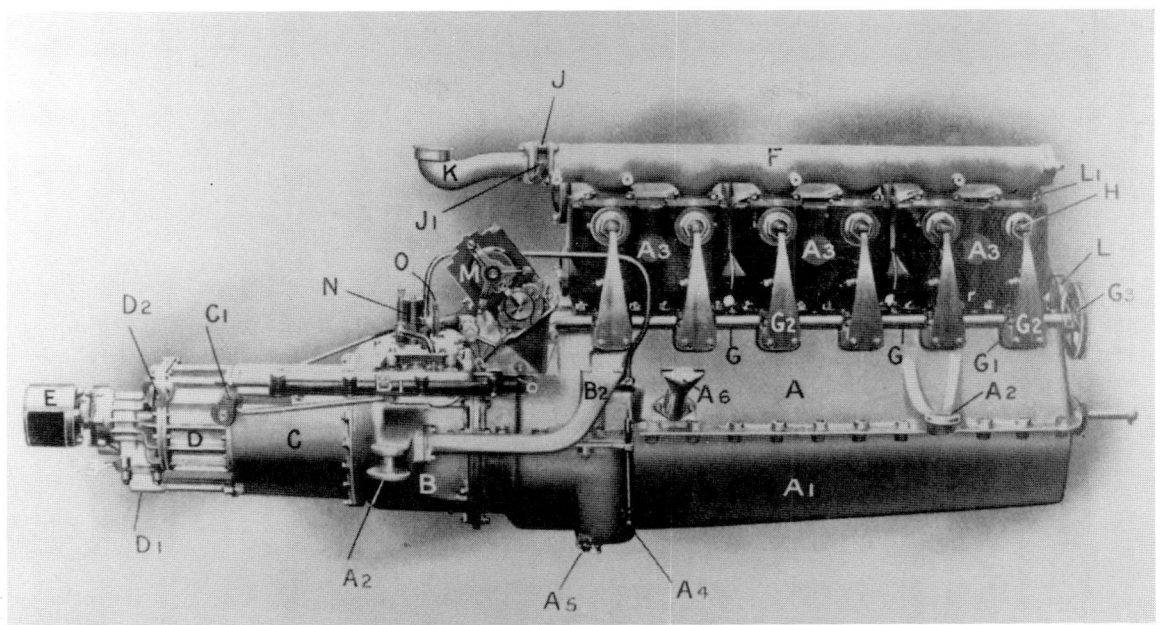

A.	Crank case.	D3.	Countershaft pump filter.
A1.	Crank pan.	E.	Driving pot or cardan coupling.
A2.	Motor support brackets.	F.	Feed pipe.
A3.	Cylinders.	G.	Cam Shaft.
A4.	Crank case oil level cock.	G1.	Cam box.
A5.	Crank case oil drain cock.	G2.	Valve springs.
A6.	Crank case oil filler.	G3.	Fan pulley.
B.	Change speed gear box.	H.	Feed valves.
B1.	Gear control box and clutch spring box.	J.	Throttle valve box.
		J1.	Throttle lever.
B2.	Countershaft oil filler.	K.	Vapour inlet pipe.
C.	Clutch casing.	L.	Water inlet.
C1.	Clutch lever.	L1.	Water outlet.
D.	Brake box.	M.	Magneto.
D1.	Countershaft oil sump.	N.	Oil pressure indicators.
D2.	Brake lever.	O.	Petrol pump driving pulley.

balanced. The cylinders were cast from a special mixture of cold-blast iron, and were said to be "as hard as it is practical to machine and are finished true by grinding". The crankshaft was machined from a solid forged billet of nickel steel and it ran in seven bearings, constructed of white metal cast on to brass shells. The flywheel was positioned at the rear of the engine and was fully enclosed.

The inlet valves, or admission valves as Lanchester called them, and the exhaust valves were positioned horizontally, opposite each other, and were operated by the two camshafts, one on either side of the engine. Although the exhaust valves were adequately water-cooled, they could be easily removed in a cage via the detachable inlet valve and seating. The flat valve springs were increased in width and length to achieve a more silent valve action.

Ignition was effected by a hand-controlled, high-tension magneto, a Bosch dual system being employed.

Lubrication of the engine and parts was by pressure-fed oil, working at 40 p.s.i. The lubricating oil fed all parts of the engine, including the small ends via a hole bored centrally up the connecting rod. Operated by this lubrication system was a tell-tale plunger on the central switch panel between the front seats, which rose visibly when there was pressure in the system and dropped down if it became insufficient. This delightful device was also used on the separate countershaft lubricating system.

The capacity of this 38.4 h.p. engine was 4,942 c.c. and its compression ratio was 4.7:1. At its normal running speed of 1,400 r.p.m. it developed 48 b.h.p. and at its maximum 2,200 r.p.m. the engine developed about 63 b.h.p.

From its first outing in 1911, it proved to be a fast engine and could power the car to 65

Every conceivable item at hand in the 38 h.p. limousine.

Luxurious individual seating in the limousine.

A 38 h.p. landaulette in the modified style, with the windscreen closer to the chauffeur.

m.p.h., returning 16 m.p.g. The Company was so pleased with all aspects of the car that the decision was made to extend its guarantee period to a full two years from the date of purchase, again with unlimited mileage. The terms of this guarantee stated that the Company undertook to "replace free of charge any material, with the exception of tyres, found to be defective". (This guarantee was to remain for each model produced by the Company until 1931.) Such an act of confidence in their products must have encouraged greater sales.

The car's all-round performance pleasantly surprised many, one of whom was the Autocar's motoring reporter, Owen John. In a previous article, he had complimented the Lanchester as a town car, implying this was its only use. He now wrote, "I wish to take it all back. I apologise. I spoke in ignorance; now I know better. Only a month or so back, after enjoying a stately progress in a magnificent Lanchester limousine body in the neighbourhood of London, I wrote that 'for the portly, middle-aged lady or gentleman, who wish to be conveyed in comfort from one place to another, there was no more suitable vehicle than a Lanchester .....' To prove I spoke in semi-ignorance, Mr. George H. Lanchester invited me and another to come out with him one fine day to see for ourselves what his cars were capable of doing in other fields, and I now rejoice that I made the initial error because it has afforded me a new experience and a joy I might otherwise never have had a chance of getting."

George took Owen John on a trip to demonstrate the car's worth and to this end no hill, either up or down, dirt track or speed limit would deter him. The reporter's comments were many, such as: "..... how fast we went, this is no place to mention ..... after Broadway came test hills. Personally I should describe them as shoots, but these were evidently what we were seeking, and Mr. Lanchester put his car up them in a manner beautiful to behold ..... never have I experienced anything like the feeling of power and ease on a car that I did on that occasion."

Owen John's companion, a mechanical expert, was equally complimentary, with such phrases as: "There is no main road hill in the Cotswolds anything like as steep as at Saintbury. At the present time (it) is in very bad condition: loose and altogether rough, but

Lanchester Exhaust Foot Warmer.

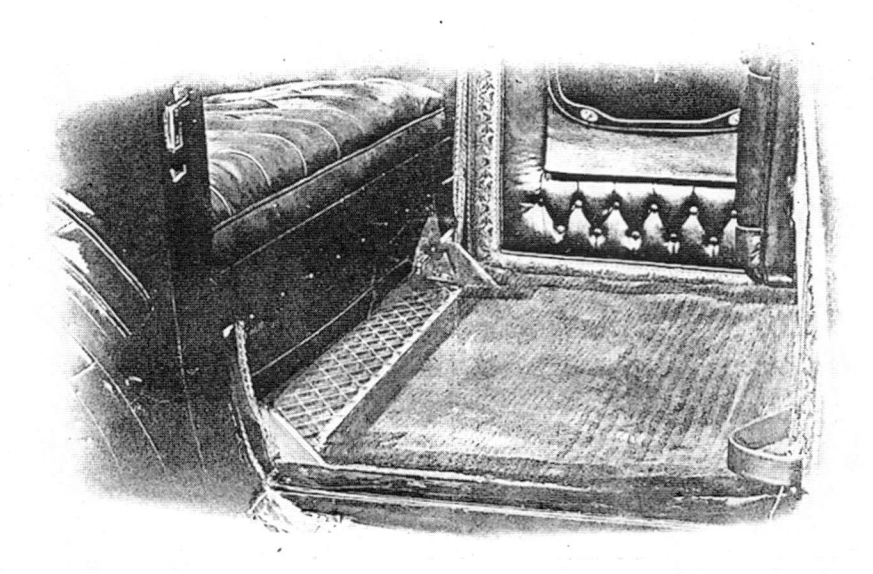

THIS has been designed to meet the requirements of those who use their Cars in all seasons.

The Photographic illustration shows the Foot Warmer fitted to the interior of a Landaulet.

The diagramatic Line Drawing clearly shows the method by which the exhaust gases are diverted, when required, to pass through a coil pipe under the foot-board. The foot warmer is operated by means of the lever below the seat on the near side.

When the foot warmer is "open" or "shut," the positions of the lever are indicated by the letters O and S.

PRICE, fitted complete at our Works or Depôts,

£7 10 0

Price of Set of Parts

£6 0 0

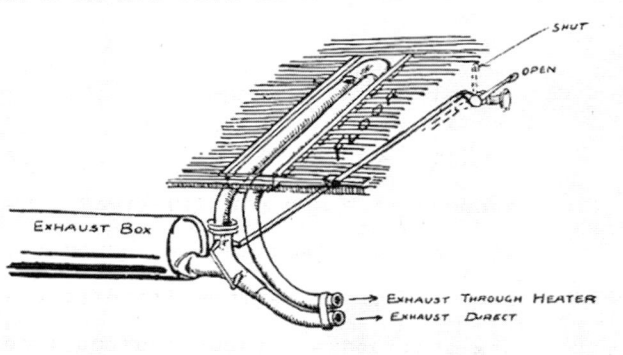

A contemporary Lanchester advertisement.

the Lanchester simply played with it ..... Nothing could have been more convincing than the way in which the Lanchester danced up this steep, horrible, gulley track, not merely as a demonstration of power, but of super-abundant life which was delightful ..... it will run down to a crawl and accelerate rapidly and then swallow up a straight, all without the least fuss or noise." As a summary he expounded, "Indeed, the 38 h.p. Lanchester is so smooth, fast, and altogether delightful a car that it is really difficult to do it justice without seeming to indulge in undue enthusiasm..."

Others to be impressed by the 38 h.p. Lanchester were from Rolls-Royce Limited at Derby. In late 1911, and with great secrecy, they purchased a rolling chassis (number 1201, with engine number 1200), via the famous west London coachbuilders, Barker and Company.

THE SIX-CYLINDER 38 H.P.

The 38 h.p. chassis which Rolls-Royce bought in 1911, for evaluation and comparison.

Working closely with Rolls-Royce, Barker provided cover to avoid the Lanchester Company realising the true owner.

After general inspection and comment the car was fitted with a "very rough" two-seater body by Barkers. It was registered with the Derby number R1261 and was driven to the south of France, in February 1912, to the home of Henry Royce, for further road trials and inspection. His subsequent comments showed that, whilst he disliked certain Lanchester features, he approved of others.

The following month, on its return to Britain, the car was dismantled for closer inspection. The Lanchester worm drive rear axle was regarded as "quiet, although not absolutely so". The epicyclic gearbox was considered more than once for introduction on the Rolls-Royce cars. Royce initially remarked, "Second, first and reverse gears very good indeed. Second gear, exceptionally silent. Nice and easy to change gear." No action was taken then, but some years later Royce again raised the possibility of its introduction when he posed the option ".... whether it is desirable to have such a gear box on the Rolls-Royce car even if it seems impossible to make it any better than the standard Lanchester."

The Rolls-Royce Company investigated thoroughly the constituents of Lanchester oils,

The eight-light, five-seater, coupe-limousine coachwork by Lanchester on the 1912 38 h.p. chassis.

38 h.p. cabriolet with artillery wheels.

maximum engine speed and a variety of other mechanical detail to use as a comparison with their own specifications. The renowned amount of passenger space in the Lanchester was compared to their own. One note to Claude Johnson, their Chief Executive, showed the smaller Lanchester car to have a surprising 15ins. of extra leg-room inside. This was borne out by the Autocar who wrote about the space given to the front passenger: "Lanchester design makes for roominess, and one notices this as one pushes a spare can of petrol forward, places a bag behind it and even then has room to stretch out one's legs to the full."

The Lanchester cantilever rear-springing, which had earlier been investigated on a 28 h.p. car by Henry Royce, was examined again on the 38 h.p. Comparisons were duly made between it and that fitted to his own 40/50 car. Although the Lanchester 38 h.p. was not intended to directly compete with the Rolls-Royce 40/50, many owners commented on the similar performance of each, even though the Rolls-Royce had a 20 per cent greater engine capacity.

In 1912, George began to make changes to the body designs and brought in more

The Maharajah of Nawanagar's tenth Lanchester, a 38 h.p. Pullman limousine with new style radiator.

Lanchester domed-roof limousine coachwork with concealed spare wheel in the top-pivoting rear panel, on a 1913 38 h.p. chassis.

rounded and flowing lines. This prompted a change of radiator shape and the 28 h.p. style was abandoned.

The Maharajah of Rewa obviously approved of these modifications, as he ordered a special state limousine which was delivered to him in May 1912. The Motor proclaimed that the car "was painted and upholstered in a most lavish style, the colour of the panels being the brightest of blues and the leather upholstery to match". The roof incorporated ventilation louvres and an electric fan inside a dome. The whole limousine top and driver's canopy were separately removable so that the car could be used as an open tourer; a folding hood was also provided. A chauffeur's seat, which could be folded-up to allow access to the front passenger seat, was another feature.

(The owners of the Ravenglass and Eskdale Railway Company were also impressed by the Lanchester's reputation, and chose an old 38 h.p. Lanchester in 1925, to use as a power-horse for moving vast quantities of stone from their quarry. They bought a 1912 open tourer, chassis number 1375, for £100, and proceeded to strip it down and graft it onto their rail bogies. The whole of the engine, countershaft, petrol tank, wick carburettor, radiator and most of the back axle were used. This Lanchester railway tractor-unit was christened Ella and gave four years' excellent service hauling 40-ton loads of crushed stone from the quarry.)

Another development was recalled by Francis Hutton-Stott and Anthony Bird. ".... remarkable was the performance of Lord Ridley's 1913 38 h.p., which retained its touring body but was modified by a slight increase in compression ratio and by re-contouring the cams to give higher lift; stronger valve springs and a low-pressure supercharger completed the job and gave the car startling acceleration and a top speed of more than 90 m.p.h., to the consternation of owners of 30/98 Vauxhalls, Bentleys and similar fast cars, who were passed and left behind by this sedate and elderly touring machine."

The touring version of the 38 h.p. was originally marketed as the five-seater phaeton deluxe and was sold at £150 above its chassis price of £885. The £1,035 total price for this model was, in fact, the cheapest that Lanchester offered. A whole range of coachwork was available, with the magnificent seven-seater Pullman limousine being the most expensive at £1,185.

The cars stayed in production until the onset of the First World War, although a few were subsequently made for the War Office. The 38 h.p. engines and chassis were used with a variety of different bodies for the War effort, the most famous of which were the armoured cars.

Chapter Fifteen

THE FOUR-CYLINDER 25 H.P.

For customers unable to afford the 38 h.p. car, the Lanchester Motor Company offered an alternative in their 25 h.p. four-cylinder model. To take account of its smaller engine, the forward chassis and coachwork were reduced in length, although this was practically indiscernible to the casual observer. The rearward chassis and coachwork, to all intents and purposes, were the same as the 38 h.p. models which had been introduced earlier in the same year of 1911.

The availability of the 25 h.p. effectively brought to an end the production life of both the 20 h.p. and 28 h.p. cars.

The square radiator denotes that this Lanchester-bodied torpedo phaeton was one of the first 25 h.p. models, in 1911.

Its introduction had not been without problems, however, as after the prototypes had been on test for a period of time, there were reports of noise and vibration from the engines. Fred had explained that the best way of retaining balance in a four-cylinder engine was to have the two end cylinders forming one pair and the two middle cylinders the other, so that the engine would be symmetrical. The 25 h.p. was designed in this way, as the earlier 20 h.p. model had been. He also pointed out that the vibration was not completely neutralised because, "The motion given to the pistons by the connecting rod is not symmetrical, and the four connecting rods are all at their maximum angularity at once. This 'secondary vibration' is very powerful in the four-cylinder engine, however carefully it is constructed." When considering the Lanchester 25 h.p. engine at maximum revs, Fred stated, "It amounts to as much as half a ton applied alternatively upwards or downwards, twice per crankshaft revolution; the importance of elimination on want of balance of this magnitude can scarcely be overstated." The problems connected with these "torque

THE FOUR-CYLINDER 25 H.P.

alternations due to piston inertia" of the engine were so unsatisfactory to Lanchester's exacting standards that the Company considered abandoning the four-cylinder car trade altogether. Fred explained, "You will understand that the Lanchester Company's business is, from a modern point of view, a comparatively small business, being only done with the clientele who will pay a high price and who are correspondingly fastidious in their demands."

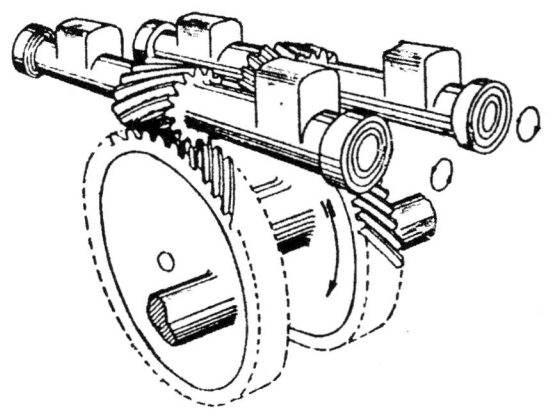

The Lanchester Harmonic Balancer. The two large crankshaft gears drive the two smaller cross-shafts and their four bob-weights at twice the crankshaft speed, in order to obtain complete dynamic balancing.

Fred's cure was ingenious and, as usual, attacked the problem from a scientific base. Drawing on his experiences with the balance weights fitted to the crankshafts of the earlier twin-cylinder engines, he calculated that the common centre of gravity of the four pistons had "a small harmonic motion of double frequency". To counteract this, he designed his 'harmonic balancer', which incorporated weights mounted on two shafts which were geared together to turn at the same speed, but in opposite directions. This device, which surprisingly absorbed no power other than by friction, was geared to run at twice the speed of the crankshaft and gave complete dynamic balance to the engine. The installation of the harmonic balancer also allowed a four-cylinder engine to run at 30% higher speeds than was previously possible, with a corresponding increase in performance.

Fred patented his harmonic balancer in the United Kingdom in 1911. Many motor companies were interested in the device but, apart from some marine and stationary engine constructors, only Vauxhall, Daimler and the Astor Engine Company took out licences in this country. To add the device to an existing engine usually meant that, due to lack of room, it was installed beneath the crankshaft, resulting in a deeper engine sump. This

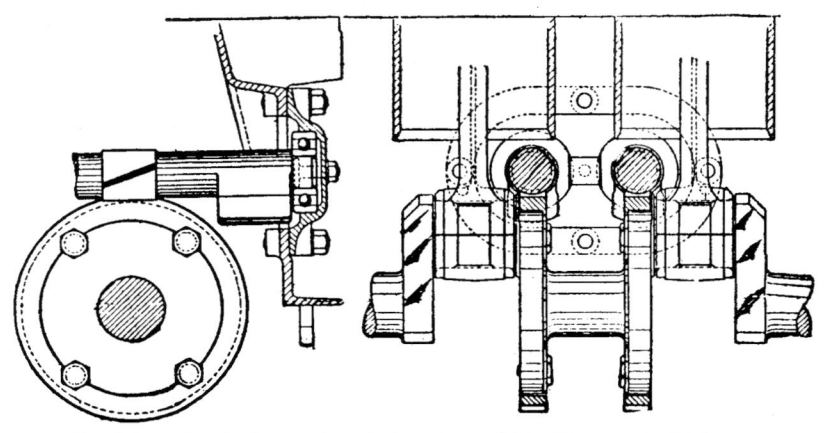

Fred Lanchester's sectional drawing of his Harmonic Balancer.

disadvantage, coupled with the adverse effects of extra weight and extra moving parts, as well as the significant expense of manufacture and payment of royalties, prevented others from taking up the option. They preferred instead to allow their engines to succumb to the secondary vibration, but to insulate them from the chassis frame by using simple rubber mountings. Lanchester's invention was actually more useful to the American market, where many of their luxury cars were powered by four-cylinder engines of large bore size. Three variations of the patent were obtained in the United States of America and a few years later, in 1915, the Willys Group bought these from Fred for £10,000. (It is interesting to note that after decades of receiving no attention, the Lanchester harmonic balancer has re-emerged as a valuable tool for engine design. Companies such as Ford, Chrysler, Colt, Dodge, Perkins, Mitsubishi, Porsche, Saab and an increasing number of motor cycle companies are taking advantage of Fred's invention, although not many give him due credit.)

A 25 h.p. fitted with a cabriolet body by Cann, in April 1912.

The harmonic balancer was quickly brought into the production of the Lanchester 25 h.p. car as an integral part of the engine, located on two internal cross shafts set above the engine's crankshaft. It fulfilled its function admirably and the Lanchester Company paid royalties to Fred for its use, although these were at minimum rates, by agreement.

The 25 h.p. engine was similar to that of the 38 h.p. six-cylinder. Its bore size of 4ins. equalled that of the stroke, and the four cylinders gave a cubic capacity of 3,295 and an R.A.C. rating of 25.6 h.p. The engine's maximum speed of 60 m.p.h. was good, as the complete car weight averaged 32 cwt. (Its potential for development became noticed some ten years later when the racing driver, Tommy Hann, startled many with his exploits at the Brooklands Motor Course in an old Lanchester 25 h.p.)

The chassis frame of the standard 25 h.p. car was one foot shorter than that of the 38 h.p. model, its 14ft. overall length giving a wheel-base of 10ft.7ins. A 9ft.7ins. wheel-base special short chassis version was also produced to order, without extra charge. The 4ft.10ins. wide track gave a remarkably good turning circle of 36ft. and, to take account of the car's lesser weight, smaller tyres of 880mm. by 120mm. were fitted.

Although many features of motor-car design derived from Lanchester innovations, Fred himself rarely felt the need to incorporate the designs of others. On the few occasions when this did occur, he utilised them quickly to gain full advantage. This was the case with the

THE FOUR-CYLINDER 25 H.P.

A 1912 25 h.p. car showing the latest-style radiator.

new 'electric self-starting and lighting set' which had been produced by the American Delco Company.

There had been a few engine starters designed previously using compressed air, including Fred's patented turbo system of 1902 and, amazingly, an experimental electric starter fitted to an Arnold car by its owner in 1896, but no company had felt the need to develop them for production use. Similarly, from 1907, electric sidelights had been fitted to Lanchesters and other makes, but these had only been powered by a non-charging accumulator/battery. The American Cadillac was the first production company to fit a complete starter and generator system, in 1912, but Lanchester gained credit for forward thinking by being the first European car producer to do so, albeit by a small margin. The October edition of Autocar announced that: "A self starter, for which an extra charge will be made, can be fitted to order. The self starter is also adoptable to all existing 25 h.p. and 38 h.p. Lanchester cars."

The electrical gear, as well as the dynamo-motor, was supplied by the Delco Company, but all the intricate linkage and gearing required was designed and produced by Lanchester themselves.

Another order for the Maharajah of Nawanagar: the 25 h.p. torpedo tourer with Lanchester coachwork.

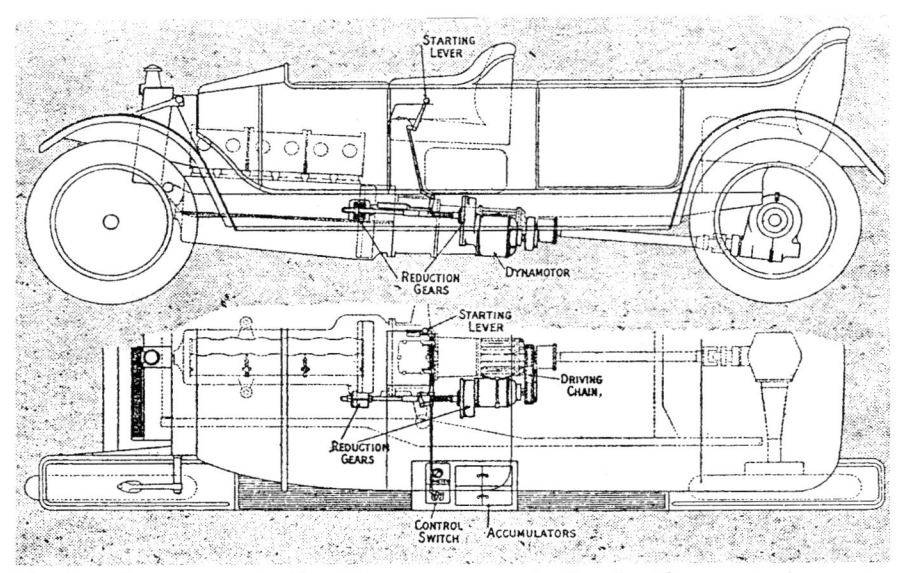

The Lanchester was the first European car to be fitted with a starter/generator installation.

The means of operation was by a short, vertical lever on the central control panel, which moved forward in its slot against a return spring. This switched the wire into its series-connection, to produce 32 volts, an effective measure when starting the big engines on a cold winter's morning. The lever also engaged the device's drive shaft and reduction gears, in order to turn the engine flywheel and so start the engine. When the starting lever was released, the drive shaft was disengaged and the electric switch immediately reverted to its normal parallel-connection position. Driven by an enclosed, lubricated chain from the car's propellor shaft, the generator section of the dynamo gave an 8 volts, 10 amperes charging rate for the four accumulators. To prevent discharge from these when running at low speeds, a cut-out relay was provided. This advanced system also allowed for automatic regulation of the current. The gauge, termed the ampere hour-meter, measured the current flowing into the accumulators, and the accompanying pointer on the device showed the

Original view of a Delco starter/generator set, as fitted to a Lanchester running board in late summer, 1912. The ampere-hour meter and gauge are seen behind the vertical control switch.

THE FOUR-CYLINDER 25 H.P.

amount of current put into and taken out from the accumulators and thus whether the system was charging or discharging.

The light switches were housed in the shaped, wooden box set above the control panel, and its quality was typical of the whole system and its layout. In fact, the whole of the electrical and mechanical device was a worthy addition to both the 25 h.p. and 38 h.p. cars.

Other improvements had been made to both models by October 1912. The gearbox oil-pump, which provided oil to the brake, clutch and gears, could be removed from its casing for cleaning with no escape of oil. Also, the brakes were enlarged by 50%, giving both greater retardation and longer periods between adjustments.

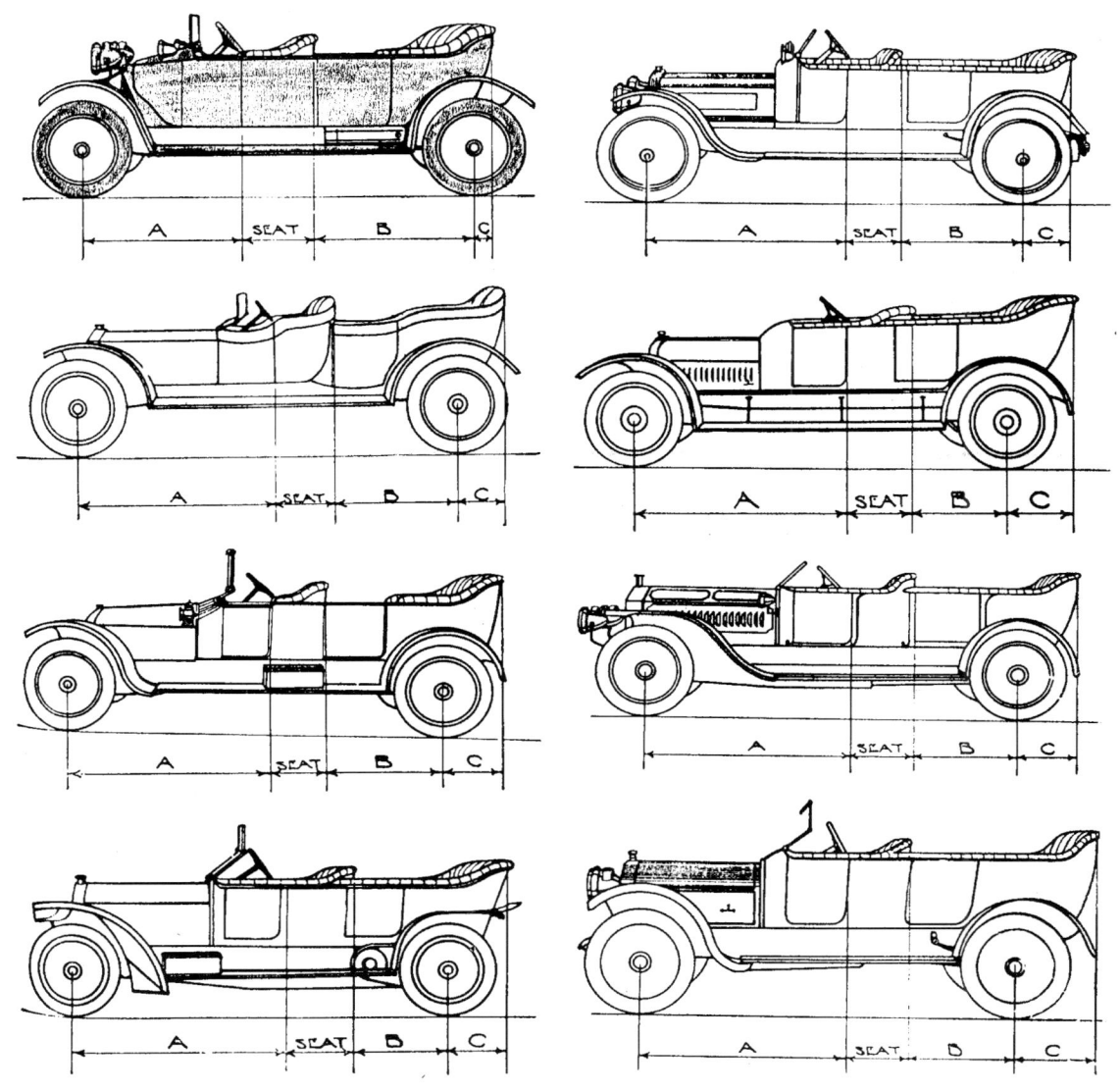

The 1912 Lanchester car, shown first, favourably compared to other leading makes for passenger space.

The first Lanchester 25 h.p. shown to the public had been at the November 1911 Olympia Motor Exhibition: a seven-seater double landaulette with canopy, priced at £800, excluding six guineas tax. The same model, probably displaying the new enlarged radiator design, was used to enhance the Lanchester stand at the Motor Show, held in February 1912.

The following month, a letter from the Lanchester Motor Company was published in the Autocar magazine, emphasising the advantages of their 'bonnetless' coachwork. The outline drawings depicting the Lanchester's competitors drew attention to the "crowding of the coach body" and the "excessive overhang behind the rear axle". It was also noted that, with competitors' saloon and limousine cars, "the overhang is still more marked". The letter concluded, "In these days of wide mudguards, large radiators and immense bonnets of no mean weight, accessibility can hardly be the excuse for retaining the engine in front of the dashboard."

Rather unimaginatively, considering the choice at their disposal, the Lanchester Motor Company turned out the same car as in the previous two years for the 1913 Motor Show, even though their torpedo tourer appeared to be more popular. In view of the limited advertising for the 25 h.p. car, it must be assumed that the Company preferred to attract the more wealthy customer for their 38 h.p. models. Although sales of the four-cylinder 25 h.p. continued for many years, this lack of enthusiasm in promoting the car was indeed an omen for the future, as the 25 h.p. was to be the very last four-cylinder model to be produced at Armourer Mills.

Chapter Sixteen

THE SIX-CYLINDER SPORTING 40

The idea of the 'Sporting 40' had been first conceived in 1913 and it was this which had led directly to Fred's departure from the Lanchester Motor Company. His hard-won success in saving both the epicyclic gearbox and the worm gear rear axle for this new car was his parting gift to George, who then took over complete control of design. George's steady progress learning from his "genius brother" over the past two decades had left him in good stead to design cars which, although different, would continue to uphold the reputation they had gained during Fred's reign.

George Lanchester in his prototype Sporting 40 h.p. of 1914.

George accepted the need for change, and therefore endorsed many ideas put forward by the increasingly dominant Directors, although he was surprised at the strength of their desire for the cars to become 'conventional'. He agreed that it would be sensible to incorporate a fashionable long bonnet and also a spray-type carburettor instead of the wick vaporiser. He did, however, fight hard to preserve the Lanchester cantilever and parallel link suspension instead of their favoured system of semi-elliptic springs. Eventually a compromise was reached, whereby semi-elliptic springs anchored to protruding chassis dumb-irons were used only on the front suspension. The measure which upset George the most was the Directors' insistence on a side-valve engine. With little awareness of engine design principles, their idea was based solely on the fact that other competitors used side-valve engines. George was aghast at this retrograde step, and informed the Board that he deplored the very thought of it and would not go ahead, unless specifically ordered to do so. Lacking the international reputation or the dogmatic forcefulness of his elder brother, however, George was unable to change their minds and was duly compelled to comply with their directive. The principles of the Sporting 40 design were set and work began.

The Sporting 40 engine of 1914.

The chassis frame had a wheel base of either 11ft.8ins. or 12ft. 8ins. and was of great strength. It was made from 0.156ins. thick plate and its depth and width measured 6ins. by 4.25ins. respectively for most of its length. To maintain its torsional rigidity and prevent twisting, especially as the supporting wick carburettor petrol tank was not employed, the frame sides were joined by two 4.5ins. diameter tubes across the centre and rear, and a

A Sporting 40 tourer with an Ace of Hearts bonnet mascot, aeroplane radiator mascot and snake's head horn.

THE SIX-CYLINDER SPORTING 40

smaller tube across the dumb-irons which also located the starting handle. The combined engine and countershaft provided additional torsional bracing.

The usual tubular front axle of the Lanchester was discarded in favour of a steel stamping of 'H' section. The swivel pin assembly was unaltered, giving a front wheel track of 4ft.10ins.

The steering mechanism was of the 'double worm and nut' type, which followed previous Lanchester practice. One worm was of left-hand and the other of right-hand construction, and both were cut into the steering column shaft. When this was rotated, a half-nut located on each worm moved up and down, with the result that the drop-arm moved in an arc, so causing the steering linkage to move the front wheels in the required direction. Although this was costly to manufacture, it gave accurate and precise steering and was long-lasting due to the great surface contact area. (This design continued in use until 1931.)

Rudge-Whitworth wheels were used, with tyres of 895mm. by 135mm. The rear, internally-expanding drum brakes were used as a reserve or parking brake, operated by a side-lever held by a spring into a vertical ratchet on the chassis. Its adjacent gear lever worked within the four-position gate change and, as before, operated the epicyclic and direct drive gears, in conjunction with the left foot pedal which operated the multi-disc clutch. This was pressure-fed with oil by a special pump situated in the gearbox, as was the foot brake, and both were housed within the countershaft.

The engine's oil pump was driven directly off the crankshaft. It supplied oil from the sump to each of the seven main bearings, through the hollow crankshaft to the big ends and upwards, via the large connecting rods, to the gudgeon pins situated within the convex top pistons. Dual ignition was provided by high-tension magneto and a coil-battery system, and could be used separately or together. The sparking plugs were screwed vertically into the top of the two monoblocks of three cylinders and ignited the petrol supplied by a Lanchester-modified, five-jet Smiths carburettor. This in turn was pressure-fed by an air pump driven off the camshaft, forcing petrol from a rear tank.

The single camshaft was set low down on the nearside of the engine. It actuated the

The Sporting 40 steering wheel folds to assist entry to the driver's seat.

vertical inlet and the adjacent exhaust valves, which were not located overhead the pistons in the most efficient position, but instead were tucked away in a space running horizontal with the top of the cylinder, thus making an inverted 'L' shape. The Lanchester Motor Company Directors proudly let it be known that the 'L head', as it was termed, was a "concession to general design principles ... in order that owner-drivers might have a car with as many orthodox features as possible".

The Automobile Engineer stated that: "The Firm have endeavoured to eliminate all Lanchester features that in the experience of the Company were not really understood, and all features against which certain prejudices may be said to have existed." In this they achieved their aim, for when the same journal commented on the engine design, it reported: "The design of the engine generally is so thoroughly orthodox that there are really few points calling for particular note."

A 1915 Sporting 40 tourer with mascot secured to the radiator's front, above the water level sight-glass, which itself was covered by a circular Lanchester nameplate. (See appendix.)

An omission which seems strange was that of the crankshaft vibration damper, which was not incorporated into this six-cylinder engine design. The benefits of the damper were appreciated by all engineers, and George was obviously more aware of its advantages than most. It must therefore be assumed that restrictions imposed by the Directors prevented George from incorporating it in the new engine. In effect he was forced to back-track a decade to the inception of the 28 h.p. engines, in which the crankshafts had to be made exceptionally robust in order to withstand torsional vibration.

One assumes that the Directors were well pleased with the results of their interference, but George was not. He described the engine as ".....about the worst form of combustion chamber conceivable, but one that was in vogue at that time. Fuel economy was poor, combustion tended to roughness and brake horse power was poor for the dimensions." George's very first design of engine, through no fault of his own, was one that he hated, and he must have felt demoralised and perhaps even embarrassed by the whole experience.

George's various designs of coachwork for the Sporting 40 chassis, however, were indeed something to be proud of, and proved worthy of all the time he had spent carving scale

THE SIX-CYLINDER SPORTING 40

A Sporting 40 with open-drive limousine coachwork.

A Sporting 40 two/three-seater phaeton with rear dickey seat.

models from rounds of Cheddar cheese. The Sporting 40's body design was set off by a more square radiator, discreetly evolved from the 38 h.p. model. The filler neck and cap resembled the earlier model, although the water level sight-gauge was moved from the neck to the top of the radiator.

The 1914 Sporting 40 received a fair trial in the press, although the Automobile Engineer began their review in sombre mood. "The fact that a Company such as Lanchester have found it impossible to educate their potential customers to an appreciation of the correctness of Lanchester principles is a matter for regret, as it proves that we still have a very large buying public not yet sufficiently educated really to appreciate a car that is designed and produced strictly by adherence to the first logical principles of design." After clarifying their position, they then gave credit where it was due and wrote praise such as: "sporting automobile de-luxe class .... refined quality and construction .... evidence of quality in every detail .... high grade design .... leaves nothing to be desired .... exceptionally well sprung .... extremely fast on the road."

Even though he had been working under duress, without his heart fully in the project, George had produced a car of considerable attributes, capable of 70 m.p.h. Making reference to one owner's letter, of both good and bad content, George later recalled, "Although I have no illusions about the defects of the 1914 40 h.p., it was stated by one owner to have averaged 50 m.p.h. from London to Liverpool. The same owner, when he did not want to pay for a fairly extensive overhaul sometime later, claimed that it was a 'dud' car. In the following law suit, when his letter was handed to the judge, the latter looked over the top of his glasses and said, 'Not a bad performance for a dud car.' Really though, I felt that to some extent the customer was right."

The defeated owner had paid £1,035 for his 6,000 c.c. Lanchester, of which £885 was the chassis price, as well as ten guineas tax. This was the same rate of tax charged for the concurrent, smaller-engined 38 h.p., but the R.A.C. rating of 38.4 h.p. meant that only the bore size of 4ins. was taken into account, not the increased stroke of 4.5ins.

Although there is no record of Fred's thoughts on this new-generation Lanchester, other owners were quite satisfied with the Sporting 40. One reported later, in the 1919 Autocar, that his 1914 car "did much good work during the War". Ironically, it was the War which put a halt to the production of the car, probably after only eight had been made. Highlighting his personal feelings for the car, George summarised: "I was most grateful that the War intervened, before more than a few of these cars had been made."

Open-drive limousine coachwork on 1924 40 h.p. chassis no. 1936. (See appendix.)

1921 40 h.p. chassis no. 1755, with Lanchester tourer coachwork. This tourer was modernised in 1925 by the Lanchester Motor Company to incorporate a braked front axle, different hubs and wheels and new headlamps. (See appendix.)

The only remaining left-hand drive Lanchester: 40 h.p., chassis no. 1968, with Lanchester tourer coachwork. (See appendix.)

1925 40 h.p. chassis no. 1965, fitted with old-style engine and four-wheel brakes. Polished aluminium two-door coachwork by Barker. (See appendix.)

A 1928 40 h.p. chassis no. 2010, with Lanchester touring body incorporating wind-up, glass windows. (See appendix.)

1924 21 h.p. Lanchester fixed-head coupe body with dickey seat, on chassis no. 3057. This has been in the owner's family since 1936. (See appendix.)

The author's 21 h.p. with Lanchester limousine coachwork, ordered by Captain Morcom in 1924 and owned by his family for forty years. (See appendix.)

The Penman landaulette coachwork utilises a sliding top on the roof and a two-position folding section at the rear, on the 1924 21 h.p. chassis no. 3123. (See appendix.)

Chapter Seventeen

THE FIRST WORLD WAR

Following prolonged tension in Europe, the First World War broke out in the summer of 1914 and Britain became embroiled on the 4th August. Except for the few motor companies producing staff or messenger cars, tender vehicles, vans, ambulances and lorries, British vehicle production gradually ceased. Many motor manufacturers received orders to produce munitions for the War effort, and the Lanchester Motor Company was initially given the alien task of producing shrapnel shells, for which their machines were quite unsuitable. However, with some 1,200 shells produced, it was realised that shrapnel was ineffective in trench warfare and new instructions were given to the Company. The new contract was actually more inappropriate than before, as the order was for the production of 4.5ins. naval gun shells.

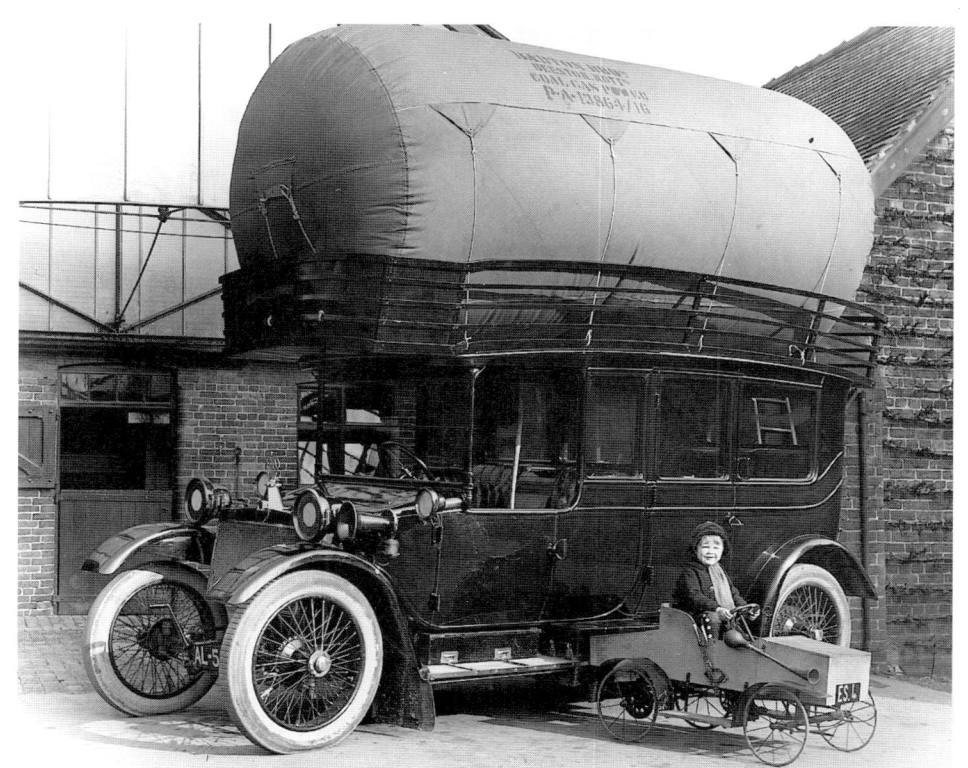

Preparing for war; coal-gas power for the 38 h.p. limousine and pedal power for the other.

Frank Lanchester was still based in London at this time and his wide circle of friends, acquaintances and business associates from key areas of industry and politics was to prove most advantageous for the Company. Frank fervently stated his case, arguing that his Company could benefit the country more by producing items for which they had a worldwide reputation; namely motor vehicles. By his sheer persistence, he won the Company a contract to build vehicles based on the 25 h.p. and 38 h.p. design, and in particular armoured cars.

One of the 25 h.p. tourers ordered by the War Office for R.N.A.S. use.

Initially an armoured super-structure was fitted to a standard 38 h.p. chassis, as a prototype. Improvements were found to be necessary, and all subsequent armoured cars produced were modified. A re-inforced chassis frame incorporated an extra set of cantilever springs for the rear, while coil springs supplemented those on the front. Due to the increased weight of the 8mm. plate armouring the body, the whole vehicle weighed 4.75 tons. To compensate for this, twin wheels were also added to the rear axle.

The endurance of tyres was of paramount importance in war time. The motoring historian, Ronald Barker, recalled the story that whilst on road trials prior to delivery, the armoured cars were testing new 'puncture-proof' tyres which were filled with liquid latex. They were not initially successful, however, as, while going through Birmingham City Centre, one of the tyres suddenly burst, covering an unfortunate woman from head to toe in the substance and resulting in the apologetic Lanchester Company having to buy her a new frock!

The sheer bulk of the Lanchester armoured car made it a formidable machine, measuring 16ft. in length, 6ft.4ins. in width and 7ft.6ins. in height. For armaments, it carried a Vickers-Maxim .303 machine gun in a fully-rotating turret, complemented by the handguns of the crew of three or four. The 38 h.p. design was deemed to be most suitable for War use as the engine and essential mechanics were centrally positioned, well protected by armour. This also enabled mechanical adjustments to be made to the engine and countershaft whilst still inside the protection of the armoured shell. From the driver's eye-slit position, the armour sloped downwards to the radiator opening, so giving extra forward vision which was particularly beneficial when travelling at its high maximum speed of about 50 m.p.h.

As with all vehicles used in the War, precise details of service are not available. It is presumed that the Lanchesters, with other makes of armoured cars, were bodied either at the dockyard in Sheerness or at the Royal Naval Air Service headquarters at the Talbot Works in Kensington, the armour plate being provided by the Beardmore Company. George Lanchester recalled that an initial order for six armoured cars was completed and sent via France to Belgium in 1915, to help protect that country's neutrality.

After their successful campaign, Winston Churchill, the First Lord of the Admiralty,

THE FIRST WORLD WAR

The prototype 38 h.p. armoured car, with a standard car chassis and single rear wheels.

ordered the armoured cars to speed back to Antwerp to help defend the city. A further fallback to the Allied lines at Ypres enabled the Lanchesters to be used to rescue pilots, shot down behind enemy lines. It later became clear that the trench warfare was really no place for 'Churchill's Motor Bandits' and the Lanchesters, with the majority of other armoured cars, were returned to England.

Whilst in France they had been complemented by a number of ancillary vehicles built by the Lanchester Motor Company. Seven 25 h.p. wagons, one 38 h.p. ambulance, and a 38 h.p. landaulette staff car were sent to France, but no details of their exploits exist. Five 38

A batch of production 38 h.p. armoured car chassis waiting to be delivered, complete with re-inforced chassis and suspension, and twin rear wheels.

The production 38 h.p. armoured car, one of forty two.

h.p. touring cars and two 25 h.p. landaulettes were commissioned for use at the headquarters in England, but their fate is similarly unrecorded.

Due to the potential shown by the armoured cars, the R.N.A.S. decided to form a more specialist force - the Royal Naval Armoured Car Division - and they contacted Lanchesters, amongst others, again. George related, "We then received a contract for thirty-six armoured cars with six high-speed tenders, six field kitchens and other equipment." The total Lanchester contingent was divided equally into squadrons 5, 6 and 15, and all, it is believed, subsequently served in France.

Presumably referring to the initial batch delivered in 1914, George reflected, "Six other (Lanchester) armoured cars were shipped to the Middle East, but were sunk in the Mediterranean."

It had meanwhile been decided to transfer the R.N.A.C.D. to the Army, but Naval

The squadron preparing to leave for Russia.

THE FIRST WORLD WAR

Commander Locker-Lampson resisted this and was permitted to keep his section within the authority of the R.N.A.S.

With three of the thirty-six Lanchesters being held back in Britain, the remainder were dispatched to assist the Czar of Russia and his army. They undertook numerous hazardous tasks on various missions throughout Russia, Persia, Rumania and Austria. The Lanchester armoured cars were the backbone of the task force, and performed admirably in atrocious conditions where often no vehicles had ever been before, and sometimes in temperatures of minus 46 degrees Fahrenheit. They were supported by numerous Lanchester tender vehicles and Seabrook lorries. Commander Locker Lampson also retained the use of one Rolls-Royce armoured car and one staff touring car for his own use, vehicles to which he had previously grown accustomed.

The Lanchesters gained a reputation for being fast, reliable and seemingly impossible to turn over. Their unique suspension and mechanical design also meant that they were most economical on tyres, which was a great benefit, especially considering historian Anthony Bird's comments about the Rolls-Royce vehicles that: "It was not unknown for them to get through a dozen tyres in a day's work." A criticism of the 38 h.p. armoured cars was, however, their comparatively low ground clearance. As the rear axle worm gear drive was of the underhung type, it did mean that the ground clearance was less than that of other armoured cars, including the solitary Rolls-Royce. It is on record that Commander Locker Lampson asked, unsuccessfully, for the Lanchesters to be substituted due to this problem. This apart, the Lanchesters were well-liked by their crews and contributed greatly to the operations of War.

In all, twenty-six Lanchester armoured cars succumbed to the fighting. The seven remaining had covered some 53,000 miles during the campaign, in such demanding conditions that tyre chains had to be used for much of the time. It seems that these armoured cars were abandoned when the task-force was returned to Britain due to the Bolshevik revolution, although there is a tale that they sank in a deep lake when its covering ice broke.

The three Lanchester armoured cars which had been left in England later helped Britain's War effort by dramatically parading in public places to encourage the sale of War Bonds. Their ultimate fate is unknown, as is that of the prototype, but they would probably have been either converted into private cars or cut up for scrap metal.

Meanwhile, back at the Works in Birmingham, changes were being made. A new Works

A 38 h.p. landaulette staff car and a 38 h.p. armoured car on patrol.

*The three 38 h.p. Lanchester armoured cars displaying
'Buy Victory Loan' posters in Birmingham's Victoria
Square, in July 1919.*

Manager, Jasper Cooper, was appointed in 1915 to replace William Caimell, and Archie Millership was appointed as his assistant.

In view of the need for more space for the War effort, George was planning another enlargement of the Armourer Mills factory, and a second galvanised building adjacent to the Radix Works was duly purchased. Both were subsequently demolished to make way for a new single-storey building, which was mainly equipped as a machine shop, but which also incorporated an enlarged metallurgy department. This move also allowed some remaining machinery at Alpha Works in Liverpool Street to be transferred, thereby providing extra storage space for finished vehicles and other War Office equipment.

Frank Lanchester's persistence with his contacts at the various Government Ministries won the Company new orders, and a contract for the construction of 450 aero-engines, of 100 h.p. rating, for the Royal Aircraft Factory was awarded. George remarked, "I was invited to be Chairman of a committee whose objects were to 'rationalise', i.e. to facilitate

Seven 25 h.p. covered wagons outside Armourer Mills and its new factory extension.

The 25 h.p. covered wagons heading down Montgomery Street with the Radix Works behind Armourer Mills in the distance. The Lanchester Company's chimney and new extension are in the foreground.

production by co-ordinating the efforts of the various contractors building the same engine. But Herbert Austin, who had a contract for four times the number of engines, protested. In the end, Rowledge of the Napier Company was appointed."

Other contracts were awarded to Lanchesters, and one included the construction of paravanes; underwater devices, secured by wires to the bow of a ship, which had jaws to slice through the cables of submerged mines, so that they could be exploded on the surface. Another contract was for the production of kite-balloon winches: wheeled winches for releasing or hauling down the observation balloons employed by the army for spotting enemy gun emplacements. The cable drum was controlled by a 38 h.p. engine and could wind in 2,000ft. of its 3,000ft.-long cable in six minutes.

Another order was for the manufacture of Constantinesco Interrupters, for synchronising an aircraft's machine-gun fire so that it avoided striking its own propeller.

The Company also played a part in the development of the V8 Sunbeam Arab Aero-engine, which developed 210 b.h.p. and which George described as "ill-starred". Again on the committee to consider standardisation between the manufacturing companies, George recalled, "I made myself unpopular with its designer, by insisting on all screws to comply with the newly created B.S.F. standard." The Sunbeam Company had its own mixture of special screws, and it was only after much persuasion that George eventually prevailed with his common-sense approach. To some of the committee, however, it may have seemed that George was promoting his own Company, as the British Standard Fine thread was, in most ways, the same as Fred Lanchester's 'M thread', which was still in use.

Fred himself was busier than ever since parting with the Lanchester Motor Company. As a proudly patriotic man, he had been deeply concerned about the readiness of his country for war, and strove constantly to assist in any way he could; normally at no financial reward whatsoever.

His work at Daimler had increased, as that Company were also busy with Government

An observation balloon winch powered by a Lanchester 38 h.p. engine.

contract work, but on a much larger and more diverse scale than that of the smaller Lanchester Motor Company. Daimlers had immediately turned over all their new and second-hand vehicles for use by the Government War Office, as Lanchesters and other companies had also done, and took immediate orders for lorries, vans, cars, ambulances, tractors and tanks. As Daimler's Consultant Engineer and Technical Adviser, Fred was deeply involved in various aspects of this urgent programme, as well as submitting new designs and ideas. The development of the aeroplane led to the need for suitable aero-engines and the Government gave a contract to Daimler, as they had to Lanchester, to produce the Royal Aircraft Factory 100 h.p. engine. They also produced the French 80 h.p. Gnome engine, which suffered many initial problems. As few drawings or design features were known of in Britain, a Gnome engine had to be stripped down for copying. The full extent of Fred's contribution to the project is not known, but his vast experience must have

A 25 h.p. general-purpose wagon.

been invaluable, as also were the many notes he had made a few years earlier when inspecting, testing and using similar engines for the White and Thompson plane project.

Most of Fred's time, however, was spent in London as a member of six Government committees set up to advise and determine policy on aeronautics in the United Kingdom, as they respected and urgently needed his unique level of practical and theoretical knowledge. He worked ceaselessly for the vitally important Advisory Committee for Aeronautics and contributed scores of reports and memoranda. He clashed with many, and indeed wrote to the Chairman regarding Percy Martin, who was not only his boss at Daimler but also a high-ranking member of the exalted Air Board (later known as the Air Ministry). In typical style, not caring one iota for reputation, he bluntly wrote, "A terrible lot of time has been wasted by a long discourse from the Chief Engineer of Daimlers and if every designer in the country were heard at that length they would not get down to work until next Christmas."

A 38 h.p. canvas-top Lanchester ambulance.

Fred visited France, in order to report on the aircraft situation there, and also frequented the National Physical Laboratory and the Royal Aircraft Establishment factories for consultation, working diligently for the good of the United Kingdom as he felt best. This involved much prompting and encouraging of the authorities, including letters to the Prime Minister, to consider many of his important strategic ideas. These included low-level flying patterns, night flying, bombing and torpedoing of submarines, ships' apparatus for 'receiving and launching aeroplanes', a four-bore 'blunderbus' to be used in flight to damage enemy aircraft, the gyroscopic artificial-horizon device for aircraft and also his comprehensive book, entitled Aircraft in Warfare.

Such was the pressure of all this work that Fred only managed to patent a total of eight other inventions during the three years up to the close of War on 11th November 1918. Many felt that Fred Lanchester should have received great credit and remuneration for his unselfish and total commitment to his country, but he received neither. He was pitifully rewarded financially and was overlooked for any sort of national honour. It is obvious that he had upset too many people by his no-nonsense talking, in his bid to simultaneously save Government money and to make more rapid technological progress. It may seem as though Fred's share of tact and diplomacy had by-passed him in the family and had supplemented that of Frank, such was the difference between them.

Feeling aggrieved at his unfair treatment, and with his War-time quest for urgent, constructive action at an end, he retreated to his London home at 41 Bedford Square, Bloomsbury, which he had earlier bought to be close to the various committees on which he served. With the importance of aircraft in warfare now obvious to all, George summarised his brother's decisive role for the United Kingdom with the words, "The rapid evolution of British aircraft during the First World War was largely due to his energy and foresight."

A 38 h.p. metal-bodied Lanchester ambulance, sign-written with the words "The gift of the residents of Westfield Road, Edgbaston".

Chapter Eighteen

THE SIX-CYLINDER 40 H.P.

At the time of the Armistice terminating the First World War, the Lanchester Motor Company was still working on various Government contracts, some of which were immediately cancelled, whilst the remainder were drastically reduced. The Company and the workforce had given everything to assist the nation during the War years, and in consequence were left with their precision machinery in dire need of renovation through its continuous heavy usage. In recognition of their contribution to the War effort, the Government awarded a settlement of £180,000 to the Company.

In a bid to resume normal operations in the forthcoming era of peace-time production, George and Frank had proposed to capitalise on their expertise and seek further aircraft-engine orders and associated work. This constructive strategy had been immediately rejected by the Board, who could foresee no future in any peace-time aircraft activity, and their decision not to diversify was ultimately to cost the Company dearly. It also made the completion of a new car, available for sale to the public, even more urgent, and design work had commenced in October 1918 when it was obvious that hostilities were nearing their end.

The Company's interim catalogue, circulated early the next year, described their forthcoming car designs in a general manner only and showed many war-time photographs of the 25 h.p. and 38 h.p. vehicles, as well as the short-lived Sporting 40. It stated: "In establishing, as we have in our pre-War 38 h.p. car, a type which is unquestionably without rival as a superb carriage, we feel that the general public have come to regard the Lanchester as a town carriage only and its qualities as a fast and serviceable touring car have been lost sight of. This is no doubt due in a measure to the somewhat 'bluff' and bulky appearance of our touring bodies, caused by setting the front seats apart, to utilise the space on either side of the engine, and thus provide a spacious after-body, the advantages of which are less apparent in a touring car than in a Town Vehicle De Luxe."

This apparent endorsement of the traditional Fred Lanchester design proved to be without substance, however, as the Board then decided not to pursue the option of a 'bonnetless' style, but instead to adopt the conventional long-bonnetted layout, such as on the Sporting 40, for all future models. Pragmatically, George conceded that the Directors were making the correct decision. He wrote: "Hence, we gave up the main feature that distinguished our cars from others, namely the engine position, which, though scientifically superior, had always caused sales resistance." Fred's opinions of this departure from the traditional Lanchester design are not recorded, although he would have endorsed the Company's statement, ".... whilst the addition of orthodox bonnet must necessarily reduce the available body space."

Certainly Fred disagreed with the Company's Managing Director about the proposed future direction of the Company. Hamilton Barnsley and his Directors decided to attack the luxury-car market, and especially the predominant position of Rolls-Royce. Fred disagreed strongly with this as the Lanchester Motor Company had always been a relatively small concern with limited funding, and Fred thought that taking on the world-renowned Rolls-Royce Company with their accomplished technical, financial and sales resources was

40 h.p. chassis number 1679, being finished at Armourer Mills.

a rash and foolhardy policy. To the Board of Directors, however, it must have been an exciting challenge to attempt to capture the Rolls-Royce slogan of 'the best car in the world' for their new Lanchester. George Lanchester and his dedicated team were not to let them down, and the 40 h.p. Lanchester began to take shape.

The pressed-steel chassis frame closely resembled that of the 1914 Sporting 40, but the side members were made even larger and the area around the engine and gearbox was enclosed to make a box section. This addition gave extra strength, although holes were drilled to eliminate unnecessary weight. An exceptionally robust structure was formed by using many cross members, one of which was a massive 6.5ins. diameter tube. The engine and gearbox bearers were bolted directly to the chassis frame and also provided additional rigidity.

The chassis front terminated in dumb-irons and heavy-duty half-elliptic springs, which were some 43ins. long and allowed great deflection; three different spring sets were available to suit different coachwork weights. The rear chassis retained the Lanchester cantilever suspension, for which five different spring sets were available.

The springs themselves were the result of the ceaseless quest for perfection, as the Lanchester brothers had never been entirely satisfied with the 'bought out' springs on all their previous models. For the first time in motoring history, the Company developed its own spring-making equipment in order to guarantee the quality which George demanded. The effort was worthwhile, as instead of the usual 20% spring leaf contact attained by the best spring-makers prior to clamping, George achieved a surface contact of an incredible 85%. On assembly, all the highly-polished spring leaves were provided with thin, perforated brass inter-leaves loaded with lubricant and finally enclosed in leather cases. The spring ends were located by means of phosphor-bronze rotating trunnions, incorporating four steel rollers in each.

The parallel-link system, allowing the rear axle to move only in a vertical path, was another Lanchester feature to be dispensed with and its absence necessitated substantial re-design work. The rise and fall of the axle was allowed to pivot radially from the massive 8ins. diameter spherical joint at the front of the Carden propeller shaft. To minimise the inevitable movement in an arc away from the correct vertical path, George made sure that the rear axle was as far from its pivot point as possible, by using an exceptionally long, enclosing torque-tube to the differential drive coupling. Diagonal tie-rods ran from the suspension spring-brackets to the front of the torque-tube, to counteract the tendency of the axle to move sideways when travelling over rough ground.

THE SIX-CYLINDER 40 H.P.

40 h.p. chassis complete, awaiting coachwork.

With the torque-tube counteracting the driving, braking and torque forces and reactions, George turned his attention to the rear axle drive. The underslung Lanchester worm and wheel assembly was available in four different ratios to suit different coachwork weights, or to customer preference: 4.375:1, 3.777:1, 3.3:1 and 3.1:1. Bolted to the housing were steel tapering casings and 'fully floating' wheel hub mountings. This system allowed the entire weight of the vehicle to be supported by the road wheels and casing. The axle shaft therefore transmitted driving torque only, and was not subjected to any side thrust or strain in supporting the weight of the vehicle. Brake drums of almost 15ins. diameter were included on the rear hubs and were of the internally expanding type, with Ferodo friction material rivetted to the cast-iron brake shoes. This reserve brake system was operated by the hand-brake lever, which engaged with a side ratchet on the chassis.

The front axle consisted of an almost straight 'H' section drop-forging, machined all over and secured on top of the springs. The front wheel hubs were mounted on roller and ball bearings, with their outer splines taking the Rudge-Whitworth detachable wire wheels in preference to George's own patented idea of using a series of wide metal strips, instead of thin spokes, to support the rims. Tyres of 895mm. by 150mm. were used, the track was 4ft.10ins. and the steering lock was 51ft. No brakes were fitted to the front axle but an

40 h.p. chassis in production at Armourer Mills.

The early 40 h.p. epicyclic gears, direct-drive clutch-plate and countershaft brake drum.

effective, foot-operated main brake was fitted into the transmission behind the gearbox. As with the clutch, this was a totally new system, and was adopted to reduce both the weight and the length of the complete power and transmission unit, in order to make room for the long torque-tube. The multi-disc-type brake used in previous models was discontinued in favour of an internally expanding pair of brake shoes, which operated against the 12ins. diameter brake drum secured to the gearbox tail shaft. The clutch was also completely re-modelled to take the form of a single steel disc revolving freely when actuated, between two similar-sized Ferodo-fibre facing discs, partially submerged in oil.

The 40 h.p. epicyclic gearbox differed little from those used by Fred before the turn of the century, which had proved themselves to be utterly foolproof and silent in operation. However, the Lanchester Motor Company was, by this time, a lone voice expounding their virtues, as all other firms used the sliding gear 'crash' boxes, which were smaller in bulk and cheaper to produce. The Company therefore felt it necessary to issue a descriptive booklet to show the merits of their "much criticised Epicyclic Gear", although to avoid any consternation, all talk of using reverse as an additional brake was resisted.

The gearbox consisted of three trains of epicyclic gear, each contained within its own drum. The handbook stated: "The outer element of the clutch forms a driving connection

The 40 h.p. epicyclic change-speed gears.

THE SIX-CYLINDER 40 H.P.

between the Engine Flywheel and the tubular shaft on which is formed the Low Gear and Reverse Gear sun-pinion; this drives alternatively the Reverse Gear, the Low Gear or the Compound Second Gear through the medium of its epicyclic train, which finally rotates the centre shaft or transmission shaft. The Single plate Clutch is in direct connection with the centre shaft and, therefore, when engaged with its corresponding elements (the engine flywheel and driving disc), drives the transmission shaft direct, carrying round the Epicyclic Gear cluster as a whole. This condition is termed the direct drive and constitutes the 'Top Gear'. The gears are actuated by arresting the motion of the drum belonging to the gear that is required. This is effected by means of a brake, or 'crab blocks', three of these being contained within the gearbox, which are brought into operation by tumbler levers engaging with the selector bars. The selector bars are operated by the selector cams."

As the top gear was a direct drive to the rear axle, the transmission ratio was 3.3:1 as standard, with 5:1 on second and 11:1 on first gear. The system was, as before, cleverly designed to permit the movement of the gear lever only when the clutch pedal was depressed, and the whole was made to resemble a conventional gearbox arrangement. George designed and patented, in 1920, an advanced method of operating these gears by using electric solenoids. Similarly, Fred designed an alternative system four years later, using hydraulics, but it seems that the Company feared these ideas too radical to adopt and neither was used. In unit construction with the clutch and gearbox was George's newly-designed engine, which was greatly influenced by aero-engine design advancements made during the War. The engine was known as the 40 h.p. although rated by the R.A.C. at 38.4 h.p. Its six cylinders were cast in two blocks of three and had a bore of 4ins. with a stroke of 5ins.

1. Valve Cover.
2. Water Circulation Pipe.
3. Camshaft.
4. Camshaft Bracket.
5. Valve Tappet Lever.
6. Cylinder.
7. Inlet Valve.
8. Camshaft Worm Wheel.
9. Distributor.
10. Vertical Shaft.
11. Vertical Shaft Casing.
12. Gudgeon Pin.
13. Connecting Rod Oil Conduit.
14. Magneto Drive Shaft Worm Wheel.
15. Piston.
16. Fan Pulley.
17. Centrifugal Water Pump.
18. Vertical Shaft Worm Wheel.
19. Oil Gauze Cover.
20. Connecting Rod Big End Bearing.
21. Oil Pipe.
22. Crankshaft Main Bearing Cap.
23. Connecting Rod.
24. Crankshaft.
25. Oil Base.
26. Crankshaft Bearing Cap Oil Way.
27. Starter.
28. Starter Drive Housing.
29. Starter Worm Wheel.
30. Starter Free Wheel Clutch.
31. Layshaft.
32. Engine Oil Pump.
33. Generator Worm Wheel.
34. Layshaft Driving Wheel.
35. Flywheel.
36. Clutch Disc.
37. Clutch Cover Plate.
38. Gear Box Centre Shaft.
39. Gear Box Oil Sump.
40. Countershaft Oil Pump.
41. Countershaft Brake Shoe.
42. Countershaft Brake Drum.
43. Torque Tube Ball.
44. Universal "Hooks" Joint.
45. Cardan Shaft.
46. Torque Tube.
47. Torque Tube Spherical Joint.
48. Oil Pressure Relief Valve.
49. Compound or Second Gear Drum.
50. Low Gear Drum.
51. Reverse Gear Drum.
52. Gear Box.
53. Fabric Clutch Disc.
54. Fabric Clutch Disc.
55. Flywheel Casing.
56. Generator.
57. Oil Pipe.
58. Cylinder Water Jacket.
59. Exhaust Port.

Near-side sectional view of an early 6,178 c.c. 40 h.p. engine.

The water jackets surrounded only the area covered by piston movement, and were connected between the cylinder blocks by rubber gaskets. The base of each cylinder stood individually on the aluminium crank case. The dome-top aluminium pistons employed three compression rings and, conforming to established Lanchester practice, an oil scraper ring at their base. Fully-floating gudgeon pins were held within the piston by conical brass washers which opened up into a groove like a present-day circlip. The connecting rods were produced from high-tensile steel and were fully machined to eliminate weight, facilitate the checking for any metal defects, and achieve identical balance.

Lubrication of the small-end of the connecting rod was pressure-fed through a small-bore pipe attached to its outside. The oil was delivered at 40 p.s.i. through the white-metalled big-ends on the 2.5ins. diameter hollow crankshaft, which was supported by seven main bearings. So robust was the chrome-nickel steel crankshaft that no Lanchester crankshaft vibration damper was needed.

Situated at the forward end of the crankshaft was a spiral pinion gear which drove a vertical shaft. Via another worm and wheel arrangement at its head, this revolved a single overhead camshaft, split to allow either half of the top of the engine to be dismantled

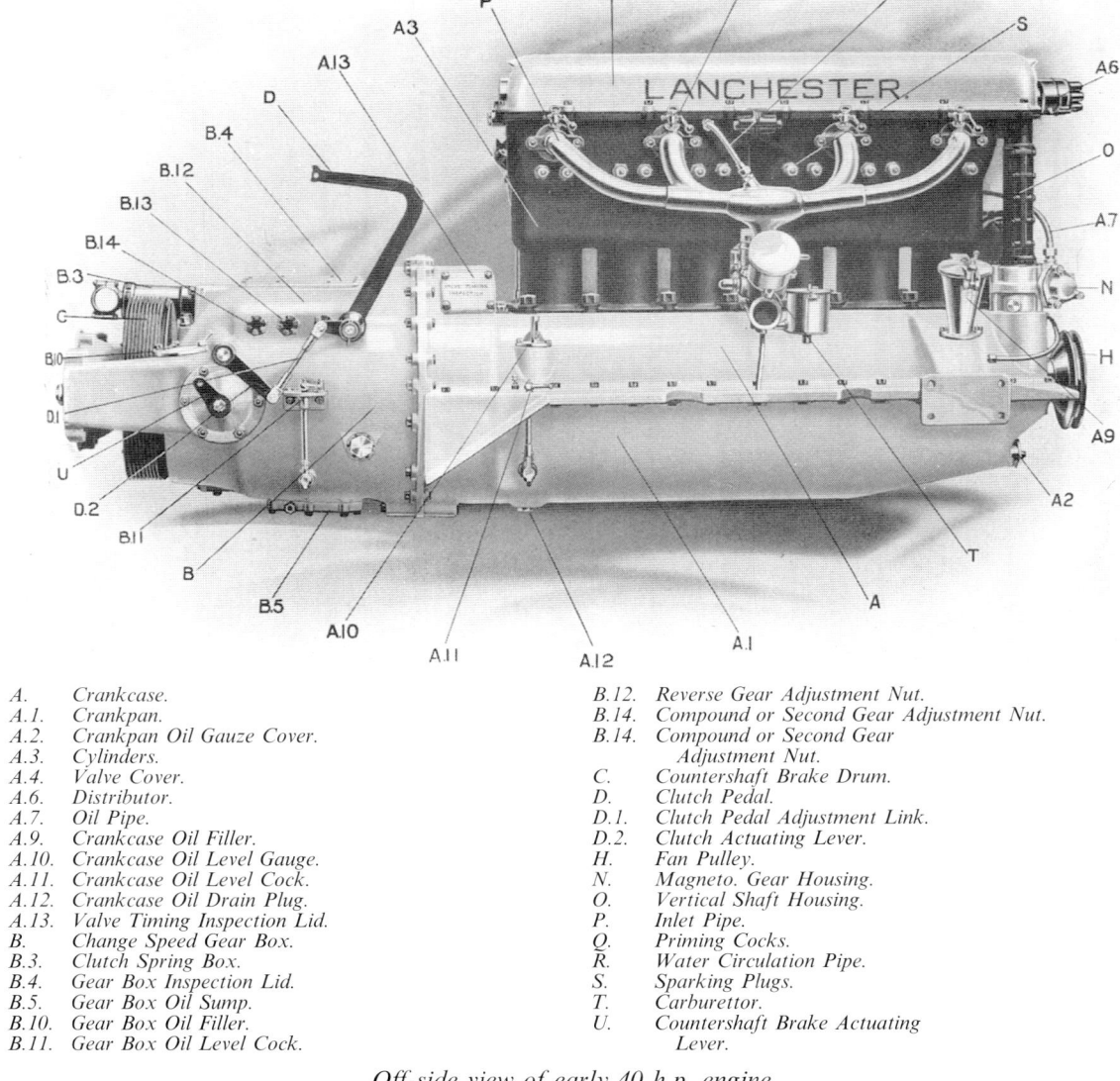

A.	Crankcase.	B.12.	Reverse Gear Adjustment Nut.
A.1.	Crankpan.	B.14.	Compound or Second Gear Adjustment Nut.
A.2.	Crankpan Oil Gauze Cover.	B.14.	Compound or Second Gear Adjustment Nut.
A.3.	Cylinders.		
A.4.	Valve Cover.	C.	Countershaft Brake Drum.
A.6.	Distributor.	D.	Clutch Pedal.
A.7.	Oil Pipe.	D.1.	Clutch Pedal Adjustment Link.
A.9.	Crankcase Oil Filler.	D.2.	Clutch Actuating Lever.
A.10.	Crankcase Oil Level Gauge.	H.	Fan Pulley.
A.11.	Crankcase Oil Level Cock.	N.	Magneto. Gear Housing.
A.12.	Crankcase Oil Drain Plug.	O.	Vertical Shaft Housing.
A.13.	Valve Timing Inspection Lid.	P.	Inlet Pipe.
B.	Change Speed Gear Box.	Q.	Priming Cocks.
B.3.	Clutch Spring Box.	R.	Water Circulation Pipe.
B.4.	Gear Box Inspection Lid.	S.	Sparking Plugs.
B.5.	Gear Box Oil Sump.	T.	Carburettor.
B.10.	Gear Box Oil Filler.	U.	Countershaft Brake Actuating Lever.
B.11.	Gear Box Oil Level Cock.		

Off-side view of early 40 h.p. engine.

George Lanchester testing the first 40 h.p. car ever built, with Jasper Cooper as passenger.

separately. The valve levers actuated the hollow, bell-shaped valves which were diametrically opposed and slightly inclined, providing the most efficient combustion chamber shape. As detachable cylinder-heads were not fitted, the tungsten-steel valves were designed to be easily removable. Each inlet valve was fitted in a detachable cage; each identical exhaust valve could be removed by dropping it onto the piston crown, when at the bottom of its stroke, and then extracting it through the aperture left by the inlet valve cage.

Two spark plugs were fitted to each cylinder, one to suit a Remy high-tension coil and battery system with its distributor located on the forward end of the camshaft, and the other to suit the magneto system. The magneto was flange-mounted and was driven from a gear on the drive for the vertical camshaft.

Large spur gears were used to drive a 14ins. long layshaft, which ran inside the crank case, parallel to and driven from the crankshaft. This had two skew gears to drive the high

Archie Millership putting the demonstration 40 h.p. tourer through its paces.

The November 1919 Motor Exhibition two-door saloon 40 h.p. on chassis no. 1634. (See appendix.)

THE SIX-CYLINDER 40 H.P.

pressure oil pump, the vertically mounted dynamo, and lastly the 'free wheel clutch' part of the worm and wheel gearing for the vertical starter. The internal electrics of the dynamo and starter were produced by the Lucas Company.

A large three-bladed cooling fan was driven by a belt from the crankshaft pulley. The water pump was driven from the front of the crankshaft by the gear driving the vertical shaft, and circulated cooling water round the engine via the large brass or nickel-silver honeycomb radiator. The Lanchester preview booklet of the 40 h.p. stated: "The radiator and bonnet will closely resemble that of our 40 h.p. car of 1914, but will be slightly narrower and higher to conform to the characteristics of the new power unit."

The engine was indeed worthy of George's insistence on replacing the earlier side-valve Sporting 40 engine. It is inconceivable that he would not have exchanged views on the project with his brother, although Fred's previous disapproval of the Company's future policy meant that he had not been officially consulted over any aspect of the new 40 h.p. design.

The petrol tank spanned the rear chassis, and George opted for an Autovac, rather than his own patented mechanical pump, to supply one updraught Smiths four-jet carburettor. For maximum efficiency, the induction manifold was located on the opposite side of the engine to the exhaust (a layout which was 're-invented' decades later, as the cross-flow head). The induction manifold resembled a ram's horns and consisted of just four copper inlet tubes; instead of the usual six-branch manifold supplying the petrol/air mixture to each port in the cylinder head, George constructed two cylinders in each block with a common port. The areas and lengths of the ports and pipes were designed so that each cylinder obtained an equal amount of petrol vapour; the exhaust system had a similar arrangement with its end ports. The reason for this unusual design was to facilitate the coring of the cylinder blocks.

Comparing well with their Rolls-Royce 40/50 competitor, which managed only 80 b.h.p. from its 7,036 c.c. engine, the first two 6,178 c.c. engines on test developed 89 b.h.p. at

Highly-polished, inlaid, walnut marquetry on the door-panelling and head-lining for the coachwork of chassis no. 1634. (See appendix.)

2,000 r.p.m. and about 96 b.h.p. at their maximum of 2,400 r.p.m. Their performance justified all the attention afforded to them, and they were duly installed in two matching chassis. By the August of 1919 these two prototypes, of 11ft.9ins. wheelbase, were clothed with a five-seater touring body of polished aluminium and a four-seater owner-driver saloon respectively.

The tourer was used for test and demonstration purposes and the leading motoring journals, eager to try out the new product from Lanchester, took full advantage of this. The Autocar motoring correspondent wrote, "Only an inspired pen could find adequate words to picture the charm of the new Lanchester. Telling the unbiased truth, though haltingly, the present writer expects to be accused of exaggeration, save by those who have had experience of this car ..... Criticism went by the board; unfettered enjoyment took its place. With the Lanchester, the earth's surface is as flat as a calm sea, all roads are good roads, and hills have no significance ..... Vehicles such as this may be high in first cost, but true refinement is as valuable as it is rare. To connect the Lanchester with purchase money is almost a sacrilege."

The journal Country Life also seemed to like what it saw. "Really I do not know where to begin in committing to paper my impressions with this super-car. They seem to crowd on me in a mass which on being peered into dissolves away into nothingness. In other words, everything about the car is so excellent that there is no outstanding impression as a result of quite an exacting little test in it, and one feels inclined to say 'The Lanchester is the finest car in the world,' and let it go at that ..... I have not found a single thing about it that can by any stretch of imagination be said to call for criticism."

The magazine entitled The Motor Owner ended its long report with the following: "Taking the new Lanchester car altogether, we should say that it is one of the two or three unmistakeably 'best' cars; it is a magnificent specimen of British design and craftsmanship, and its performance is all the tribute that the skill of Mr. George Lanchester, the designer, requires."

The Motor started its report with similar unbounded enthusiasm: "To compare the latest 40 h.p. Lanchester with ordinary cars would be akin to comparing a Cunard liner with a yacht, for ..... this super car is immediately recognised when seen on the road as one of the world's highest class automobiles."

All these glowing reports, and many more besides, were made by different reporters at different times, although the same prototype tourer was used throughout, each time driven by the Lanchester Midland Sales Manager and Chief Demonstrator, Archie Millership. The test reports follow the same theme of exhilaration gained by the correspondents, placed in the hands of their fearless driver, who revelled in driving the car through 'impossible places'.

Each reporter was surprised by the flexibility and torque of the engine, "at 30 m.p.h. one hardly seems to be moving". Each received the 'Millership treatment', when he would set the Lanchester at "two or three miles an hour on the top speed" and then get out of the car and walk alongside the shocked occupants. When he was back behind the wheel, he would then depress the accelerator and power up to its maximum speed of 78 m.p.h. Referring to the car's ability to accelerate from 10 m.p.h. to 30 m.p.h. in just over eight seconds, the Autocar remarked, "The comparatively heavy car is as lively as some of the small sporting models, which appear to have terrific acceleration."

The suspension received high praise, with reports of the car being deliberately driven into the deepest pot holes or up and down kerbs 9ins. high, whereby "sitting back in the seat, the only sensation to which we can compare the movements, is that of a large liner in a gentle swell".

As always with a Lanchester, the operation of the epicyclic gears stunned the correspondents, with Archie Millership casually stating at 50 m.p.h., "We are on our third" followed immediately, and silently, by, "We are now in reverse". One expert later commented, "Third to first, second to reverse, reverse to third, this impious youth plays tunes, and does counterpoint, on this wonderful epicyclic gearset."

With their superlatives becoming grander with each report, one wrote of his experience as, ".....being drawn through the air by a piece of extra-fine elastic thread, bound in silk." Maybe the correspondent for The Motor summed up the feelings of his journalist colleagues when he stated, "We have never ridden in or driven a finer car than this new Lanchester."

The other completed prototype was shown at the Society of Motor Manufacturers and Traders' Olympia Motor Exhibition in November 1919, and was described as "a saloon limousine, seating four persons, painted and upholstered in grey and black". The two-door coachwork was an original design by George and its construction portrayed lavish craftsmen's skills in the passenger compartment, with inlaid walnut marquetry door-panelling and head-lining, silver fittings and silk blinds. The purchase price was £2,400, making it the most expensive car in the entire show.

As President of the S.M.M.T., Frank Lanchester was that day host to King George V, who was inspecting the exhibits. After studying this vehicle the King, who was renowned for his reserved approach to motor carriage presentation, remarked to Frank, "Very fine, Mr. Lanchester, but more suited to a prostitute than a Prince, don't you think?"

Landaulette coachwork by J.Blake & Company of Liverpool on a circa 1920 40 h.p. chassis.

Other visitors to the Lanchester stand, which also included a 40 h.p. rolling chassis priced at £1,500, obviously did not agree with the King's remark, and the press talked later of the Lanchesters "attracting a never-ending throng of connoisseurs ... admired by thousands". The show report in the Daily Telegraph allowed the reporter to dream: "The car which would take my order is the 40 h.p. Lanchester. Its design from end to end is a pleasure to examine ..... as real a work of art as any picture or statue to be found in the most treasured galleries of Europe."

Following this successful Motor Exhibition, production of 40 h.p. cars began in earnest. Armourer Mills and its adjacent Lanchester factories in Montgomery Street, Sparkbrook, began turning out completed chassis, although only at the rate of a few per week to start with.

Problems did surface, however, caused by the desire to break free from Fred's influence, which still dominated the Company even in his absence. The use of his 'M' thread was discontinued due to adequate stocks of the very similar British Standard Fine thread becoming available throughout the United Kingdom; most other motor and engineering firms were also beginning to adopt its use. Another one of the foundations of Fred's engineering principles dating back to the turn of the century was his theory and practice of clearances, limits and tolerances in components. For some twenty years the Company had prospered on its good engineering record, but now this reliable system was relaxed. Fred recalled that, "An attempt was made to widen the limits with a false idea of economy. The old limits had to be restored in a very short space of time after something like £5,000 had been lost, and the reputation of the firm had suffered. Retribution came in the assembly shop, the test shop and in free repairs, as I advised the management would inevitably be the case." All components in doubt on the cars already sold must have been quickly changed as not one complaint against the Company was received by the motoring press, where great harm could have been done. The business of production was restored.

Lanchester cabriolet coachwork on a 1920 40 h.p. chassis, unusually fitted with artillery wheels.

By the time of the Olympia Motor Exhibition in November 1920, with many open and closed 40 h.p. cars already sold, a few minor alterations were put into effect. To accommodate the large, heavy coachwork required by many customers, a longer wheel-base chassis of 12ft.6ins. was included as an alternative to the original, allowing the correspondingly longer overall length of 17ft.4ins. The 4ft.10ins. track remained the same. The tyre size was also altered, coming down to 895mm. by 135mm. The Lanchester stand displayed a complete rolling chassis and a dark blue, seven-seater, owner-driver, saloon limousine. The accompanying sales photo showed the car as a 'V' windscreen, enclosed-drive limousine of Lanchester coachwork.

Before the close of 1921 more changes were made. The use of a Smiths carburettor was discontinued in favour of an S.U. and, although this returned over 15 m.p.g., it may not have been quite as compatible with the engine, as the decision was reversed in 1923. Other minor alterations included the fitting of a worm drive arrangement for the new Watford magneto, the use of a Watford clock and speedometer in place of Smiths instruments, an increase to twenty gallons in the capacity of the petrol tank, and the fitting of a water temperature thermostat. The cooling fan was relocated from the engine front to a cowling on the rear of the radiator. This carefully-designed fan incorporated a turned-up end on each blade and, combined with its spider-shaped cowling, proved most efficient, as well as eliminating any possibility of the fan developing resonance. An automatically-operating Sylphon thermostat was installed to regulate cooling water temperature. The most visible alteration, however, was the 3ins. increase in the height of the radiator, bonnet top and coachwork which, to many, gave a more balanced impression to the car. There were eight advertised styles of coachwork to choose from, apart from those made to special order.

The 1921 Motor Exhibition stand was bigger than before, with three exhibits. The grey,

THE SIX-CYLINDER 40 H.P.

1920 sedanca-de-ville coachwork on a 40 h.p. chassis, wrapped up for the winter.

five-seater tourer, priced at £2,750, included the fashionable rear passenger screen, whereas the dark green, seven-seater limousine was £3,000. By comparison, the Thrupp and Maberly Coachwork stand showed off their own treatment of a 40 h.p. chassis, displaying a four-seater limousine body with a roll up extension roof over the driver, which sold for £2,746. Lanchester's third exhibit was their traditional presentation of a rolling chassis, which included the price tag of £1,950 for either length. This was a new reduction from the summer price of £2,200 and showed the Company's determination to be attractively-priced against its competitors in the luxury car market.

Comparing the 6,178 c.c. Lanchester with other top-quality British chassis, it is interesting to note that the 7,000 c.c. Leyland Eight sold for £2,500, the 7,777 c.c. Sheffield Simplex sold for £2,250, the 6,105 c.c. 40-50 h.p. Napier sold for £2,100, and the 7,112 c.c. Daimler 45 Special sold for an amazingly cheap £1,450. There is no doubt, though, that the Lanchester Board of Directors had set their sights on beating the 7,428 c.c. Rolls-Royce, which sold for £2,100. However, whereas Lanchester's product was counted merely in its hundreds, the Rolls-Royce Company had produced thousands of their 40-50 models, thus making their capital expenditure and subsequent costs considerably lower. Although Lanchester cars attracted much glory and praise, the Directors' policy of undercutting Rolls-Royce did mean that their profit margin was lessened.

Apart from a good customer base in the United Kingdom, sales were also made to many countries overseas. In fact, in 1921, the Lanchester Motor Company became the first British manufacturer to produce left-hand-drive cars for export, when it embarked on a sales campaign in the United States of America.

As part of this, George travelled to America in order to arrange a business deal. An agreement was reached with Brewster, the famous coachbuilding company of New York, to act as Lanchester agents in the United States, although the first car sold was mainly due to George's influence. He had persuaded a great personal friend and previous customer to buy another Lanchester, the customer's name being Henry Ford. Both his car and the second car produced had Lanchester coachwork, but the remainder were all bodied by Brewster themselves. At least six of these Americanised Lanchesters were provided with left-hand-drive steering and a central gear shift. At this time there appeared to be a potential market developing in the United States for high quality European cars and sales hopes rose

Lanchester landaulette coachwork on a circa 1921 40 h.p. chassis.

for the partnership, but after about twenty cars were sold this optimism was dashed, as Rolls-Royce began negotiations to take over Brewster and Company Incorporated. This was a bitter blow to the Lanchester Company; no other suitable agencies could be found, and from 1924 onwards Lanchester cars exported to the United States of America were sold on an individual basis.

Market sales on the home front, as with all luxury car manufacturers, had been affected by the post-war depression. Lanchesters tried to counteract this by undercutting the price of the Rolls-Royce 40-50 chassis by £50. A price of £1,800 was shown on the chassis displayed at the 1922 November Motor Exhibition at Olympia, accompanied by a three-quarter landaulette and a touring car. Their prices also reflected the chassis reduction and were £2,600 and £2,300 respectively.

During 1923, the Company decided to concentrate their advertising solely on the long chassis of 12ft.6ins. wheelbase, as so few customers were ordering the shorter 11ft.9ins. wheelbase version; most people ordering the 40 h.p. wanted large capacity coachwork. Another consideration was that, from this year, the new Lanchester six-cylinder 21 h.p. of 10ft.9ins. wheelbase was available. Indeed, at that year's November Motor Exhibition, the 40 h.p. rolling chassis and its stable-mate, a seven-seater, clematis-coloured limousine priced at £2,588.10s. had to share the exhibition stand with their new 'little sister'.

One of the items on the new 21 h.p. model which aroused great interest was its four-wheel braking system, in contrast to the 40 h.p., which had no front brakes. The latter's system, at this time, was modified by enlarging the rear wheel brakes to 20ins. diameter, in order to use them as the main brake and to demote the transmission brake to the reserve hand-brake. However, even this improvement was regarded as old-fashioned. It seemed strange to many that the heavier, faster, larger, and more expensive, 40 h.p. should have less braking capabilities than a car smaller in every respect. The author, Anthony Bird, recalled that a loyal customer had questioned the Managing Director about this peculiar oddity. "His memorable reply was to the effect that the 21 was an owner-driver's car and owner-drivers often lacked skill: but the 40 was a gentleman's car and gentlemen had

The 1922 40 h.p. with five-seater landaulette coachwork by Lanchester bought by Lionel Rapson, the tyre producer and Brooklands racer.

chauffeurs who knew better than to go dashing about, relying on their brakes to get them out of trouble."

This delightful story did not give the full picture, however, as George's patented design for the four-wheel braking system had actually been installed for test on a 40 h.p. chassis in 1922. George's official Company letter, in October, to Fred, who had been temporarily brought back to the Company to act as a part-time consultant, gave details of the chassis' first road test. Describing the brakes as "satisfactory and promising" he proclaimed that the "pull-up on the front and rear brakes together is very sweet and silky". It was to be the new 21 h.p., however, which was given priority for their introduction.

When they eventually arrived, the brakes for the 40 h.p. were ostensibly a more robust version of those used on the 21 h.p. model, although George may have considered using his patented invention of a road wheel where the inside rim was machined to be used as a brake drum. In order to counteract the torque generated by its front-wheel brakes, the front axle was altered from the 'H' section to a tubular design, reviving memories of that style used in pre-war days.

The rod-operated brake shoes were lined with Ferodo fibre which acted on a steel liner riveted to each 20ins. diameter aluminium drum, ribbed to quickly dissipate the heat generated. The rear drums were of double width in order to take a separate pair of shoes in each for actuation by the hand brake; the earlier transmission brake was discontinued. Although the brakes were effective, the foot brake was heavy. To alleviate this, George designed a system whereby the pressure exerted by the foot pedal was servo-assisted. Hydraulic power from a cylinder and ram, secured to the outside of the gearbox, was fed by high-pressure oil from the gearbox pump and was automatically brought into effect by foot-pedal operation.

The whole braking system was available not a moment too soon, as Lanchester's competitors, and especially their arch-rival, Rolls-Royce, were instigating four-wheel brakes on their models also. A 40 h.p. chassis was displayed at the April 1924 British Empire

Polished aluminium saloon coachwork by Lanchester on a 40 h.p. chassis of March 1923.

Exhibition, but apart from the introduction of a Delco coil ignition system driven from a re-designed camshaft, little other alteration was made. Customers had to wait until the Olympia Exhibition in November to inspect the new "internal expanding brakes on all four wheels".

The chosen chassis was clothed in a purple-lake seven-seater limousine body, tastefully upholstered with mulberry-red leather front seats and the rear interior in "antique, morocco-grain, velvet calf leather", with folding tables, triplex glass and silver fittings throughout. The price was £2,689, and it is interesting to note as a comparison that the weekly wage for one of the Company's skilled craftsmen was £3. The unashamed luxury attracted many wealthy and famous people to the joys of Lanchester motoring and letters of praise were sent to Armourer Mills by their delighted owners. One of the first received was from a loyal customer. "I feel I must write to tell you how absolutely delighted I am with my new Lanchester. When I had a '38' I thought I possessed a super-excellent Car and one that even time and experience could not improve, but I am afraid it will now have to play 'second fiddle', for your new model is vastly superior in every detail, except the suspension, and that is quite as good, which means a lot when you are talking of Lanchester suspension. If I were to write 'reams' in praise of the Car, I could not adequately give expression to my feelings, so I had better not try to paint the lily, but just leave the subject by saying there is only one word for the new 40 h.p. Lanchester and that is 'PERFECTION'. I congratulate you on your achievement in producing the finest Car in the world today."

Another letter remarked, "Now that I have had my 40 h.p. Lanchester for nearly two years, and having covered over 17,000 miles under the most trying conditions (driving in London and up and down the Surrey hills, to boot) I am pleased to tell you that I have never had an involuntary stop and my machine has not cost me one halfpenny for repairs. I would further say that it seems to improve with wear, and that I am well content with my choice. Only those who have driven the 'Lanchester' can appreciate the supreme achievement in modern automobile engineering."

An owner new to Lanchester wrote, "I have motored without break since 1896, and was the first owner of any kind of mechanical carriage in these northern counties. Since those distant days, I have owned and driven dozens of Cars, and have been generally regarded in this district (Sunderland) at all events, as one who invariably seeks and obtains the very best car procurable. I was the first owner of a six-cylinder Napier, and Mr. Edge will remember many four-cylinders of that make which previously passed through my hands.

The Prince of Wales leaving his Lanchester 40 h.p. tourer whilst on a visit to the Midlands in 1923.

Twelve months ago, after exhaustive tests and trials, I bought a new open Lanchester. I had, shall I say, reluctantly come to the conclusion that there was nothing better, but being a new model I was somewhat shy of it, in spite of the fame of its designer. A month ago, having sold a Panhard-Knight all-weather, and being wishful to replace it, I set out to find if there was any car on the market that could equal the almost incredibly brilliant running of my Lanchester. One does not spend £3,000 without great care, so that every late model car with a good name, such as Rolls-Royce, Napier, Hispano, Daimler, etc., was seen and tried and afterwards I went straightaway, without further hesitation and bought another Lanchester, thoroughly convinced in my own mind that there is nothing on the road today to equal it. I know I am making a tall assertion, but actions speak louder than words, and the fact of having bought a second Lanchester must convince the unprejudiced that I honestly think this make is absolutely the world's best car."

A letter from Wales also remarked on rival makes. "Wherever we have gone the Lanchester has been peerless and her supremacy is unquestioned. She was challenged from Bettws-y-Coed to Peny-gwrd by a Rolls-Royce touring car and won the trial with distinction and still something to spare. The owner admitted afterwards that it was their first defeat and thought it must be a foreign car, but I was pleased to tell him it was a 'Lanchester' all British, and when they saw it they were full of admiration."

The cars were exported in their huge wooden packing cases around the world, and again letters were received. "It is extremely agreeable to deal with people who make such a splendid car and treat their customers so well", said one from Brazil, whereas another from France remarked about his European tour, "During my 6,000 miles of travel the Lanchester car has been eminently satisfactory." The governor-general of Australia, Lord Somers, imported a number of 40 h.p. cars for official use and, indeed, that country, with the United States of America and India, were the main overseas markets.

With India then being part of the British Empire, the Maharajahs naturally bought most of their cars from British companies and many favoured Lanchester. This was partly due to the enterprising efforts of Frank, who visited a great number of their opulent palaces,

Four-light, two-door, drop-head cabriolet coachwork by Lanchester on a 40 h.p. chassis circa 1923.

entertained them on their visits to England, and regularly corresponded to keep his Company prominent in their car-buying programmes. Most had more than one Lanchester, with many having models going back to the early 10 h.p. and 12 h.p. cars. Amongst others, the Maharajahs of Rajkot, Kolapur and Rewa indulged themselves with the luxurious quality; the latter's assistant wrote to the Company in praise of his latest Lanchester: the 40 h.p. with chassis number 1785. The motor-superintendent to the Maharajah of Nawanager wrote: "The three 40 h.p. Lanchester cars which His Highness bought, two in 1922 and one in 1924, have been run over 60,000 miles under all sorts of conditions in India and England. It would interest you to know that I have a fleet of over 60 high-grade cars of different makes, including Lanchesters, etc. therefore I am in a fair position to judge their merits. The Lanchester is easily the best car I have under my care."

The Maharajah of Rewa purchased another 40 h.p., chassis number 1915, in February 1924, and the Autocar reported that the car was: "What must surely rank as one of the most completely equipped and luxurious carriages in the world." The enclosed-drive limousine was painted light blue with gold coach-lines and incorporated a double roof with louvres, and an electrically-operated ventilation-fan housed in a dome on the roof; features in common with other Lanchesters belonging to the Maharajah. This car was probably the first vehicle ever to make use of dark-tinted glass, ordered so that the Maharajah would not be defiled by others gazing in on him. The Motor remarked: "In the rear compartment is a mahogany locker containing a silver picnic outfit, and a red morocco-leather attache case with writing materials. At the sides of the body are ladies and gentlemen's companion cases." Amongst other items, such as a silver cigarette box and cigar lighter, was an electric bell and speaking tube, and a direction indicator resembling a ship's telegraph. The car was designed to carry eleven passengers, with three on the rear seat, four on the occasional seats, three on the front seat, and an external saddle, which incorporated a back and arm rest, for the servant.

Arguably the most amazing vehicle design of any was that built for the Maharajah of Alwar in 1924, to his own general design. He had already purchased a 21 h.p. "baby Lanchester", complete with solid gold fittings and gold-plated radiator, but his 40 h.p. state carriage was superior to all. Its extended chassis of 14ft.10ins., number 1925, was fitted with the unique style of a horse-drawn, double-opening landau coach body, designed as a replica of a Victorian state-carriage and finished in royal-blue with gold, double coach-lines. Large 'C'-shaped springs complemented the usual suspension and The Motor commented that: "Those who have ridden in the coach say that this suspension is the most luxurious thing they have ever tried, and is also free from roll." It employed a special, low-ratio rear axle gear to maintain pace more easily with elephants in processions, although a top speed of 55 m.p.h. was still available. It was the most expensive of carriages, with blue and gold silk brocade used for the interior upholstery, as well as for the blinds covering the dark-tinted glass windows. Many fittings were produced in solid silver, while the remainder were lavishly finished in hand-engraved silver-plate.

Alf Martin worked at Serck Radiators, as well as assisting "that genius Fred" on various private projects with his routine maths calculations. He recalled the cars and the process whereby the Maharajah's radiator would have been of brass finish, plated in copper, nickel and finally silver. The water level sight-glass was a feature that all owners became accustomed to and, indeed, the Maharajah of Rewa's many chauffeurs were so used to this on his fleet of Lanchesters that, when he ordered a 1926 silver-plated Daimler 45 h.p., he insisted that its radiator should be Lanchester-shaped and that it included a similar sight-glass also.

The Lanchester honeycomb radiator, which was finished as a complete unit rather than a core and separate shell, was of the same general style as previous models. Alf Martin recalled that George initially produced sketches of his desired radiator shape, from which a brass prototype was hand-made. It was duly checked on a chassis for style and proportion before joint discussions with Sercks about its frontal surface area, type of core, size of header tank, etc. and how effective it would be compared to the size and performance of the engine. After ensuring that all the criteria were fulfilled, a master-pressing was made from which all others were subsequently produced. The Lanchester radiators were of thicker metals than those destined for most other companies. By 1924 most radiators were being produced from nickel-silver, as the fashion for all brass fittings was diminishing due to the problems of keeping brass clean and shiny. (For this same reason, nickel itself went out of vogue when chromium plating was developed in 1928.)

With the realisation that the model was in need of some modernisation, the coachwork was completely re-vamped and again George showed his skill as a body designer. The wheelbase, track and bonnet-line stayed the same as before, the coachwork was 11ins. longer and the higher door, roofline and headlamps, coupled with a 2ins. increase in the width, gave the car a very solid and impressive appearance. The scuttle-mounted air-scoops, looking like miniature ships' ventilators, were retained due to their popularity, but were peculiar to the 40 h.p. model.

The main mechanical alteration was to the engine layout. At this time silence was, for many people, the most important feature of a car. Eric Sisman, who worked at the Company for many years, recalled that there were some nineteen gears in mesh within the engine and that slight noise was unavoidable after great usage, although the engine performance and efficiency were not impaired in any way. The modifications consisted of the complete removal of the layshaft and the horizontal re-location of both the starter motor and dynamo. The former was positioned to drive the flywheel via a Bendix gear, while the flange-mounted dynamo was rotated by an extended magneto drive-shaft, which itself was operated by the front, vertical camshaft drive. A re-designed camshaft, still produced in two halves, also improved efficiency, and this now drove the horizontally-mounted distributor protruding through the centre of the valve cover. The all-aluminium pistons were discontinued as there was a tendency to a slight amount of piston-slap. In the quest for a totally quiet engine, the Company adopted an expensive two-piece piston design with its domed crown remaining in aluminium, secured to a steel skirt; a combination of two optimum features. The oil pump, developing 45 p.s.i. pressure, was situated at the rear of

Two views of the Maharajah of Alwar's 1924 State-landau, built on a special, 14ft.10ins. long chassis, no. 1925. The Maharajah ordered this unique car when he was in London, initially to purchase a Rolls-Royce. Rather shabbily dressed, he entered their showroom but was soon evicted. He immediately visited the Lanchester showrooms in New Bond Street and ordered the landau, and other Lanchester cars. The motor trade soon learnt of this story and a senior Rolls-Royce official was sent to the Maharajah to apologise. He subsequently ordered two Rolls-Royce cars, but on their arrival in India he gained his revenge, it seems, by personally destroying their interiors with an axe and demanding they should be used as refuse vehicles! (See appendix.)

the engine and was driven by a worm drive from the crankshaft. A housing was secured to the bulkhead in the engine compartment for a by-pass filter, which cleaned about 10% of the oil in circulation and was designed to be replaced as a unit after 8,000 miles. Combined with all this, the old copper-tube inlet manifold was disbanded in favour of a cast aluminium manifold, which was water-heated. The cooling fan was returned to the engine from its previous location on the radiator, and was driven by a pulley secured to a

The Lanchester craftsmanship on the Maharajah of Alwar's State-landau, with external mirror-finish and gold silk brocade interior. (See appendix.)

Lanchester crankshaft vibration damper. This was the first time one of Fred's dampers had been used on a Lanchester since the pre-war 38 h.p. design, and it enabled the flywheel to be drastically reduced in its size and mass. To take advantage of these engine alterations, the gearbox ratios were changed to 3.56:1, 6.5:1 and 12:1, allowing 38 m.p.h. to be reached in second gear.

The result of this re-design was an engine which was more responsive and powerful than the old one, achieving a better acceleration and about 10 m.p.h. increase in the maximum speed. Although the Lanchester engine was 1490 c.c. smaller than the new Rolls-Royce engine of 7,668 c.c., it bettered this Phantom 1 model by 5 m.p.h. It was certainly quieter than previously, matching the Rolls-Royce as well as the sleeve-valve Daimler, which Fred was still involved with. The engine retained its good fuel economy of 14 m.p.g., despite the car weighing between 48 and 53 cwt. depending on the type of coachwork fitted.

After this new style was displayed in 1925, the Duke of York, later to be King George VI, bought his first Lanchester; the baby Princess Elizabeth made her first public appearance in this 40 h.p. limousine. Frank had earlier offered the Duke and Duchess a trial run in the 40 h.p., and Archie Millership was sent to collect them at Buckingham Palace in order to give them access to the car in total privacy. They were obviously impressed with their trip to Cambridge, for the Duke subsequently stayed loyal to the Lanchester Motor Company throughout his life.

Due to advancements in tyre design, subsequent changes were made by Lanchester. The beaded-edge tyres, which had been in common use since the turn of the century, were essentially high-pressure tyres. With cord woven into their rubber casings it had been essential to keep the correct pressure, which for the 40 h.p. had been 65 p.s.i., in order to avoid overheating and possible tyre blow-out. However, as there was little suspension comfort from them, other types were developed.

Around 1924, the Dunlop Company joined with Lanchester to develop an idea on very-low-pressure tyres, with the former making the tyres and the latter the wheels. The tyres

184 THE LANCHESTER LEGACY

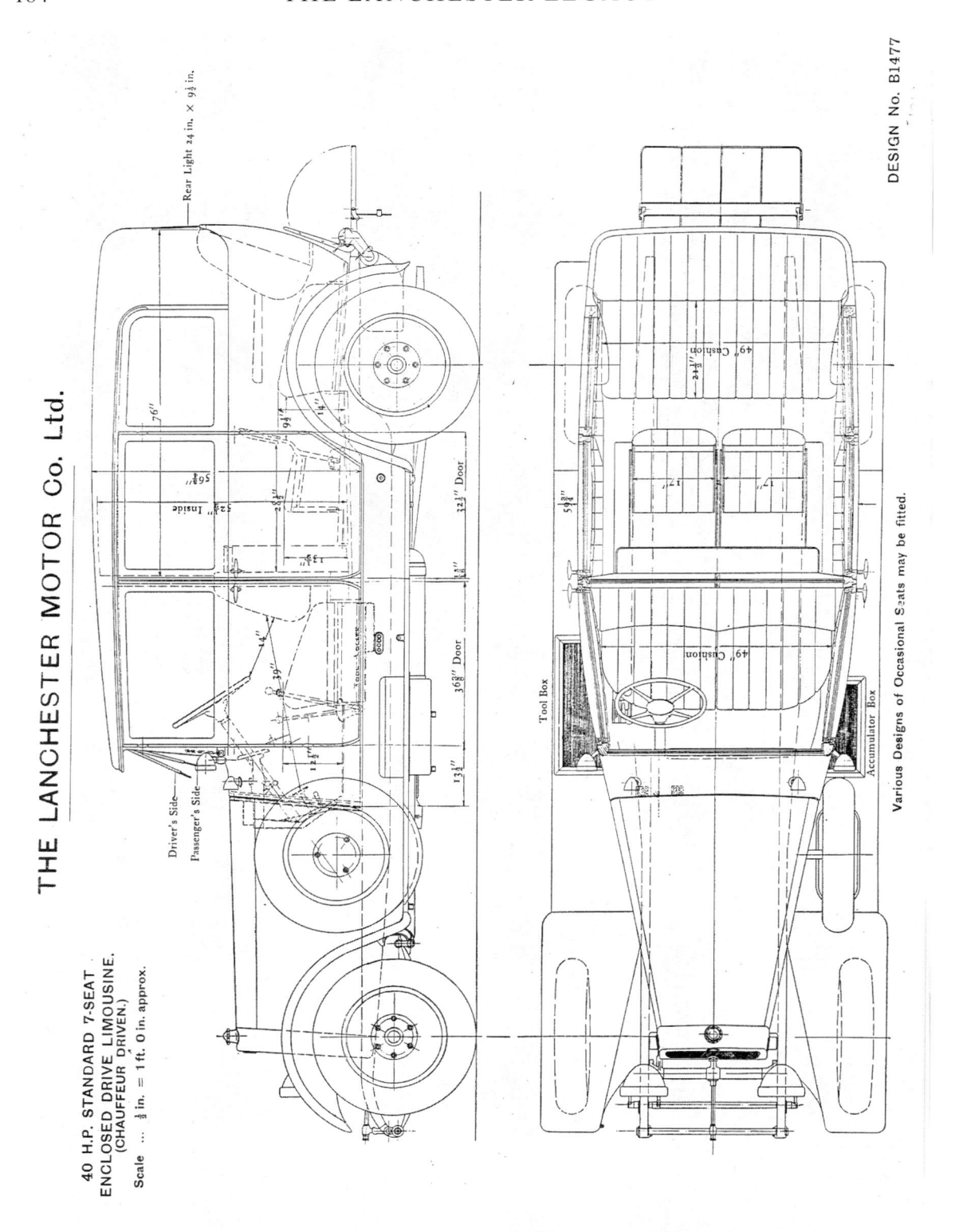

Detailed plan for the 40 h.p. limousine.

Lanchester drop-head coupe with dickey seat, on 21 h.p. chassis no. 3180. (See appendix.)

1925 21 h.p. saloon with coachwork by Caffyn of Eastbourne. (See appendix.)

Lanchester touring coachwork on a 1925 chassis, no. 3334, owned by the same family since 1954. (See appendix.)

Re-bodied in 1942 by F.Moyle Coachbuilders, after removing 18ins. from 1927 21 h.p. chassis frame no. 3390. (See appendix.)

Owned by the original owner from 1927 and then by his chauffeur until 1963. Maythorn limousine coachwork on 21 h.p. chassis no. 3497. (See appendix.)

1928 chassis no. 3574, with 1913-style Hooper landaulette coachwork for Lady Morrison, who wanted to replace her 1913 Lanchester. (See appendix.)

Chassis 3552 was ordered with low-ratio back axle gear so that a loyal customer's daughter only needed to use third and top gears! Lanchester sports-tourer coachwork. (See appendix.)

Connaught saloon coachwork built to the Weyman-fabric style on the 1926 21 h.p., chassis no. 3445.

THE SIX-CYLINDER 40 H.P.

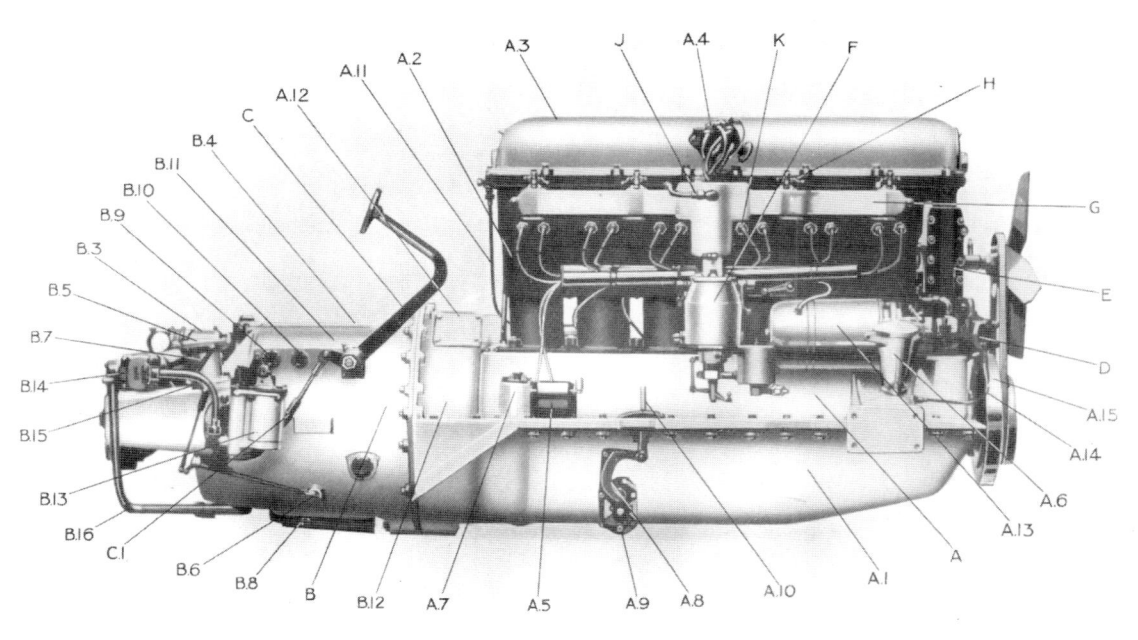

A.	Gearbox.	B.6.	Gearbox Oil Overflow Cock.
A.1.	Oil Base.	B.7.	Gearbox Oil Level Actuation.
A.2.	Cylinders.	B.8.	Gearbox Oil Sump.
A.3.	Valve Cover.	B.9.	Compound or 2nd Gear Adjustment Nut.
A.4.	Distributor.	B.10.	Low Gear Adjustment Nut.
A.5.	Coil.	B.11.	Reverse Gear Adjustment Nut.
A.6.	Engine Oil Filler.	B.12.	Flywheel Casing.
A.7.	Engine Oil Level Gauge.	B.13.	Hydraulic Brake Cylinder.
A.8.	Engine Oil Level Cock.	B.14.	Hydraulic Brake Accumulator Cylinder.
A.9.	Engine Oil Drain.	B.15.	Hydraulic Brake Cylinder Delivery Pipe.
A.10.	Engine Oil Level Cock Handle.	B.16.	Hydraulic Brake Feed Pipe.
A.11.	Camshaft Oil Delivery Pipe.	C.	Clutch Pedal.
A.12.	Valve Timing Inspection Lid.	C.1.	Clutch Pedal Adjustment Link.
A.13.	Dynamo.	D.	Magneto and Dynamo Gear Housing.
A.14.	Oil Return Pipe.	E.	Vertical Shaft Casing.
A.15.	Crankshaft Damper and Fan Driving Pulley.	F.	Carburettor.
B.	Gearbox.	G.	Inlet Pipe.
B.3.	Clutch Spring Box.	H.	Priming Cocks.
B.4.	Gearbox Inspection Lid.	J.	Water Circulation Pipe.
B.5.	Gearbox Oil Filler.	K.	Sparking Plugs.

The 6,178 c.c. modified 40 h.p. engine of 1925

were of 8ins. tread width and were inflated to just 12-15 p.s.i. There were many problems, and George drove a 40 h.p. fitted with these tyres to Fred's nearby home to ask his opinion. The brothers found that when the tyres were at Dunlop's intended pressure, the 40 h.p. could not be driven above 35 m.p.h. The scheme was eventually dropped as these tyres rendered the car virtually uncontrollable.

When the straight-sided tyre arrived in 1925 it was designed to suit a new profile of wheel, whereby a removable rim flange secured the tyre firmly in position. However, as the pressure had to remain as high in order to secure the flange in position, its ride was still hard. This 33ins. diameter by 5ins. wide cord tyre was fitted to the pale grey limousine of that year's October Motor Exhibition.

New developments in 1926 brought out 'balloon tyres' which, in the case of the 40 h.p., were reinforced 33ins. by 6.75ins. Dunlop cords. These low-pressure tyres made for a more cushioned ride so, to allow for easier steering, the size of the 40 h.p. steering wheel was increased by 1.75ins. to 19.75ins. in diameter.

Unfortunately, these tyres initially brought with them massive problems of wheel wobble, shimmy and tramp, and were cursed by all the motor manufacturers. Fred Lanchester, whilst trying to advocate the use of his 'shimmy damper' device, highlighted the general problem when he exclaimed that, "Surgeons have told me that they dare not drive their

Lanchester landaulette coachwork on the circa 1926 40 h.p. chassis.

own cars when they are going to carry out an operation, because of the unsteadying effect on their hands, of the steering wheel."

However, the problems were overcome and balloon tyres were used throughout 1926 by the Lanchester Company. Indeed, their drawings of January 1926 recorded, "Wheel track increased 0.5ins., width of tyres altered." They were shown at their Olympia stand on their six-seater, enclosed-drive cabriolet, a style which incorporated a fixed roof over the rear seat passengers and a folding extension over the driving seat. The coachwork, which surprisingly was built by the Hooper coachbuilding company, was finished in buff and black with a brown interior, and the whole sold for £2,725.

The 1926 Olympia Motor Exhibition Lanchester stand, featuring a 40 h.p. Hooper-bodied cabriolet.

THE SIX-CYLINDER 40 H.P.

The 40 h.p. Lanchester with Hooper cabriolet body, no. 6611, exhibited at the 1926 Motor Exhibition.

Circa 1928 Lanchester landaulette coachwork on the 40 h.p. chassis.

The 40 h.p. car, chassis no. 2014, with State-limousine coachwork by Lanchester, ordered by His Imperial Highness Prince Chichibu of Japan. It was finished in January 1929 and shipped to Japan in July. The car was destroyed by fire during the Second World War.

As good as the previous Lanchester 40 h.p. cars were, Rolls-Royce had raised the stakes slightly by the introduction of their Phantom 1 model. This incorporated their new, overhead valve engine of 7,668 c.c. in a re-vamped 40-50 chassis, and the Lanchester Board of Directors felt that they should respond accordingly.

In early 1926, therefore, George worked out the design for a new engine. It had a bore of 4.25ins. and a stroke of 5ins., with a capacity of just under 7,000 c.c. Unlike the standard 40 h.p. engine, detachable heads were used, and they incorporated two inlet and two exhaust overhead valves for each of the six cylinders. A test engine was produced and was used successfully in a 40 h.p. chassis, complete with test body. In August of that year, this was used by Arthur Bird to drive Joe Woods, who worked in the road test department, and Eric Sisman, from the drawing office, down to the Brooklands Motor Course to watch the Grand Prix race. The engine performed well, but the order for full production was never given. The reason for this was probably due to the general economic climate of the time, for the depression of the late 1920s had begun and the Company had, for the very first time since its inception, failed to make a profit.

Production of the standard 40 h.p. car carried on as before, but only sixteen were sold during the whole of 1926. No modifications were recorded for the 40 h.p. before the 1927 Olympia Exhibition, where its last public showing featured a Lanchester-bodied "enclosed-drive limousine seven-seater body, painted black with ivory-white lines; front seats upholstered in natural grain black leather; interior in silk Damask; two folding chair occasional seats, central armrests to rear seat; woodwork polished mahogany inlaid; two corner electric lights; screen frame and all other plated fittings electro-silvered; thick sheepskin rug". Its price complete was just over £2,692.

Sales of the 40 h.p. declined further in 1928, with just ten cars sold. Minor modifications took place, such as the fitting of a C.A.V. coil, as well as one item of particular interest: the adoption of the popular Dewandre brake vacuum-servo unit operated from the induction manifold.

THE SIX-CYLINDER 40 H.P.

Anthony Bird and Francis Hutton-Stott wrote that George Lanchester "liked this (Dewandre system) for its simplicity, but disliked it for its lack of 'feel'. The salesman undertook to have the defect eradicated and Lanchester undertook to place an order once he was satisfied. The Dewandre mechanism was duly modified, and Lanchester duly fitted the 40 with vacuum-servo in the interests of economy."

The adoption of the new servo-system was also taken in the interests of standardisation, as the new 30 h.p. Straight-Eight Lanchester was making its debut. This model did not use the Lanchester epicyclic gearbox, so there was no oil pressure to drive George's own hydraulic ram system to assist the brakes. Accordingly, from 1928, all the 40 h.p. cars were fitted with the Dewandre vacuum-servo unit.

The new Straight-Eight model, in effect, provided even more competition for the 40 h.p. Fully in vogue with the times, it undoubtedly took sales from the 40 h.p. in the ever-decreasing luxury car market, and only three or four 40 h.p. cars were sold in 1929. One of these was an exquisite dark blue and black, V-windscreen limousine ordered by His Imperial Highness, Prince Chichibu of Japan. The rear compartment, complete with occasional seats and loose cushions, was finished in grey, silk brocade. The rear, corner recesses and other woodwork were of polished mahogany and all metal fittings were silver plated.

All advertising for the 40 h.p. model was stopped later in 1929, although some were built to special order after that date. From the first cars in 1919 there had been fluctuating periods of high sales, but these figures averaged out at under one 40 h.p. sold per week; a small amount when compared to the large Rolls-Royce models which had sales of many thousands between them. This reflected more on the Lanchester Company's history of financial hardship than on the quality of the 40 h.p. car. The company's advertising had been limited and the lack of sales and spares outlets throughout the world must have given potential customers cause to purchase more accessible cars. It must have seemed, in hindsight, that Fred Lanchester's initial worries about competing against the hugely-organised and well-financed Rolls-Royce Company had been justified.

Chapter Nineteen

THE VINTAGE YEARS

The Lanchester coachwork for the 40 h.p. and other models was produced in the works adjoining Armourer Mills. Although convenient in many respects, this may have caused problems in the engineering works, with sawdust drifting through the linking doorways and contaminating the precision machinery. In view of this, most of the coachbuilding operation had been transferred, immediately after the First World War, to the Alliance Works in Foremans Road, Greet, the neighbouring village to Sparkbrook. This single-storey factory had formerly been the site for Rudge-Whitworth's wheel-making operation, and the move was obviously sponsored by the Pugh brothers, who were directors of both companies.

The Greet factory started producing coachwork to George's design, and continued the traditional Lanchester practice of well-balanced, conservative coachwork, rarely bordering on the racy or sporting style. The construction of the coachwork was years ahead of its time and set standards which were very advanced compared to those of the other, mostly independent, coachbuilders; Rolls-Royce and the majority of other motor manufacturers never considered making their own bodies.

Many of the Lanchester bodies and fittings had slight differences from each other in order to meet every whim of the customer; indeed, even a commode, complete with silver-plated fittings and curtaining, was included in the limousine specifications for one customer. Under the control of Works Manager, Albert Lee, who had been with Lanchesters since 1897, most work was done in-house, but orders were occasionally put out to specialist suppliers. Some seat squab supports in the early 1920s, for instance, were supplied by a small local family firm named Ashen, and were made from best quality cane rather than the usual webbing. Most essential raw items were delivered by the rail network to the nearby Great Western Railway marshalling yard, whereas items such as timber travelled by barge on the Birmingham and Warwick Canal to a nearby timber yard, and then completed their journey to the Works by horse and cart.

*Lanchester 40 h.p. tourers awaiting delivery from
The Alpha Works, Liverpool Street, in Birmingham.*

At the Alliance Works, the paint-undercoats were applied in their own paintshop. The car bodies were then taken to the Alpha Works, in Liverpool Street, where a large paintshop had been located to produce the finished product. After some sixteen coats of paint had been applied to each car, with an overnight drying-time between each, the varnish was applied. One ex-employee recalled that, "This was brushed on liberally and then 'laid flat and even' by one man, to create the most perfect finish. Any deviation from perfection was unacceptable and the whole panel would be rubbed down and re-finished." The resulting surface, though initially immaculate, remained a relatively soft finish; even vigorous cleaning of the car with a chamois was likely to cause scratching. After all this, re-varnishing a car after two or three years was not uncommon. Final assembly of body-fittings and trimming also took place at Alpha Works, which had space for up to thirty cars to be parked there at any one time, awaiting delivery to their customers.

In March 1923, coinciding with the introduction of the new 21 h.p. car, the main coachbuilding operation was transferred yet again, the Alliance Works being taken over by Joseph Lucas Limited, the car electrics company. Lanchester's coachbuilding arm, and also the servicing department, were moved to premises purchased from Alldays and Onions, the cycle, motor-cycle, car and commercial vehicle manufacturer who had just been declared bankrupt, in Fallows Road, just a short distance from Montgomery Street. It was a typical Victorian building mostly of two floors, occupying a ground floor area of 120ft. square and was originally known as The Matchless Works, a name signifying the Alldays Matchless motor-cycles built between 1903 and 1914.

The great problem of the paint's drying-time meant that this operation, and therefore the subsequent trimming and finishing, had to be continued at Alpha Works. Any colour of paint imaginable could be chosen by customers for their cars, although the majority were dark. One feature which Lanchester carried on from the Edwardian era was the alternate three or four inches wide vertical stripes of paint in varying colours, such as yellow and black or two shades of green. This tradition, however, was abruptly halted when the American Company, DuPont, perfected a new cellulose-based paint which was found to be unsuitable for the Lanchester striping. The advantages of adopting the paint far outweighed this small drawback, for it drastically reduced the time spent on finish-painting a car. From a previous total painting time of many weeks, a car was subsequently out of the paint shop in a fraction of that time, as each coat of the new 'Duco' paint-lacquer "dried in minutes". The paint also gave a very hard finish and was scratch, stain and weather resistant. By 1925 most American firms were using this product and George, always up-to-date with new technology, gave orders for its trial use on his cars in the same year, possibly becoming the first firm outside America to use other than traditional coachpaint enamel and varnish, which had been in use for over a hundred years. A disadvantage did arise, however, as this finish was not quite as good in the short-term compared to the enamel paint, although after some months' service the situation was reversed. It was therefore only applied when requested by the customer; one such customer was the Duke of York. Improvements led to its general use the following year and from then on, all the final painting, trimming and finishing was conducted at Fallows Road, enabling all the coachbuilding operations to be housed under the same roof.

One of the sixty-three craftsmen employed there, Dennis Butler (now named Dennis Corben), recalls that the Works had room for up to twelve car bodies to be constructed at any one time. The first floor could accommodate six in the coachbuilding and cabinet-making department, two in the panel-beating shop, two in the paint shop and one in the trimming shop. With access to the ground floor via a large hand-operated lift, there was room for one car body in the finishing shop.

The Lanchester process of producing a car body started in the coachbuilding shop with the fettling and cleaning of the body base frame, and also the door pillars, screen pillars, scuttle and bulkhead. This body frame was regarded as vital by Lanchesters to ensure standardisation of parts. In these post-war years, duralumin castings were adopted rather

One of the first Lanchester cars to be finished in cellulose paint: the Duke and Duchess of York's 1925 40 h.p. limousine. Princess Elizabeth made her first public appearance in this car with her parents.

THE VINTAGE YEARS

1926 Lanchester 40 h.p. finished in "wood-grain cellulose paint".

than the steel used previously. The Daimler Company, prompted by Fred, was the only other coachbuilder to adopt this "application of Engineering to Body-Building", as George termed it, although they still used the heavier steel frames.

Depending on the type of car chassis required, the body frame was secured to a jig, which had identical location points fixing it securely in its correct position, so that any tendency to flex or bend was avoided.

Ash was used, in view of its combined strength and lightness, to construct the remainder of the body frame. Sandwiched in the vertical pillars, however, the ash acted only as 'fillers' into which tacks and screws could be driven, the door hinges and other items being secured into the surrounding duralumin pillars. Also in this department, the cabinet makers applied their craft to the various fittings ordered by customers, such as writing tables, ladies' powder sets, drinks cabinets and so on. George designed many of these features and patented special items, such as his automatically-locking window-levers and his occasional seats, which would "fold flat, flush into the rear floor".

When complete, the body skeleton was wheeled on a truck into the adjacent panel-beating shop, where it would be secured to another matching jig. With felt stuck onto the timber batons with Shellac, to avoid any risk of vibration, the thickness of aluminium for the panelling would be selected to suit the type of coachwork required. Normally this would

A Lanchester's duralumin and timber body frame for a touring body.

Duralumin and timber body frame for a limousine body.

be 18 gauge, which is 0.048ins. thick, but if the body was to be highly-polished instead of painted, 0.064ins. thick 16 gauge aluminium sheeting would be used. After the panel-beating hammers, dollies and wheel-rolling machines had done their work, the panels were checked for size and shape against ready-made templates. Where separate panels butted against each other, George decided not to follow tradition by securing to the timbers beneath with screws and pins, but instead instigated a new and highly-skilled method of welding the aluminium panels together by oxy-acetylene gas equipment (the first such usage in motoring history). Only the edges of the large, shaped panels were then secured with pins to the timber-fillets beneath. For polished aluminium bodywork, aluminium screws were used to secure the panels before their heads were filed down flush and polished to ensure they were unnoticeable.

The paint shop was the next port of call, where the body received numerous undercoats and top coats, together with meticulous rubbing-down sessions, until no blemishes remained.

In the trimming shop, the leather roof-covering and interior design would be produced exactly to the customer's requirements with regard to the type and colour of cloth, leather, veneer, lighting, ventilators, heaters and so on.

A trip to the ground floor in the platform lift brought the body to the finishing shop, where its rolling chassis had arrived, tested and correct, from Armourer Mills.

Previously constructed mudguards, which were later known as wings, had already been made to order from either steel or aluminium in the adjacent sheet-metal shop; the front pair and the chassis were painted after being finally bolted together. Four men were needed to lift the car body onto the chassis; large circular rubber blocks were sandwiched on special protruding brackets, to ensure there was no contact at all between the body and the frame. Although a few other coachbuilding firms laid felt or rubber strips between the body and chassis to act as insulation, their securing bolts were still in metallic contact with both, so negating the idea. The Lanchester's securing bolts were surrounded throughout their length in rubber, to eliminate any transmission of vibration. The accuracy of construction of both

THE VINTAGE YEARS

The interior fittings of one customer's 21 h.p. limousine included a leg-rest and extending table, both of which were able to be concealed under the folding occasional seats.

Luxurious silk finish provided by the trimming shop for His Imperial Highness, Prince Chichibu of Japan on his 1929 40 h.p. limousine.

The customary Lanchester pattern sewn into the door pocket-flaps of the Maharajah of Nawanager's 21 h.p. hunting car.

the body and chassis meant that no problems were experienced lining the two up, and the whole operation would take less than twenty minutes.

Aluminium sheet was shaped to encase the rear wheel-arch body structure and the rear mudguards were secured, to finish the bodywork. The fitting of the electrics concluded the complete operation, allowing the final checks to commence.

Each departmental inspection throughout this process had demanded the very best, so that only occasionally was the finished work not found to be perfect. Dennis Butler remembers that, "On the rare occasion that items were not up to standard, great pains were taken to eliminate the faults. It once took two days to hang one door."

Most cars received one further process before delivery, and this age-old skill was accomplished by the 'master', Harry Tommerson. He coachlined wings and body panels on the Lanchesters using a brush, continually turning the 9ins. long bristles to make a perfect line. Black tape, and then a tool like a glass-cutter to make a slight indentation, held the bristle point true. Between coachlining jobs, he used his skill for the painting of new and second-hand vehicles which locals brought to Fallows Road, revelling in his application of artistic bowls of fruit, loaves of bread and the like.

Possibly his most interesting job, and certainly his most lucrative, was the monogram he hand-painted on a solid silver fretwork base mounted on the outside of the doors of the Maharajah of Alwar's 40 h.p. State Coach. The Maharajah would not accept this, or any carriage, as finished, until his Royal monogram was displayed. In possession of the Maharajah's signet ring in order to perfectly reproduce the monogram, and permanently flanked by one of his security guards, Harry Tommerson spent three days painting the intricate design as he had to allow for each of the three colours to harden before the next one was applied. He was rewarded by the delighted Maharajah with a sovereign.

The Works were not restricted to Lanchester chassis for the application of their coachbuilding skills. Owners of other makes such as Rolls-Royce, Daimler, Austin and Sunbeam, had various Lanchester bodies constructed and fitted to their vehicles. The Company would readily undertake other work also; an order from the Daimler Motor Company for a design of bonnet-catch to lock their three-piece bonnet design was just one such example.

There is no memory among ex-Lanchester employees of any complaints received from customers on the standard of work. A happy atmosphere prevailed at the Works, conducive

Cloth interior of a 1929 30 h.p. landaulette, with swivelling occasional seats and polished walnut-burr fittings.

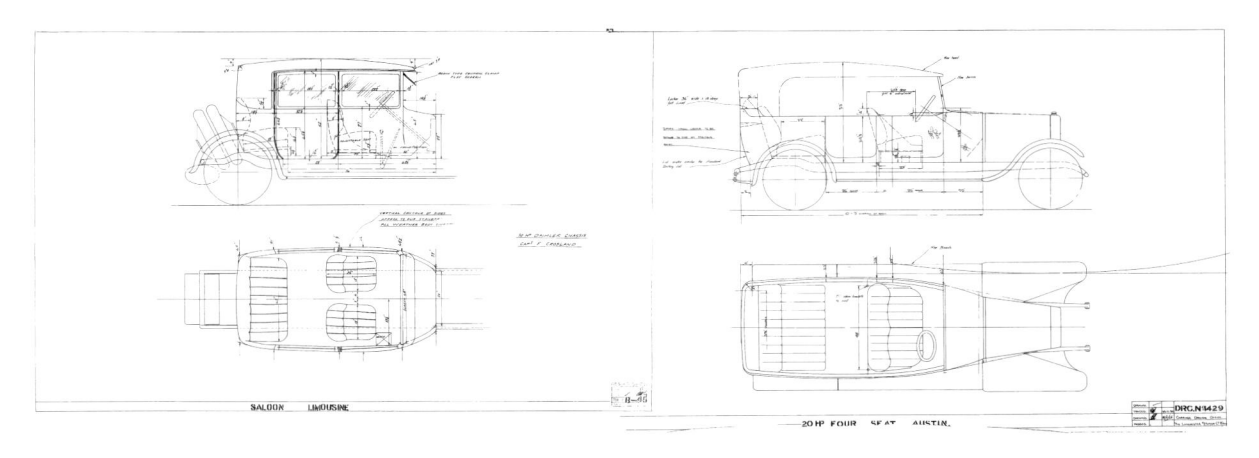

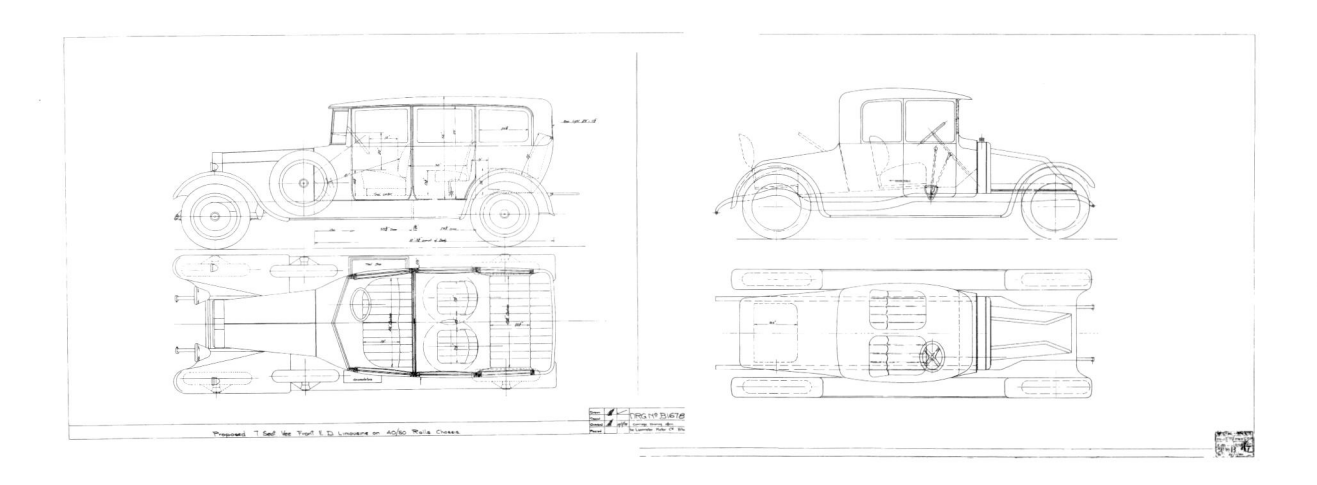

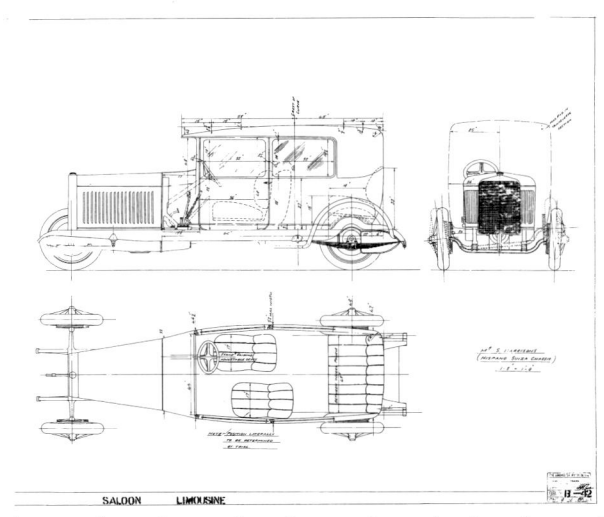

Some coachwork designs for other make of cars, from the Lanchester Drawing Office.

to the high quality of their product on which the employees prided themselves. There was even a suggestion-box in which workers could post their ideas; an enlightened and progressive move which contributed to the success of the coachbuilding department.

Lanchester customers were strongly encouraged to have the combined package of chassis and coachwork, but the sale of only the chassis was allowed so that outside coachbuilt bodies could be added. Lanchester's body-construction method proved to be most beneficial in this respect as on the complete chassis, which they would send to companies such as Mulliner, Hooper or Windover, a body base-frame would be mounted, on which the coachwork could be secured. This system made it unnecessary for anyone to drill into the Lanchester chassis frames, for fixtures or brackets.

However, some coachbuilders objected to the body base-frame on the grounds that it "hampered their designs". George strongly contradicted this by arguing that, "There is ample latitude for any variations that are reasonable for the size of chassis for which the body is designed." He also pointed out that it was desirable to use the Lanchester scuttle and door pillars, but not essential, and remained adamant that the base-frame was in everyone's best interests.

George also observed that, "Many instances of bodies built by firms of the highest repute have been spoilt by slavish adherence to traditional coachbuilder's craft." He cited one case where he "..... computed that the bottom, sides and edge plates employed weighed 160 to 180 lbs. in timber and steel, where 60 or 80 lbs. could have been saved without diminishing the strength".

Another case in point was the prevalent practice of using two uprights and two or more horizontal rails to stiffen a door, where the use of just two diagonals reduced the components and also the number of joints to five, from at least twelve formerly. His use of aluminium corner pieces instead of wooden blocks again made savings, not just in weight but also in a craftsman's time, and therefore cost.

George "emphatically condemned" the fact that first-class coachbuilders would happily include sheet metal, plywood or wood panels in a roof structure without a care for their resonance. He continued, "It is difficult to conceive any structure better calculated to transmit noise than these large sound-boards." His coachwork used instead a flexible roof of leather, reinforced with wire gauze to prevent sagging.

So serious was the problem of resonance in other bodies that many of the cars' owners turned to the Lanchester Motor Company for assistance. George explained, "In some cases it has been impossible to satisfy customers of a nervous or sensitive temperament without re-building the roof, mounting the body on rubber buffers and softening the trimming in every way."

The Lanchester coachwork was considered to be technologically-advanced as well as comfortable to the occupants. It was practical, too, as the Company continued their tradition since early in the century of "hanging the doors with their hinges to the front, so that when the car starts, the door will slam automatically should it be left open". Customers did countermand this occasionally, by ordering doors hinged at the rear. Typically, the Company never capitalized on their coachbuilding prowess by advertising in motoring journals, or by separate display stands at exhibitions, nor was Lanchester coachwork specifically shown on a different make of chassis. It can be safely assumed that any of these measures would have helped sales and enabled more motorists to have sampled Lanchester's "scientific principles of coachbuilding".

One-third of the Fallows Road coachbuilding works' ground floor area was devoted to the service department, managed by Enoch Aston. About a dozen mechanics worked there, maintaining and repairing Lanchester cars in a most detailed fashion; even the 40 h.p.'s silencers were serviced by cleaning out the internal tubes. It was from here that the new cars would be delivered by Company personnel, or collected by the owner or his driver.

Often with new customers, Lanchesters would also be asked to provide a chauffeur. Groups of potential drivers were tested for temperament, general health and eye-sight,

THE VINTAGE YEARS

The Service Department's next challenge?

before being put through their paces for seven or fourteen days of tuition. Apart from learning about the car's design and maintenance procedures, they would undergo driving tests in all conditions. Only when proficient would they be permitted to take up duties with their new employer, and frequently had to move house to do so.

Also required to be flexible in their living arrangements were the engineers, who would travel out to various parts of the country and abroad to work on customers' Lanchesters. These service engineers were necessary as, although the Company had been producing cars for two decades, there were still only three major sales and servicing outlets in the whole of Great Britain. Invariably making the return trip by train, servicing engineers usually stayed in the servants' quarters at the customer's house, or at a local guest house, for the duration of the repair. Although they received no extra pay for being away from home, their board and lodging was paid by the Company when necessary. Between calls they worked in the service department at the Fallows Road works, although their priority was for outside repair work.

The Company had retained their northern repair centre and showrooms at 88 Deansgate, in Manchester, whereas the south was still served by the London showroom at 95 New Bond Street, in Mayfair. A separate repair centre nearby, managed by George Upton, was at 107 Pimlico Road, Fulham, before it transferred to Star Road in West Kensington, about 1922. Other than these and the service centre in Fallows Road, there were only about twenty regional agents spread throughout the United Kingdom. Many customers relied instead on their chauffeur for repairs and maintenance, or tried out the local garage who would then apply to Lanchester for spare parts. Here the Company excelled, as parts would be despatched to customers through the parcel delivery service on the same day as receipt of a telegram order.

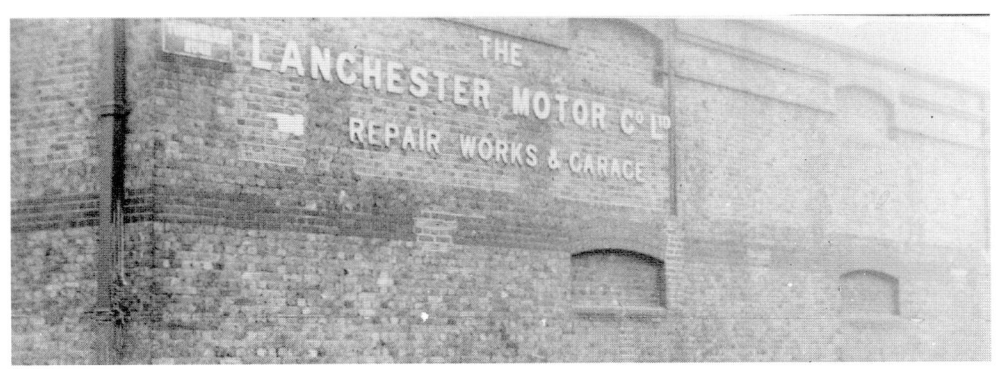

Directions to the Star Road repair centre in West Kensington, on a factory wall in adjacent Turneville Road.

> CONTRACTORS TO THE ADMIRALTY, WAR OFFICE & AIR MINISTRY.
>
> **THE LANCHESTER MOTOR CO. LTD**
>
> 95, NEW BOND STREET,
> LONDON, W.I.
>
> 25th October 1924.
>
> F. W. Hutton-Stott Esq.
> 52, Queen's Gate Terrace S.W.
>
> Dear Sir,
>
> Please accept our thanks for your esteemed order given to our representative last evening at the Olympia Show for a 21 h.p. Lanchester similar to the one shown on our Stand; and as promised we have pleasure in enclosing detailed specification in duplicate, together with an order form, which we would ask you to kindly sign, and also sign the "Works Copy" of the specification and return both to us with your cheque of £100. as deposit. The Client's copy we would ask you to kindly retain for your own reference.
>
> Regarding the question of a driver, we are using our best endeavours to find you a really suitable man, and do not think there will be the least trouble in this respect, by the time the Car is ready for delivery.
>
> With regard to delivery; we shall do our very utmost to let you have the Car before Xmas; and we are to-day passing your esteemed order on to our Works so that the Body can be commenced immediately.
>
> Should there be any other small details that you may think of within the next few days, we shall be only too pleased to incorporate them in the Car, if you will kindly call at the above address; in fact, we shall be glad if you will kindly call in any case, as there is one other small item we should like to discuss with you.
>
> Again thanking you, and we feel sure that you will be pleased in every way with the Car upon completion.
>
> Yours faithfully,
> THE LANCHESTER MOTOR Co., Ltd
>
> London Manager.

Confirmation of order for car and driver, from the Mayfair showroom, relates to 1924 21 h.p. chassis no. 3108. (See appendix.)

40 h.p. engines on test, with the white-coated foreman making adjustments.

There was little contact between the coachbuilding works and the Armourer Mills works except for the factory runabout, which carried stores, equipment, parcels and communications between them. From about 1910 to 1921 this runabout had been an early 20 h.p. landaulette which was continuously seen in and around the Birmingham streets. This transport was somewhat modernised in 1921 by the use of a slightly newer 38 h.p. with van body.

Also to be seen on the roads were the chassis under test from Armourer Mills which, unlike other companies', were never left bare and open with just a driver's seat. They were instead finished with a simple, two-seater test body to which a windscreen, spare wheel and toolbox were secured. Weights were added to imitate the potential loading of the final body complete with passengers, although occasionally other workers would be taken to serve as 'ballast'. This was a great treat, as many employees who had been with the Company all their working lives had never travelled in any car at all, let alone a Lanchester.

The chassis were tested initially by a run of about a hundred miles, under all conditions, through towns, up hill and down dale. A few of the test drivers, like Alfredo Bezani, Bill Copestake and Joe Wood, had their own favourite routes through Stratford-upon-Avon, Wales or the Malvern Hills. Certain sections were also popular with test drivers from other companies, and Stan Powers recalls the infamous, long and winding Fish Hill, near Broadway in the Cotswolds. "There was a lady who lived in the cottage near the top of the hill, where many cars would be spluttering and boiling over. She used to sell the test drivers a bucket of water for tuppence; the Lanchesters, of course, never needed her water."

On return to Armourer Mills a record sheet would be hung from each radiator filler spout, detailing any deficiencies in engine power, brake retardation or any other failings. Shorter road runs would then confirm that the fault had been rectified, whereupon Arthur Bird, of the Road Test department, would complete his final checks on each engine, using the doctor's stethoscope he used to wear round his neck to check their quietness. Final oil, water and tyre inspections allowed the car to be driven to the coachworks for its body to be fitted.

The Engine Test foreman, Freddie Chase, and his staff would already have 'run-in' the engines under their own compression, immediately they were received. They were then run at full throttle, as a complete unit with clutch and gearbox attached. One propeller shaft from each drove one of the four Heenan and Froude dynamometers in the test shop, loaded down to a fixed r.p.m. by the water-brake, to show maximum horsepower available. Under test, only the petrol and cooling water supplies were not to be part of the finished unit, the water being taken from overhead tanks and emptying via an open gulley in the floor, through the wall and into the canal at the rear of the building.

When confident of each engine's performance, ease of running on coil or magneto, and brake horsepower obtained, the engines would be left running for twenty-four hours in the care of an astute apprentice: one who knew where the 'off' switch was! They were set at 500 r.p.m. for this duration, a timespan which would probably have seized many other engines, as little or no lubricating oil would have been splashed up the cylinder bores at these low revs. As all Lanchester engines utilised oil-feed pipes to the piston gudgeon-pins, this problem was completely avoided.

Although four engines could be tested at any one time, it was normally only two, with one of the remaining test beds generally reserved for experimental work. This was usually undertaken at the instigation of George Lanchester, assisted by Arthur Bird and others, to investigate various ideas on improvements. Robert Boden was one of this team, and recalls that many modifications were tried, including lengths of piston skirts, the number of piston rings, cam lift, manifold designs, the length of valves and their guides and so on. He especially remembered Mr. Skinner, of Skinner Unit Carburettors, attending the works, trying to persuade George to use his S.U. carburettors. There was, however, no experimental department as such, more a continuous flow of ideas under consideration. A keen apprentice was invariably allowed to assist in the experiments and he would indubitably have gained an invaluable source of experience from this.

The few apprentices accepted at the Company were from three distinct backgrounds. The most common group were in their mid-twenties, mostly ex-Royal Flying Corps and, later, Royal Air Force personnel who were keen to learn and progress. The second were the younger intake, normally straight from school, whose parents or sponsors paid a £25 annual premium to the Company for a three-year course. The third group of apprentices, generally regarded as a lazy bunch as they never appeared to exert themselves, were the "well-heeled children of Lanchester owners" and, as a rule, were not highly regarded by the rest of the workforce.

All apprentices were required to attend technical college for one afternoon and evening each week, and for all their efforts their pay averaged five shillings a week, less six pence for insurance; the same weekly wage as had been earned by the Lanchester apprentices before the turn of the century, although this was to double by the end of the decade. During their three-year apprenticeship they were monitored by each foreman, who was expected to write a report on their progress through each department, to be inspected by the Works Manager.

Throughout the 1920s, the Company had its own printing and publishing department on the top floor of Armourer Mills, for its own and other firms' literature, but this was outside the scope of the apprentices. In the interests of safety, not all were allowed to work in the machine shop, where one had to treat the ranks of mills, lathes, shapers, grinders, etc. with their myriad of flapping and twisting drive belts, with the greatest of respect.

George Uzzell, one of the top-rate grinders at the Company, well remembers the machining limits that had to be adhered to: half of one thousandth (0.0005) of an inch tolerance was all that was allowed on valves, swivel-pins and numerous other parts. If a machined part was outside these limits, the inspectors in the view room would scrap it, and if this scrap exceeded 5% of work, the perpetrator would face the sack. This was perhaps a justified policy, as the cost of parts was very high; three valves for instance, would cost more than a skilled worker would earn in a week. An expensive crankshaft was therefore afforded their very best attention during machining with a glass-like grinding wheel of 3.5ft. diameter, in order to minimise the risk of flats and blemishes which were possible with a smaller grinding wheel.

The workers in the machine shop took all their detail from the blueprint drawings, which were copied from the linen originals produced in the drawing office. George paid tribute to

THE VINTAGE YEARS

Machining begins on the 40 h.p. crankshaft and crankcase.

the team and especially to Miles Watkins, the Chief Draughtsman, who "personally checked all drawings and by his vigilance practically 98% of those issued to the works were free from errors; no mean achievement".

There were three others in the small design-drawing office. George and Miles Watkins regularly spent time with them as ideas developed and high standards of innovative work resulted. The whole of the workforce was a dedicated and contented team, although the wages could be bettered elsewhere in all grades. In view of this, some workers left for greener pastures, but many of them came back. One who returned commented, "Go to the

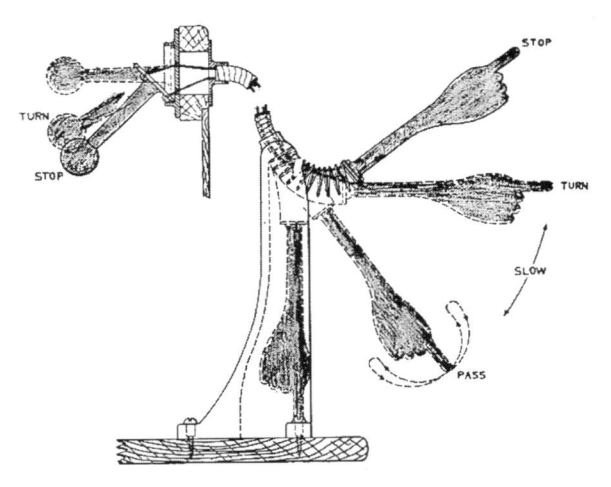

George Lanchester's patented invention of 1927 for any external fitting such as a spot-lamp or mechanical hand-signal, cable-operated from the dashboard.

Austin and you will appreciate the Lanchester; there's real pressure at the Austin and great noise."

Many ex-employees of the Lanchester Motor Company have quite independently spoken of the Lanchester brothers in glowing terms, and recalled the consideration they showed for their employees. One told of George Lanchester taking food parcels to workers who were ill, and another told of him personally taking an injured man to hospital, waiting for him to be treated and then driving him home. Dennis Butler spoke of the time when he was caught doing some work for home, a possible sacking offence. He remembers, "I was startled by a heavy hand grasping my shoulder. Turning, I saw Fred and George Lanchester who had, as I was informed later by a workmate, been watching me for two or three minutes. Still holding my shoulder, George said gravely, 'I have no idea what you are making, young man, but whatever it is make a good job of it. Otherwise you will be wasting your time, as well as the Company's.' At that they both smiled and left."

Although Frank rarely visited the factory floor, Fred did tour the Works with George on several occasions. He was an ever-welcome guest, still a shareholder and, during the early 1920s, must have been more than interested in the outcome of the designs on which he had cast his opinion. He had an obviously deep feeling for the Company itself, and he did not let his lack of authority there deter him from speaking out if he saw something amiss. One such instance involved Fred Wall, who was admonished by Fred for "dubious engineering" when reaming swivel-pin bushes. The crestfallen fitter recalled that Fred and George appeared unexpectedly the next day, approved his revised method and left, confident that the swivel-pin bushes for the 40 h.p. and 21 h.p. would be nothing short of perfection from then on.

Chapter Twenty

THE BROOKLANDS RACERS

The Lanchester Motor Company's insistence on mechanical perfection had earlier benefitted Lord Ridley, who had found no problems when developing and supercharging his 38 h.p. tourer. This demonstrated the obvious potential for increasing the speed of the Lanchester cars, but Fred had shown no interest in the spectacle of motor racing, or of using the sport as a means of testing his cars to the limit of their capabilities. Indeed, he felt that experimental work should be undertaken on the drawing board rather than on the race track. George explained Fred's attitude thus: "Lanchester had no need to have his theories proved on the race track: he knew, and time has proved, that his theories were correct. They were there on record for those interested to put into practice, and had the racing fraternity made more use of the scientific research available, and less on trial and error, many a brave man's life would have been saved."

With sliding roof open, Tommy Hann impresses the crowd in 1922 with his 1911 Lanchester 25 h.p. saloon-racer, aptly entitled Hoieh-Wayaryeh-Gointoo.

After the First World War, Tommy Hann decided to test Fred's theories for himself on the race track. He bought a 1911 Lanchester 25 h.p., which had been in use as a luxurious landaulette shortly before, and developed it for racing. He left the car mechanically as standard, apart from the addition of front shock absorbers, and built a single-seater racing body for it which, due to its lack of vision, earned the car its name: Hoieh-Wayaryeh-Gointoo. This was only the second saloon car ever raced at Brooklands and in Tommy Hann's words, "The general design embodied an engine space insulated from the driver. The cooling air, after passing through the radiator, left the engine space through a duct on either side of the seat, and issued through an opening in the tail of the car, virtually forming at this point a streamline 'tail' composed of air. This enabled a shorter tail to be used than would otherwise be necessary to avoid air drag. Access to the interior of the body was gained through the centre sections of the roof-plating, which were hinged along the elbow-rail on the off-side and fastened down by spring clips on the inside, within easy reach of the driver. A valve was fitted on the main petrol pipe-line near the steering wheel as a precaution against an extensive fire. An opening covered by a slide at the driver's elbow gave access to the gear lever, situated in a horizontal position along the chassis frame. A slot in one panel of the pointed Triplex windscreen provided full vision should oil or water vapour accidentally cloud the glass, and admitted sufficient cool air to the interior. After a considerable amount of experimental work, the car won her third race and justified the months of hard work her 'crew' had devoted to her." This description, told to the historian Bill Boddy, shows Hann's delight at winning in this old 1911 car; the year of the race was, incredibly, 1922.

He had already achieved a third place in the autumn meeting at Brooklands with a lap speed of 72.81 m.p.h., racing against the new Bentley 3-litre and other such makes, and actually increased his lap speed later to 76.39 m.p.h. Hoieh-Wayaryeh-Gointoo performed well as a saloon but Tommy Hann complained, "The enclosed body, apart from the almost deafening noise of the reverberations of the exhaust amplified through the open tail, disclosed certain disadvantages...." He decided to replace the rear body with an open, single-seater style.

Tommy Hann in his 25 h.p. racer, remodelled for the 1923 season and re-named Softly-Catch-Monkey.

The re-modelled car, which first showed itself in 1923, was christened Softly-Catch-Monkey, after the naval expression for 'proceed with caution', which may have seemed sensible as the car was by then twelve years old. It retained the standard mechanics of its saloon predecessor, but was now clothed in more orthodox racing coachwork. Bill Boddy remembers the car: "The body was very narrow, with a wedge-shaped tail, the side panels of which were tapered horizontally to present the smallest possible side area. The very distinctive radiator cowling was retained, likewise the Smith's Elland shock-absorbers connected to the front axle by their long piston arms. The epicyclic gearbox, cantilever suspension and Lanchester radius arms recalled the ancestry of this odd car. The driver had a bucket seat behind the large steering wheel, and the fuel tank was in the tail ...... The body was painted in 16ins. wide vertical bands, alternately orange and black, to provide easy recognition when the Lanchester was running amongst a bunch of other cars, or on the far side of the Track, which dazzle-painting Hann adopted for all his cars. Wire wheels carried 32ins. by 4ins. tyres and Hann used touring Lodge plugs and Shell Super Light oil ....."

Softly-Catch-Monkey's first run out was at the Whitsun Brooklands meeting. Tommy Hann entrusted his car to Tony Bellingham-Smith, who secured a best lap time of 73.67 m.p.h. This was but a preliminary for Bellingham-Smith as, out in the car for the midsummer meeting, he won at a lap speed of 78.31 m.p.h. beating a Frazer Nash, a Bugatti and an Aston Martin, on handicap. It was about this time that Hann fitted fairings over the front shock absorbers and rear axle casings to achieve better streamlining. Soon after this the 25 h.p. was entered for another race, but this time it came third to a Bugatti and Barnarto's Wolseley Moth, although Softly-Catch-Monkey lapped at its best-ever time of 80.59 m.p.h. The autumn meeting saw Tommy Hann trying to improve on Tony Bellingham-Smith's remarkable time, but his lap speed was 0.26 m.p.h. slower.

Softly-Catch-Monkey in the Brooklands pits, still a winner at thirteen years old.

Malcolm Campbell's Star pushed the Lanchester into second place at the Easter 1924 meeting. Bill Boddy recalled that, during that year, Tommy Hann lightened the car considerably until "..... it actually became quite skittish, indulging in a considerable skid on the Byfleet banking. Many of the fairings were removed, the steering column was further raked to bring the wheel inside the cockpit instead of in the driver's line of vision, and the tail was shortened, while the car was lowered slightly by mounting the front anchorage of the rear cantilever springs lower on the side-members. Hartford shock-absorbers, mounted Frazer-Nash fashion, replaced the Ellands, and the magneto now protruded from the off-side of the bonnet."

Although never beating Tony Bellingham-Smith's record time, these improvements kept the thirteen-year-old car competitive and resulted in a thrilling climax at the August meeting between Softly-Catch-Monkey, driven by Tommy Hann, and an Austin. "As they swept up the Finishing Straight there was seemingly nothing in it, but the Lanchester, its tappets now set to 6 and 8-thou. clearance, lapping at 79.94 m.p.h., got over the line a length ahead ......" Tommy Hann decided against further improvements for the Lanchester and retired

The 1921 40 h.p. test car, called Winni Praps Praps due to its exhaust note.

it; its fate is unknown. However, this was not the end of Lanchesters at Brooklands, as a different car had also been making the headlines.

This 40 h.p. car was initially intended not for racing but as a test vehicle, from which George could make observations to complement his research into the problems of wheel-wobble and wheel-bounce at high speed, the study of which was being jointly undertaken with Fred, on behalf of both the Lanchester and Daimler companies.

George and Arthur Bird prepared the car in Armourer Mills simply by taking a completed chassis from the production area and installing a light test body on it. Together with a standard 40 h.p. car, they drove it to the Weybridge race circuit, where they "..... had an interesting day, driving the car alternately, during which time they discovered much about the need for shock-absorbers on the Track." Bill Boddy wrote that, on returning to the paddock, Arthur Bird was informed that something had earlier fallen from the car. He discovered later that seven ballast weights had come adrift. He had also lost a rear tyre, and only recovered that after offering a reward.

A new and slightly more elegant body was fitted in time for the autumn 1921 race meeting. Caught up in the excitement of Brooklands, they allowed the well-known racing driver, C.A.Bird, to race the car which, due to the sound of the exhaust, had been given the onomatopoeic name of Winni Praps Praps. Against such cars as Count Zborowski's 23,092 c.c. Chitty Bang Bang 1, K.Lee Guinness' 18,322 c.c. twelve-cylinder Sunbeam and a 15,095 c.c. Lorraine Dietrich, the car did not fare well, which led the Company to complain about its handicap rating. The official response was that, "A new car, seen to be fast in practice, was hard to assess, but the Lanchester Company would no doubt benefit from having lapped at 96.71 m.p.h. with a standard car."

In August 1922, Winni Praps Praps was again at Brooklands, still being encouraged to give wheel-wobble so that its effects could be monitored. In order to gain greater speed for an attempt at the Twelve Hour World Record, however, the engine was modified slightly and was also given an S.U. carburettor incorporating leather-bellows. A new exhaust system, together with these modifications, resulted in a different exhaust note, causing Winni Praps Praps to lose its name. The 35 cwt. car was fitted with a rear axle worm-drive of 3:1 ratio and a narrow, purposeful single-seater body.

The famous motoring pioneer, correspondent and racing driver, S.F.Edge, had been asked to race the car in 1922, after previously driving a standard 40 h.p. car in order to acquaint himself with Lanchester engineering. He obviously approved of the road-going car and wrote, "I have had occasion to drive some hundreds of miles recently in a 40 horse, 6-cylinder Lanchester open touring car. I have not been on one for about a year, and the feature that impresses me very forcibly is the suspension. There seems little doubt that the Lanchester 40 is the best suspended car in the world today." This experience was to stand him in good stead for the notoriously bumpy surface of Brooklands Motor Course and the ongoing investigations into the causes of wheel-wobble.

THE BROOKLANDS RACERS

S.F.Edge, well-attired for his attempt at the Twelve-Hour Brooklands record in Winni Praps Praps.

On the race track in the car, which had been converted to a single-seater, S.F.Edge reported that the 'tramp' he experienced and had been asked to maintain, ".....was so effective at 80 m.p.h. that the front tyres would not last an hour, whereas the rear tyres under the same circumstances lasted four hours".

29th August 1922. George Lanchester at the wheel of the 40 h.p. in the pits at Brooklands, while Arthur Bird tries to straighten the steering and S.F.Edge (in the racing helmet) watches. To his left are Doug Ambrose, Bob Clegg and Godfrey Barnsley, all from the Lanchester Works.

After a pit stop to assess the reasons and conditions for wheel-wobble, S.F.Edge began his attempt on the Twelve Hour Record. He obviously enjoyed the drive: "Now this car on Brooklands literally gave one no shock, or suspicion of shock, even when lapping at over 100 m.p.h. I took it on the track merely for the purpose of driving at 85 or 86 miles an hour, with the idea of doing a thousand-mile single-handed drive inside twelve hours; but in my preliminary runs it was so easy to drive, and so comfortable, that I gradually worked the speed up to lapping at over a hundred miles an hour without the least discomfort. This is a very remarkable tribute to the Lanchester design of suspension. The biggest bump on Brooklands was swallowed up by gentle motion and no bump of any sort was felt. I have driven all kinds of cars on Brooklands, and racers of various descriptions, but nothing ever went round the track with such comfort and smoothness as this Lanchester. It seems difficult to believe that a firm who specialise in luxurious private cars should be able to produce practically their normal car capable of such a wonderful performance."

Archie Millership and 40 h.p. at the Shelsley Walsh Hill Climb on 8th September 1923.

During the high-speed run, S.F.Edge broke many World Records for 'Class G' cars from 4,998 c.c. to 7,784 c.c., even though the 40 h.p. was some 1,667 c.c. short of the maximum allowed engine capacity. The historian, Anthony Bird, listed the achievements:-

Record achieved	Detail
2 hour	191 miles & 1,610 yards at 75.48 m.p.h.
3 hour	273 miles & 1,213 yards at 75.72 m.p.h.
4 hour	361 miles & 1,422 yards at 80.28 m.p.h.
5 hour	442 miles & 1,415 yards at 80.40 m.p.h.
200 miles	at 77.77 m.p.h.
300 miles	at 81.33 m.p.h.
400 miles	at 80.37 m.p.h.

and all corresponding kilometre records.

S.F.Edge never achieved the twelve-hour record, however, as a recurrence of wheel-wobble, which was undetected by him but observed by George, caused the latter to direct the car into the pits and to abandon the record attempt.

In view of the car's earlier and deliberate four-hour pounding under test, there was a fear of fatigue fracture of its components. A little later, after full consideration, S.F.Edge

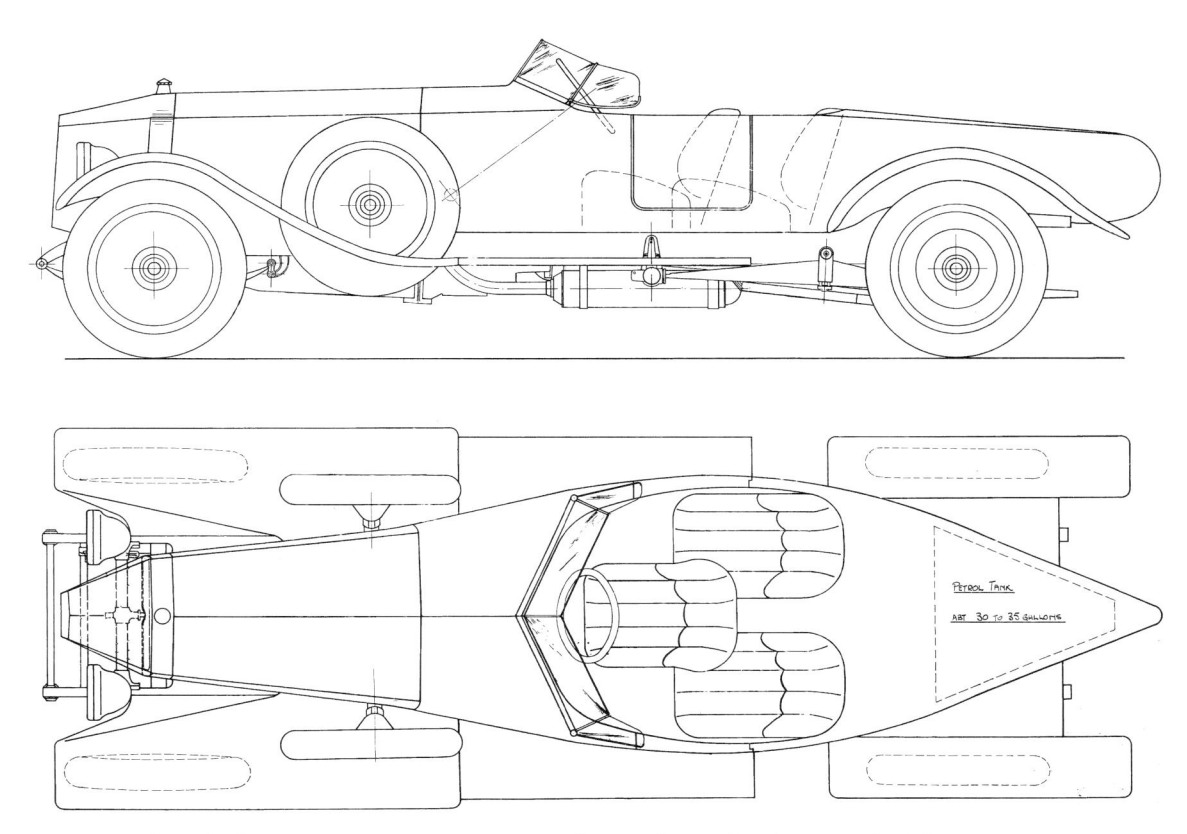

Details for conversion to a sports car for road use, for the retired 40 h.p. racer.

took the 40 h.p. back onto the circuit and drove it at its critical speed in anticipation of a collapse. This did happen, when the near-side steering arm broke, luckily near the pits. The car swerved violently and S.F.Edge controlled the car to decelerate and come safely to a stop. As the front wheels were splayed in different directions, he had to reverse into the pits.

With the primary objective of wheel-wobble investigations completed, the car did not return to Brooklands in 1923, although it did race at the September hill climb meeting at Shelsley Walsh, with Archie Millership at the wheel. Nothing more was heard of the racer for eighteen months, until George signed the order for its conversion to a road-going sports car. Its fate from then onwards is unrecorded.

The tyre manufacturer, Lionel Rapson, had been aware of the Lanchester's record-breaking activities at Brooklands, as well as the merits of their road-going cars. In addition to owning two earlier Lanchesters, he also owned a 40 h.p. five-seater landaulette and a 40 h.p. two-seater tourer. In early 1924, he placed an order with the Company to supply a 40 h.p. racer, to undertake the long-distance testing of his own Rapson double-tread tyres. He chose a Lanchester in preference to the Rolls-Royce cars he had previously used, in view of the way the Lanchester cantilever suspension system could easily cope with Brooklands' infamous uneven surface.

George recalled that a standard 40 h.p. was taken from a production batch at Armourer Mills and duly modified. The engine compression ratio was raised from the standard 4.8:1 to about 6.8:1 and the valve timing was altered to suit the car's high-speed use. A larger induction pipe with twin Zenith carburettors, and a bigger exhaust pipe were fitted. Although the steering-column mechanism was unmodified, the column itself was shortened and moved to a central position mounted above the gearbox. The unnecessary cooling fan was removed and a tapering cowling was added, protruding some 18ins. from the radiator to allow only one-third of the usual radiator surface area. George described this nose-piece and the remainder of the new, single-seater body as ".....of a design inspired by aeroplane fusilages of the period", although the science of aerodynamics was not applied to the large tool box protruding from the off-side rear chassis. A hand-pump was fitted for pressurising the new petrol tank mounted in the tail. Hartford shock absorbers were fitted to the rear axle, whose worm gear ratio was altered to 3.3:1. All this work was supervised by George and carried out by Arthur Bird and his team at Armourer Mills. When complete with road-going equipment, the car was handed over to Lionel Rapson and it was used regularly for long-distance, high-speed tyre-testing. Lionel Rapson was so impressed that, when it was not otherwise required, he entered the 40 h.p. for races at Brooklands.

The accomplished J.P.Parry Thomas was usually nominated as the driver and he obviously suited the car, as in his first race in 1924 the large Rapson 40 h.p. came in second, behind a Vauxhall OE 30/98. He made amends at the Easter meeting when, after lapping at nearly 105 m.p.h., he beat Lionel Martin's 16-valve Aston Martin and Count Zborowski's Indianapolis 2-litre Bugatti. A second placing gained later capped a fine day. The Whitsun 1924 meeting resulted in a respectable third place for Parry Thomas and the Lanchester, followed by a win in the Surbiton Lightning Handicap. The last race entered for the year, in August, produced another third placing. Also in August, Parry Thomas made an attempt on some World Records for Class G rated cars, and successfully captured the following:

250 km record at 159.77 k.p.h. average speed
300 km record at 160.53 k.p.h. " "
400 km record at 157.00 k.p.h. " "
500 km record at 157.76 k.p.h. " "
300 miles record at 97.95 m.p.h. " "
400 miles record at 98.32 m.p.h. " "

as well as the distance award for:
2 hours covering 208 miles & 150 yards at 104.08 m.p.h. average speed
3 hours covering 293 miles & 1272 yards at 97.91 m.p.h. " "
4 hours covering 393 miles & 651 yards at 98.34 m.p.h. " "

The 1924 40 h.p. Lanchester ordered by Lionel Rapson for high-speed tyre-testing work at Brooklands.

The Rapson 40 h.p., still on its original Dunlop tyres. The streamlining of the racing body did not extend to the tool-box!

These admirable results proved that the 40 h.p. was capable of making an attempt on the World's long distance records and plans were made accordingly.

The Rapson 40 h.p. was returned to the works for a full check and service, as well as for modifications to be made. The most important of these was the changing of the standard inlet manifold and Smiths carburettor for a larger, horizontal copper tube running close to the inlet ports, with a Zenith carburettor fitted to each end. Stronger valve springs were used and a new four-branch exhaust system with the new regulation Brooklands silencer was installed. Coupled with slight changes to the front shock absorbers and rear spring mountings, these alterations enabled the Rapson 40 h.p. to be ready for an attempt on the World Record.

George Lanchester, Arthur Bird and four Lanchester engineers subsequently drove the Rapson car and a 40 h.p. tender vehicle down to Weybridge on 31st August 1924, to make the necessary preparations. Staying at a nearby hotel and living on sandwiches during the four, fifteen-hour working days at Brooklands, they put the car through many test runs whilst final adjustments were made. Pit stop service rehearsals were carried out, involving every aspect of maintenance, such as magneto changes and rear axle oil top-ups.

Lionel Rapson announced that he was to drive, alternating with Parry Thomas and George Duller, another famous racer. When all was ready, the entire Brooklands Motor Course was reserved for the attempt, on 2nd September 1924. Parry Thomas drove first,

Arthur Bird leaves Birmingham for Brooklands with the re-vamped Rapson 40 h.p. racer, in August 1924.

and the watching George Lanchester recalled, "During the first hour and forty-five minutes the car averaged no more than 80 to 90 m.p.h., owing to tyre troubles. Rapson had been over-confident about the merits of a new design." Lionel Rapson obviously realised that the sixteen stops for tyre changes were spoiling their chances for the twelve-hour record attempt. When reliable tyres were used, the speed increased and the first record was gained after three hours, at 97.91 m.p.h. During the pit-stops, the Rapson team looked after tyre-changing and George Lanchester's mechanics worked with Arthur Bird to keep the car full of lubricants and fuel, and carry out any necessary maintenance. Bill Boddy wrote, "After Duller had taken over, a mysterious misfiring slowed the car. It was brought in and the magneto, twelve plugs, and all the wheels were changed and the petrol and oil topped up,

While the Rapson 40 h.p. was reaching 120 m.p.h. at Brooklands, the pit crew wait patiently. To the left of Lionel Rapson's waist-coated nephew, Mr.Gardner, are Billy Poole, Doug Ambrose and Eric Sisman, all from the Lanchester Works.

THE BROOKLANDS RACERS

A push-start after a pit-stop, during the World Long-Distance records, for the the Rapson 40 h.p. racer. Lanchester's Joe Woods (by the silencer) gives maximum effort.

in the short space of three-and-a-half minutes. Afterwards, Duller admitted that he had, in error, fully retarded the ignition lever instead of advancing it." One of the mechanics there, Eric Sisman, recalls a similar instance: "We were worried in the pits by backfiring whenever Duller was driving and whenever he passed the pits. He was a jockey and quite small and a bump between us and the Vicker's sheds was jerking his foot off the accelerator. A cushion at his back soon cured this."

With this problem solved the Lanchester kept a constant pace for a further ten hours. It broke thirty World records along the way and, as a finale, the last one hundred miles were covered at an average speed of 104 m.p.h.

The highest speed attained by the Rapson 40 h.p. racer was 120 m.p.h. and the Brooklands Gazette reported that the car covered 1,148 miles, 843 yards in twelve hours, at an average speed of 95.66 m.p.h. Parry Thomas drove for six hours, with Lionel Rapson and George Duller driving for three each.

They also exceeded the Class G 13-hour, 14-hour and 15-hour distance records, as well as all records from 500 metres to 1,000 metres and from 700 kilometres to 1,800 kilometres.

J.P.Parry Thomas in 1925 posing with the Rapson 40 h.p., complete with experimental wheel discs.

There was a delightful sequel to this challenge for, in a light-hearted moment, Lionel Rapson had offered his co-drivers £10 for every record the Lanchester broke, only to find that Parry Thomas's bill amounted to about £200 and George Duller's to about half that sum.

Coupled with August's gains, the Rapson Lanchester collected no fewer than forty-seven World Class Records; an amazing tally for a car of relatively slight modification. However, whereas other companies would have hailed this success and advertised the 40 h.p. car's merits, there was, typically, no Lanchester fanfare of publicity to encourage future sales. As Eric Sisman recalls, "It was all very low key, with no public rejoicing."

The car was rested all winter and then brought out again by Parry Thomas for the 1925 Easter meeting, at which he exceeded 100 m.p.h. but was not placed.

The drive for the Whitsun meeting went to D.R.W.Gedge, who secured a creditable third place. Parry Thomas returned to the Lanchester with success at the summer meeting, with an Alfa Romeo and Barnato's Bentley as runners-up. Yet another first place was secured at the August meeting followed by a later second place, again ahead of Barnato's Bentley.

1926 produced a welcome win at the Easter gathering. The summer meeting was, however, the last occasion that the Rapson Lanchester 40 h.p. would take the chequered flag, when it averaged 100.78 m.p.h. with Parry Thomas at the wheel. After a third place in its last race at the autumn 1926 meeting, the car was retired and stored at Lionel Rapson's home as a reminder of its successful years. (The Lanchester was unsuccessfully offered for sale during the Second World War and was finally bought by a scrap-dealer for £5 and broken up.) Although George was quoted as saying that "racing was not the golden gate to fame", it is surely a pity that the Lanchester Company's publicity and advertising managers did not capitalize on the great successes of this outer-circuit Brooklands racer, in order to encourage sales of the road-going 40 h.p. cars.

The Rapson 40 h.p., with Lionel Rapson by the tool-box. The Brooklands club-house and test hill are shown in the background.

Lanchester limousine coachwork on 1928 chassis no. 3558. (See appendix.)

Sports-tourer coachwork by Corsica, removed from a vintage Bentley, fitted to the author's 21 h.p. chassis no. 3574. (See appendix.)

This 1930 Maythorn-bodied landaulette, on 21 h.p. chassis no. 3642, was owned by the family of the original owner until 1965. It includes the taller radiator and metal dashboard. (See appendix.)

The Lanchester wooden car - Mark Seven.

Side view of the Mark Eight, four-cylinder, horizontally-opposed engine, with its direct-coupled electric motor.

Dr. Lanchester's horizontally-opposed 10 h.p. engine.

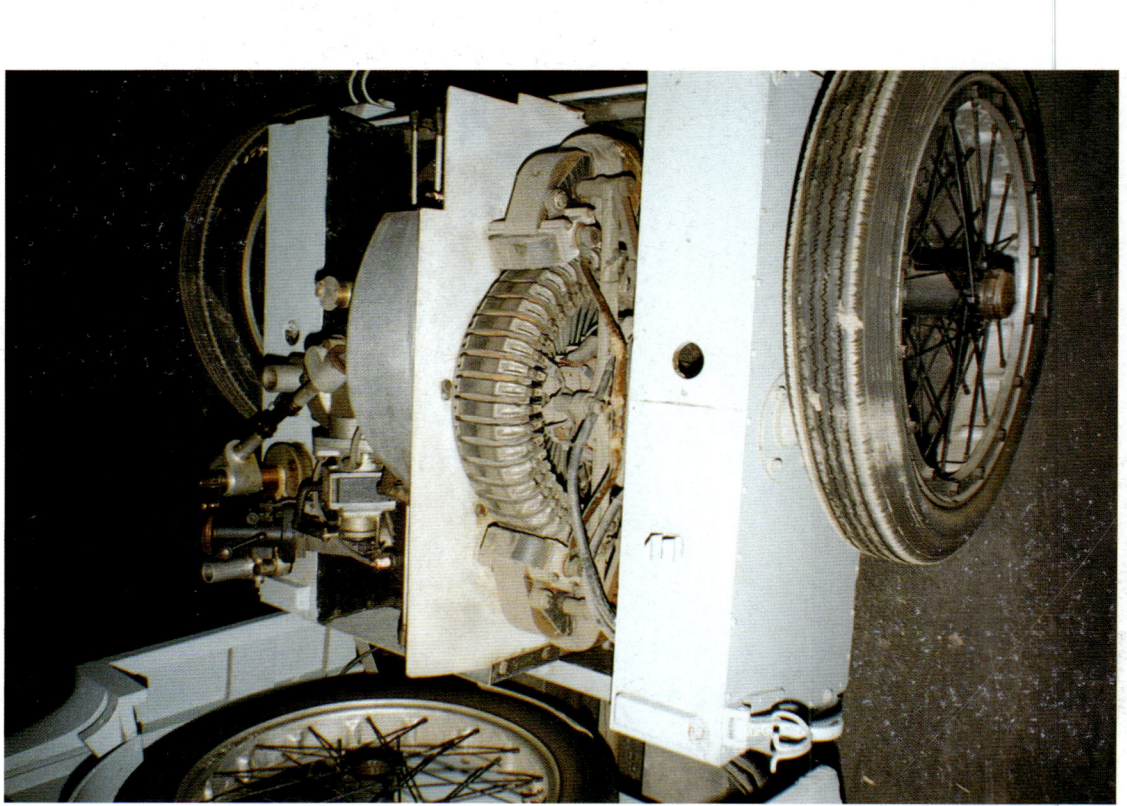

The electric motor of the Mark Seven wooden car.

The only surviving Lanchester armoured car, this 1931 40 h.p. Mark Two model is shown in the colours of the 12th Royal Lancers.

The 1929 Motor Exhibition 30 h.p., chassis no. 8046. Original Windover sedanca-de-ville coachwork replaced by two-door sports-tourer body. (See appendix.)

Sports-tourer coachwork on this 1930 30 h.p. chassis no. 8079. (See appendix.)

Touring coachwork on the 30 h.p. chassis no. 8094. (See appendix.)

Fabric-covered Lanchester Southport 30 h.p. saloon on 1929 chassis no. 8019. (See appendix.)

Chapter Twenty-One

THE SIX-CYLINDER 21 H.P.

As Fred had earlier predicted, in 1918, the Company's sales policy of exclusively marketing the expensive 40 h.p. was not a wise one, and sales of all luxury cars were unexpectedly low. With this in mind, George and Frank, early in 1921, persuaded the Directors to allow them to embark on a development programme for a smaller car and, after due consideration of estimated costings, the go-ahead was given.

George immediately began roughing-out designs and sought approval for his proposals. With so much investment at stake for the Board, Hamilton Barnsley decided to invite Fred back to the fold for a short time, as Consultant Engineer, to advise on various issues of design as they appeared. It is highly unlikely that George, as Chief Engineer, would have felt slighted at this invitation to his brother, as he had the highest respect for Fred in every way; indeed, they, and Frank, always signed letters to each other as "your affectionate brother".

Fred's reply to the Managing Director revealed his enduring deep concern for the Company. After clarifying the design principles expected of the new car, he wrote, "You will be well advised to call the engine a 20, if it is to be anywhere about 20 h.p. The old Lanchester was in its day quite a popular article, there is a certain goodwill attached to the name and if you were to call it less, it immediately suggests something not quite in our class."

The engine George designed and built was actually of a bigger bore than Fred had anticipated although the stroke remained as planned, being 2.937ins. and 4.5ins. respectively. The 2,932 c.c. engine had a treasury rating of 20.6 h.p. and the car was christened the 21 h.p. Its cylinder head was integral with the cylinder block and incorporated valves in

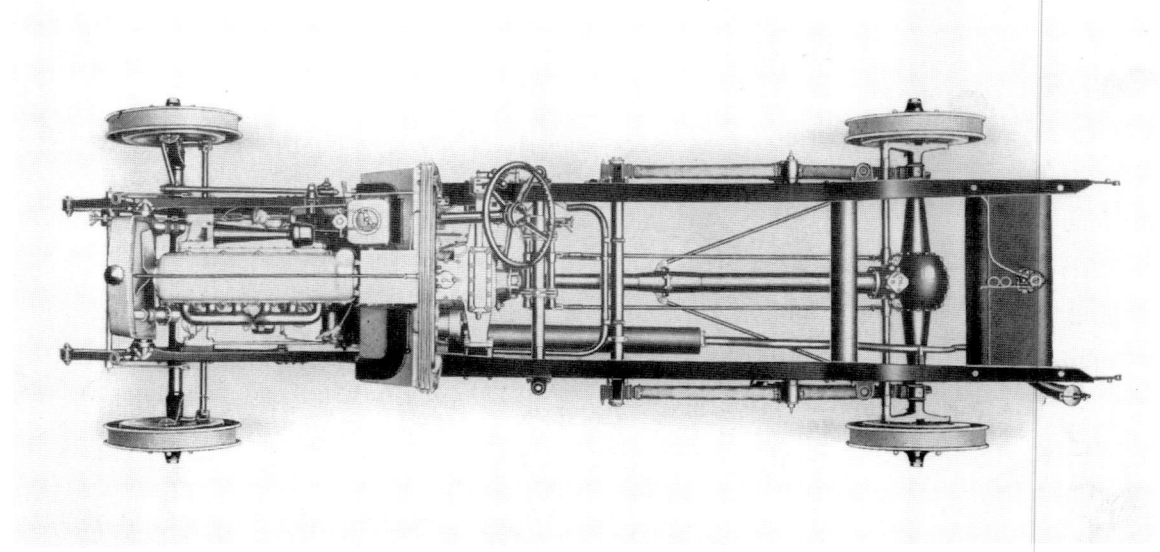

Plan view of the first 21 h.p. chassis.

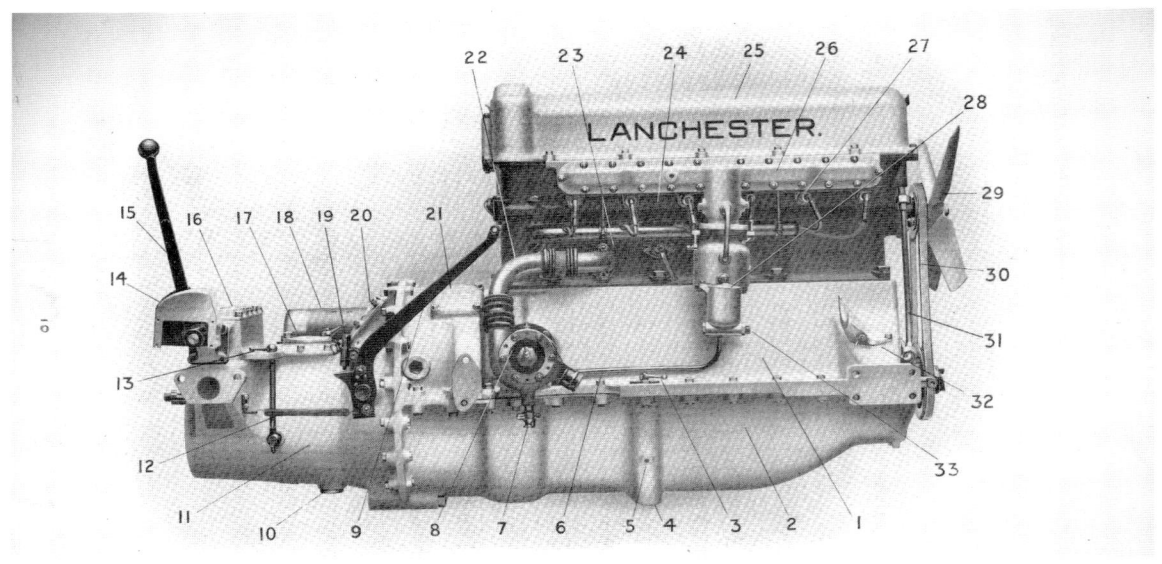

1	Crankcase.	17	Gear Box Oil Filler.
2	Oil Base.	18	Engine Starter.
3	Crankcase Oil Level Cock Handle.	19	Clutch Stop.
4	Crankcase Oil Drain.	20	Inspection Cover.
5	Crankcase Oil Level Cock.	21	Valve Timing Inspection Lid.
6	Carburettor Jacket Water Circulation Pipe.	22	Water Circulation Pipe.
7	Water Drain Tap.	23	Cylinders.
8	Water Pump.	24	Detachable Cylinder Head.
9	Clutch Pedal.	25	Valve Cover.
10	Gear Box Oil Drain Plug.	26	Induction Manifold.
11	Gear Box.	27	Sparking Plugs.
12	Clutch Pedal Spring.	28	Carburettor.
13	Gear Box Oil Level Cock.	29	Cooling Fan.
14	Gear Change Gate.	30	Fan Driving Belt.
15	Gear Change Lever.	31	Oil Return Pipe.
16	Gear Change Selector Box.	32	Electric Horn Bracket.
		33	Petrol Union.

Inlet side of the 21 h.p. engine.

detachable cages, as in the 40 h.p. engine. The power obtained from this engine, however, was disappointing and the design was quickly altered to include a detachable cylinder head, which allowed for valve and porting improvements.

Another major worry was the crankshaft George had designed, which was 1.5ins. in diameter. In an early test engine one broke, and caused much rapid re-design work. George instigated the machining of new crankshafts with sizes as great as possible without necessitating the re-boring of the crank case, and Fred fully endorsed his brother's decision. The revised diameter of the bearing surfaces gave an extra 0.186ins.

The gear box was a bone of contention between George and his Directors, as George wanted to maintain tradition and keep the epicyclic gearbox. Furthermore, he hoped to include in that gearbox four forward speeds, instead of the normal three, to be more compatible with the smaller-output engine. In pre-War years Fred had won a reprieve for the epicyclic gearbox, but now the Directors were adamant that the car should have a four-speed, sliding-gear, 'crash' box which, apart from being cheaper and smaller, was the same system as used by every other car manufacturer.

Such a gearbox had not been produced by the Company before, and George put a lot of thought into its design. He decided to make use of the experience and machining capabilities of his good friend, Harry Orcutt, owner of the Gear Grinding Company, who was also an associate of Fred's. Indeed, Fred had earlier introduced him at an I.A.E. meeting where the latter read a paper on gear design. Discussions followed between the three of them, regarding the standardisation of certain components for both the Lanchester and the Daimler gearboxes. Harry Orcutt standardised the gears' pitch pressure angle, number of teeth, the splined shaft sizes and bearing sizes. Although his general design was accepted by the British Engineering Standards Association, trouble loomed on the horizon. On both the Lanchester and the Daimler cars which used his gearbox layout, the gears tended to come out of mesh. On investigation, Fred attributed this solely to the design recommended by Harry Orcutt. George wrote to Fred, "I have taken the matter up emphatically with the Gear Grinding Company, who seem to be very tenacious of their opinion to the contrary." (This discourse did not sour their friendly relationship and indeed, some five years later, Orcutt and George took out a joint patent for their 'Variable Speed Gears' invention.) The two brothers resolved the problem by experimenting themselves, and informed Harry Orcutt that they had altered the "sliding assemblage" so that the gears tended to wind into rather than out of mesh.

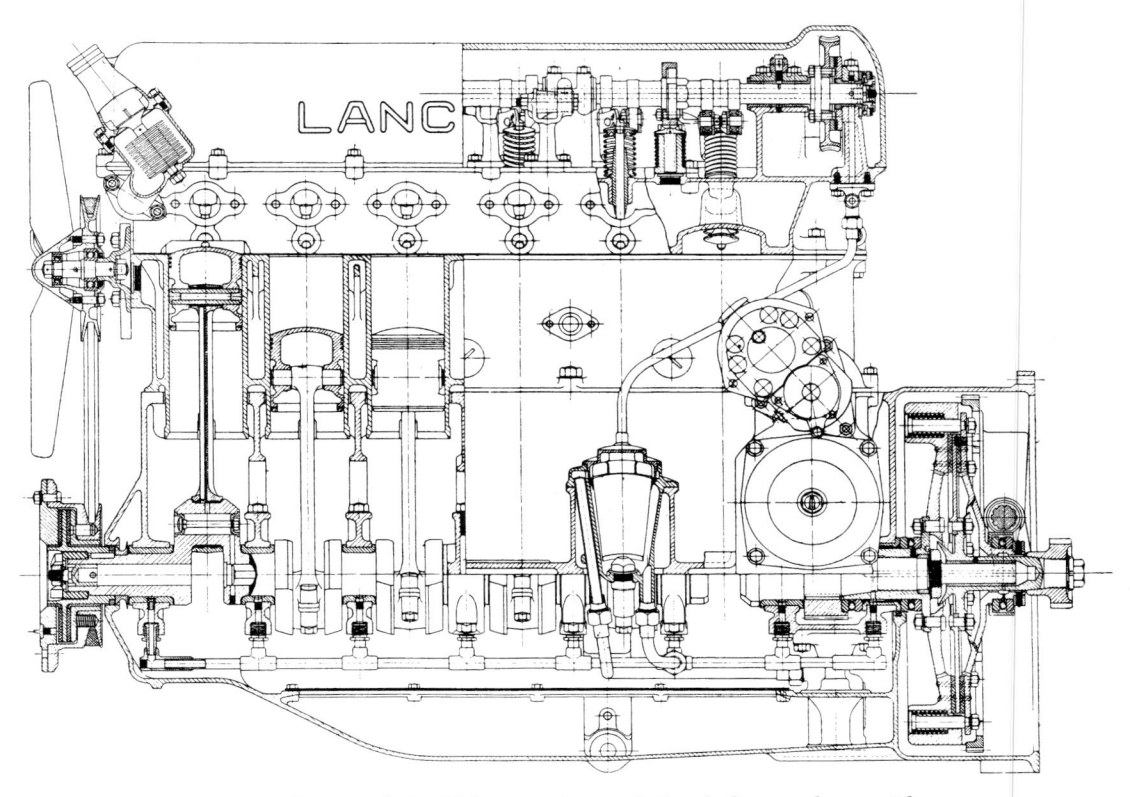

Sectional view of the 21 h.p. engine and clutch from exhaust side.

Hamilton Barnsley asked Fred's opinion on items ranging from the width of seat cushions to rear axle ratios, and George was also keeping his brother informed of the 21 h.p.'s progress. George arranged for Arthur Bird, foreman of the Road Test department, to put the car through initial trials and on 10th March 1923 he sent a telegram to Fred announcing: "TWO SHORT RUNS 21. BIRD TESTING OVER WEEKEND. REPORT MONDAY." This satisfying occasion prompted George to complain to the Managing Director about a proposed change in bodywork requirements. He had originally been asked to design a car intended to be "a comparatively light, four seat machine" and he commented that, "We seem to be drifting away from this policy ... it now appears that there should be ample accommodation for six passengers." He reminded Hamilton Barnsley that he had predicted this situation would arise and so, in order to provide for an ample, comfortable, six-seater town carriage, he had wished originally to design a 25 h.p. car instead. If this bigger bodywork was to be used, he continued ".....either the chassis should be geared lower ... or better still, the size of the engine should be increased to carry the larger body with the present gear". In compromise, George concluded that, "If we increase the bore of the engine to 3.125ins. we shall have approximately 12% more power, which should be adequate to carry the heavy load." However, although the Board's permission for this measure was not granted, George had the foresight to design an engine which could be adapted with very little effort should there later be a change of policy.

Hamilton Barnsley wrote to Fred and, after confirming that his initial desire had been for the new car to be "a very fast, very beautiful, very silent Model-De-Luxe car; the best in the world, but a small four-seat at the most", he excused his change of mind by saying that ".... it was your brother, Frank, who said that there would be no sales unless it was possible to put on a six-seat body". As Hamilton Barnsley went on to make special mention of the fact that the Daimler and Rolls-Royce 20 h.p. cars were capable of carrying seven people, it indicated that he had come to agree with Frank. He asked Fred to come into the Works to participate in the ongoing discussions.

1924 21 h.p. landaulette with coachwork by Lanchester.

The Company superficially inspected a Daimler for comparison with the 21 h.p. Lanchester, but a Rolls-Royce was afforded a more meticulous examination. Eric Sisman, who worked in the experimental department at the time, recalls that the whole team were very impressed by the quality of its components, but that the general consensus of opinion was that the Rolls-Royce 20 h.p. was a "gutless wonder".

Nine prototype 21 h.p. Lanchesters were duly built in 1923 and, after extensive testing and examination by the Company, were made available to the motoring magazines for trials and test runs. The correspondents discovered that the engine's six cylinders differed from the cylinders of the 40 h.p. in that they were cast monobloc. The dome-headed, aluminium pistons matched the profile in the detachable-head's combustion chambers, and incorporated three compression rings and one oil-scraper ring. Their fully-floating, tubular, gudgeon pins topped the machined and accurately-balanced high-tensile steel connecting rods. White-metal bearing surfaces were used against the hollow crankshaft, which itself was supported in the aluminium crankcase by eight main bearings. The most rearward bearing carried the loading of the flywheel as well as the crankshaft's worm gear, engaging with two drive shafts running at 1.5 times the engine speed. One of these was horizontally located and drove the water pump on the engine's off-side and the Lucas dynamo on the near, with a duplex chain driving a Watford magneto above. The other shaft was vertical

THE SIX-CYLINDER 21 H.P.

Vanden Plas town coupe coachwork on 1924 21 h.p. chassis.

and ran in plain bearings in the cylinder block. It incorporated a worm gear to mesh with the single overhead camshaft driving wheel.

On either side of the camshaft were inclined valves, which had one exhaust and one inlet per cylinder. They were of identical size, with bell-shaped, large-diameter heads. These were operated by long valve levers, which were adjustable for tappet clearance by means of an eccentric pin device. The levers ran underneath the camshaft and carried rollers to follow the contour of the cam lobes; these operating rollers were housed in an oil bath.

General oil lubrication was pressure-fed by a pump, which was driven from the base of

The Prince of Wales visiting South Africa in a 21 h.p. with Lanchester coachwork.

the vertical drive shaft inside the sump and was protected by a gauze strainer. It fed directly into an easily-cleaned conical gauze filter, whose housing incorporated a pressure-relief valve, a branch to feed the overhead camshaft bearings and another to feed each individual crankshaft bearing. Oil could pass through the crankshaft to feed the big ends before being forced up small pipes attached to the connecting rods to the gudgeon pins. Oil was also forced along the crankshaft extension to feed the clutch spigot thrust race. This oil was contained in the housing, resulting in the two Ferodo friction plates being partially submerged, before the oil was returned to the main engine sump via an internal channel.

The cooling system utilised a large honeycomb radiator, with an integral casing of brass or nickel-silver and, of course, the traditional, water-level sight gauge. Driven by a centrifugal pump, the water was cooled by a four-bladed aluminium fan operated by a belt from the front crankshaft pulley. An automatically-controlled thermostat maintained the correct temperature.

The petrol supply was brought from the twelve-gallon rear tank by a bulkhead-mounted Autovac, secured in a Lanchester aluminium housing, before gravity feeding to one automatic carburettor. A Smiths four-jet carburettor was used, in preference to either the twin Zenith or the twin Claudel carburettors tested previously, and gave a fuel consumption of 18 m.p.g. The carefully-designed aluminium, water-heated, inlet manifold incorporated angled sections opposite each cylinder port, to obtain the best vapour distribution. The Autocar summed up the visual aspect with the words: "The engine is such a picture as really to justify the use of glass for the bonnet panels. It is a sin to hide such a power-plant."

A two-door, four-seater cabriolet on 21 h.p. chassis.

George's gearbox layout, which was much shorter than other makes, terminated in an automatically-lubricated spherical joint, allowing movement to the long torque tube which the Lanchester literature poignantly described as "approximately equivalent of the old Lanchester Parallel Link Motion". The enclosed propeller shaft drove the underslung Lanchester worm and wheel gears. On Fred's advice, the standard ratio used was initially 8:35 (4.375:1) but an alternative was to be offered at 7:34 ratio (4.857:1). The axles were, unlike the 40 h.p., of the semi-floating type, having one large ball-type bearing mounted on the axle casing, so that it and the axle shaft fully supported the weight of the car. The bearing was mounted in line with the centre of the wheel track. The half shafts were secured

THE SIX-CYLINDER 21 H.P.

The boat-deck sports-tourer designed by George Lanchester and built by Mulliner for their 1925 Motor Exhibition stand. The car incorporated teak decking and the bonnet top was scumbled to match. Earlier examples had scumbled wings also.

"Of the many so-called sports cars, very few really meet the requirements of the sporting motorist. Here is one specially designed to fill the want. The body is mounted on the famous Lanchester 21 h.p. six-cylinder chassis, and follows in design the beautiful lines of a racing boat."

to the wheel hubs by means of splines and a cone, a similar system to that of the 40 h.p. model. The steel disc wheels included, for a while, six cut-outs near the 820mm. by 120mm. beaded-edge tyres, so that a wheel could be changed without handling the dirty tyre. These wheels were dished so that the tyre centres were in alignment with the swivel pins.

George also designed the front wheel brake mechanism to be directly in line with the swivel pins, a universally rotatable cam allowing full steering movement of the wheels. The brake-controlling rods were located on front axle fixtures to avoid any disparity of movement.

The 14.5ins. diameter brake drums were of aluminium with a cast iron liner, against which a single pair of brake shoes acted, for both foot and hand brake. No brake adjustment was provided on the four road wheels but both the front and rear systems, leading out from a central compensator device, were each easily adjusted by their own turnscrew.

Although the Argyll car was the first to adopt four-wheel brakes in 1912, it was not until they were standardised in the post-War years by Austin in 1922 and Lanchester and Daimler in 1923, that the system began to make headway.

Long cantilever springs were produced by the Company and were used only on the rear axle. Semi-elliptic springs were employed on the front axle, in conjunction with Andre Hartford friction shock absorbers.

The front tubular axle of 4ft.6ins. track gave a turning circle of 44ft. The chassis, supplied by Rubery Owen to George's design, had a wheelbase of 10ft.9ins. and an overall length of 15ft. Its sections were wide and deep, and were connected by tubular cross members. As expected, great rigidity and torsional stiffness were provided.

The Press gave glowing reports. The Motor stated, "This car is one of the most arresting propositions in the automobile world today, incorporating as it does unparalleled experience in the art of building automobiles with super fine workmanship." The Autocar, who tried out a tourer version in the winter of 1923, complete with the latest fashion of all-weather side screens, praised the car, stating, "The application of the brakes is very sweet ... few gear changes are so easy to handle ... the gears are well-nigh perfect ... for the running of the engine there is nothing but praise ... it is not easy to tell whether the engine is running or not ... the back axle is noiseless ... there is no car on the road today that is better sprung." Their road test on the tourer, which weighed 35 cwt. unladen, showed that its maximum speed with two passengers was 63 m.p.h. The engine gave 60 b.h.p., some 7 b.h.p. more than the Rolls-Royce 20 h.p. competitor, and its range was from 160 to 3,400 r.p.m.; using only top gear it would accelerate from 10 to 30 m.p.h. in 13 seconds.

The Company proudly displayed their 21 h.p. at the November 1923 Motor Exhibition at Olympia, clothed in an eye-catching, ivory-painted four-seater tourer coachwork with red upholstery. The car was priced at £1,250 and a chassis price was quoted as £950. A similar price was asked for the chassis displayed in June at the British Empire Exhibition.

Regular production had started in the Spring of 1924 and a variety of coachwork was offered, including a heavy, six-seater, enclosed-drive limousine weighing nearly 41 cwt; George's earlier fears had been proved correct. To take account of this extra weight, the standard rear axle ratio was changed to 7:34 (4.857:1). To compensate for this change, the gearbox's third speed was altered to 1:6 from 1:715. The other speeds remained the same, with first gear at 4:1, second at 2.4:1, reverse at 3.4:1 and top gear being 1:1 on direct drive.

The car attracted many new customers who wanted to try this famous marque, but did not want to buy the 40 h.p. due to either its size or its cost. The price difference between the two cars was about £1,000, with similar styles of coachwork offered. At the 1924 Olympia Motor Show, a 21 h.p. six-seater 'V' front landaulette, painted royal blue with two relieving lines on its mouldings and black mudguards, was displayed for sale at a mere £1,747. The 21 h.p. chassis exhibited had risen in price to £1,000, still maintaining the price differential with the Rolls-Royce 20 h.p. chassis. This had risen to £1,100, and had been modified to include four-wheel brakes and also a four-speed gearbox with right-hand gear change, thus offering greater competition to Lanchester's successful design.

Hooper-bodied landaulette on 1926 21 h.p. chassis, no. 3355. (See appendix.)

The Lanchester Motor Company's hierarchy had changed since the debut of the 21 h.p., as Arthur Gibson, Chairman since Charles Pugh's death in 1921, and Joseph Taylor, the Vice-Chairman, both of whom had been with the Company since its formation, had died in 1923. Hamilton Barnsley, the Managing Director, had increased his power base by also becoming Chairman. In June 1924, after nearly thirty years' involvement with the Lanchester brothers, James Whitfield also died. Hamilton Barnsley agreed to George and Frank becoming Directors, on the condition that his son, Godfrey, and Frank Berresford should also join the Board; presumably to enable him to maintain the balance of power. Possibly as a result of George and Frank becoming Directors, together with the fact that they were supported by influential shareholders, such as Fred and the Works Manager, Jasper Cooper, new ideas were now considered.

Invigorated by his recent joint paper with Fred on epicyclic gears, George again considered such a gearbox for the 21 h.p. As the traditional Lanchester epicyclic gearbox with cast iron brake shoes was considered too bulky and heavy, he explored the possibilities of using plate clutches. A great deal of time was spent on the project but (as with the 1954 Lanchester Sprite plate-clutch gearbox) it was not made to work completely satisfactorily, and the scheme was later abandoned.

George was finally able to convince the Board that the 21 h.p. car should have a more powerful engine, and in 1925 it was increased to 3.1ins. bore, which gave a capacity of 3,340 c.c. and a treasury rating of 23.1 h.p. Due to his foresight at the time of the design in 1922, new engine castings were not required. It had previously been surmised by the motoring journals that the great distance between the cylinders in the original 21 h.p. engine had been to ensure that no leakage of any sort could track between the cylinders. In fact, George had based his initial design on his forecast of more power being required, and so the only alteration now necessary was to bore out the cylinders by an extra 0.163ins. diameter and fit new pistons. The top speed of the 'new 21 h.p.' was increased accordingly, although two correspondents on separate test runs, made differing claims. The Auto Motor Journal stated "close upon 75 m.p.h.", whereas the Autocar reported "on top gear 66 m.p.h. was comfortably reached".

21 h.p. army truck.

Whilst this up-rating was being achieved, the opportunity was taken to make other improvements. The engine was altered in a number of other ways, the main difference being that a Lanchester crankshaft vibration damper was added. Commenting on the engine's vibrationless running, the Motor Owner magazine wrote of the damper's introduction, "Applied to an already smooth running engine, the result is extremely good."

As well as a bigger diameter camshaft, an extra bearing and friction disc was fitted near the phosphor-bronze driving wheel, to resist end thrust. The camshaft's sprung-loaded, hexagonal compensating cam remained as before, eliminating any tendency for the camshaft to become noisy when rotating at low idling speeds. A horizontal drive was given to the camshaft for the provision of a coil ignition if desired, as an addition to the existing magneto system. The addition of the Delco coil system, which became a standard fitting in 1930, cost an extra £21, but did somewhat spoil the glorious aluminium valve cover which proudly displayed the name 'LANCHESTER'. Lubrication of the camshaft and the valve gear was via the hollow camshaft instead of the previous system of overhead oil pipes

Lanchester fixed-head coupe coachwork on 21 h.p. chassis. Built to the customer's design is a strange folding hood over the dickey seat.

squirting onto the cam rollers. The revised engine was quieter than previously and led one reporter to admit, "Indeed, one has to plead guilty to attempting to start the engine when the unit was already in motion."

The clutch's operating motion was slightly re-designed, but there was little noticeable difference to the driver. The gear ratios were altered to suit the extra power of the engine and became 3.955:1 for first gear, 2.401:1 for second, 3.4:1 for reverse and 1:1 for top gear on direct drive; the third gear of 1.606:1 allowed a maximum speed of 40 m.p.h. The rear axle ratio remained at 4.856:1.

21 h.p. engine showing the circular marks left by a rotary felt-bob, which was a feature of all the aluminium castings.

The wheels were altered to take advantage of new developments in that industry, and steel disc wheels with detachable rim-flanges were adopted, with 32ins. by 4.5ins., or sometimes 31ins. by 4.5ins., straight-sided tyres being fitted. An extra length of chassis was also produced, longer than previously by 4ins., making 11ft.1in. overall. An additional £50 was added to the shorter chassis price of £1,000.

The revisions of 1925, which included the change from Jaeger to Watford dashboard instruments, as well as various minor items, were displayed at the October 1925 Motor Exhibition on a long-wheelbase car. Its coachwork was by Hooper, and internal fittings of silver and ivory were included in the price of £1,825.

The operation of the brake shoes by the rotation of large square cams was retained. The system was altered, however, to enable individual adjustment of the brake shoes whereby any wearing of the brake linings was taken up by a rotating assembly, secured by two bolts in the key-hole adjustment slots. This updating of the brakes convinced Star Engineering Limited to adopt George's patented system for their cars from mid-1926, in order to acquire four-wheel brakes. (About 1,500 Star cars were produced over the years, each with a front axle showing a brass plate inscribed 'Manufactured under licence from The Lanchester Motor Company, 200875/1922'.)

The new Lanchester was generally referred to by George and the employees as the 23 h.p. but, probably to keep in line with the small Rolls-Royce, Daimler and other competitors, it confusingly kept its model name of 21 h.p. for advertising and sales purposes. On the road the revised car performed well; the engine's b.h.p. was recorded at 57 with an engine speed of 2,800 r.p.m. increasing to 3,400 r.p.m. at its maximum.

With the Lanchester Motor Company's profits slowly declining from the £26,985 net profit shown for the period from 1915 to 1923, Frank and George felt that their product range was still catering for too expensive a market, and that they should offer a cheaper car to the public. At that time, vehicles of around 16 h.p. were very popular, so a high-quality car of that engine size, still retaining the individual Lanchester style, was designed and submitted to the Board. Frank and George realised that the Board's approval was of the utmost importance, as at the end of the Company's financial year, in July 1926, the profit had dropped yet again to a mere £8,043. Sadly, the design was turned down by the other Directors in view of their reluctance to provide the necessary capital outlay, the recurring theme of under-confidence that so dogged the Company.

1928 Lanchester-bodied four-door tourer on 21 h.p. chassis no. 3583. (See appendix.)

The accounts for 1927 were a low point in the Company's history as, for the first time, a net loss was recorded: £3,100. Anticipating this, George had updated his 16 h.p. design in order to again offer the Company a new opportunity for greater sales; it was once more rejected. This pattern was repeated when the following July's accounts were announced, showing a net loss of £7,926. Indeed, the 16 h.p. design was updated each subsequent year by George to keep abreast of technological innovations, but the Board's overall decision was always an emphatic refusal. It was even recorded that one of the Directors was so thoughtless that he proudly drove around in an Austro-Daimler which he had bought, seemingly oblivious to the Company's slowly sinking fortunes.

Archie Millership had tried to revive these in the snow of February 1927, when he set off for a fortnight's trip over the Alps to show the merits of the new Lanchester 21 h.p. to the motoring correspondents of the Autocar. The report told more of the Alps than of the car's performance but, needless to say, the reporters were impressed. They summarised the trip by remarking, "The tool bag was not opened, the Lanchester behaved in exemplary fashion; its engine and gears always up to the work and first gear never called upon. It would be difficult to imagine greater comfort in road travel." To supplement this high praise, they were startled by the average petrol consumption of around 20 m.p.g.

The 21 h.p. continued in production giving steady sales, and was more or less the same car as that designed in 1922 and 1923. Various modifications had kept it in the forefront of technical developments and these continued during 1927, with a change to a rotary magneto produced by the B.T.H. Company. The chassis length was standardised to 11ft.1in.,

THE SIX-CYLINDER 21 H.P.

A well-laden 21 h.p. on its annual trip to Switzerland. Lanchester limousine coachwork in lemon-yellow with black mudguards.

as orders for the shorter chassis were by this time few and far between. Rudge-Whitworth centre-lock wire wheels, using tyres of 21ins. diameter by either 5.25ins. or 6ins. width, were offered as an option to the usual steel disc type.

From late 1928, other alterations were made, the main one being to the front axle arrangement. Although it remained in tubular form, its profile was straight until once past the chassis side members, where it was angled upwards to take the wheel hubs. This was all in contrast to the earlier axle which curved gently beneath the base of the engine sump. The new axle did not need to take this into account, as it was set lower than previously, due to the front springs being located on top of the axle, rather than underneath as before. The semi-elliptic springs were much flatter in profile than with the previous design, and were noticeably bigger in section than those used on most other makes. They were anchored at their front ends to the chassis dumb-irons, not with shackles but with phosphor-bronze trunnions, similar to the rear axle arrangement. The rear ends of the front springs also terminated in phosphor-bronze trunnions, but these offered both sliding and rotational movement. Two grades of front springs were offered to suit different weights of bodywork, and both employed Lanchester-produced polished spring leaves with interleaves of brass strip. Luvax vein-type shock absorbers replaced the previous Hartford friction-plate type.

A fabric-bodied 21 h.p. saloon.

The track of the 21 h.p. was increased by 2ins. to 4ft.8ins. although the turning circle remained at 44ft.

The brakes had been modified by using aluminium shoes, adjusted by a rotating worm gear on each wheel, all enclosed in a flat-sided brake drum. Foot pedal actuation was, as an option, assisted by a Dewandre vacuum servo unit, although this became a standard fitting in 1930. The handbrake lever was altered from its side-ratchet arrangement to one at its base, operated by a locking lever at its head. The rear axle was re-designed to employ the three-quarter floating system, whereby the bearing on which the hub was mounted was carried in the axle casing.

The main alteration to the engine was that of the pistons, which were changed from the all-aluminium type to a composite construction, having an aluminium domed head secured to a steel skirt. This complicated design was to avoid any tendency of a piston's side to 'slap' on the cylinder walls. It was machined all over and incorporated three compression rings and one oil-scraper ring. A fully-floating gudgeon pin was enclosed within the steel skirt.

A by-pass oil filter, for the cleaning of the engine lubricating oil, was mounted on the bulkhead. Yet another delightfully engineered system of telling the level of oil in the engine was devised; this time of a floating level indicator design. There had been various types employed on the 21 h.p. over the years, for George would not consider the use of "that primitive instrument: the dipstick". Possibly he came to feel that his clientele disagreed with him, for he later remarked wryly of his very first system, "I once made the stupendous blunder of thinking that motorists disliked getting their hands and shirt cuffs dirty, and under this illusion provided an oil-level cock and drain cock, instead of dipstick and drain plug. Each had a handle extended to a clean and accessible position, allowing the owner to test the oil level or drain the sump without collecting on his cuffs samples of black oil, grime, and dead flies."

A tube, open to the atmosphere, had earlier been secured on top of the front of the aluminium valve cover, to lessen the chance of condensation forming. This must have been beneficial, as a similar engine-breather was fitted at this time to the rear also.

The other major modifications were the adoption of a dash-mounted Nivex petrol gauge, telling the contents of the sixteen-gallon rear tank, and an A.T. speedometer and time-clock to replace the Watford type.

The additions were proudly shown at the various Motor Exhibitions as they were included on the chassis. Lanchester's stands were always eye-catching and, unusually, often included light-coloured cars. The 1927 show had featured a cream, three-seater, fixed-head coupe with chassis and mudguards finished in green, selling for £1,660, as well as a Hooper-bodied, six-seater, crimson limousine for £300 extra. The coachbuilding firm of Penman also showed a Lanchester 21 h.p., clothed with their own black and white two-seater coupe coachwork, priced at £1,495. The following year saw a blue limousine accompanied by a cream, Hooper cabriolet body, with two front seats, two sideways-facing rear seats and a dickey containing one extra seat. A 21 h.p., four-door saloon of grand specifications by the coachbuilding firm of Brainsby, was displayed on their stand for the extremely competitive price of £1,500. The shows proved extra beneficial to small companies such as Lanchester as so many people could readily inspect their products; in view of the relatively small numbers made there were not many Lanchesters to see in normal daily life.

An earlier private showing of a Lanchester 21 h.p. chassis had been arranged for the Duke of York. He subsequently bought two, one coachbuilt as a shooting-brake for informal occasions, and the other a fabric, Weymann-pattern saloon, bodied by the coachbuilders, Gurney Nutting and Company. It was reported that the Duke of York later damaged his wrist and afterwards found gear-changing difficult on the 21 h.p., so he sold the cars and replaced them with his second 40 h.p. Lanchester landaulette, which employed epicyclic gears. Another Royal personage to purchase simultaneously two 21 h.p. cars was the Maharajah of Nawanager. His pair were absolutely identical, designed specifically for use on hunting trips, and were fitted with wheels of greater diameter to increase ground clearance.

An inducement for customers to purchase a Lanchester was the Company's record for service and maintenance. Godfrey Barnsley told of one customer who approached Rolls-Royce and enquired about their procedures. He was informed that a car would be expected to be sent in for maintenance once a year, to which he protested that his cars had to be on the road day-in and day-out. He subsequently bought a 21 h.p. Lanchester sedanca-de-ville and ran it every day for five years, covering 200,000 miles.

1930 21 h.p. with fixed-head coupe coachwork by Kellner of Paris.

1929 saw just one 21 h.p. on Lanchester's Olympia stand, coachbuilt by Maythorn and Son. The landaulette body was painted red with black "mud-wings" and valances, and among its fine fittings were a cloth interior, revolving occasional seats, velvet carpeting and a dictaphone to the driver. The whole was advertised at £1,700, while a chassis prior to coachwork remained at the price of £1,050.

Before the end of the year, various options were made available to customers. These included a taller type of radiator, which was fitted in by lowering the chassis cross-tube on which it rested and shaping the radiator base around the central starting handle aperture. Chromium-plated, vertical, radiator ventilation slats, automatically opened and closed by a thermostat, were offered as an option, to complement the chrome-plated radiator surround. Chromium-plated bumpers, as well as a metal dashboard, were also available if desired. Tyre sizes were altered again, and were standardised at 6.00ins. by 20ins.

Two 21 h.p. cars were exhibited at the October 1930 Olympia Motor Exhibition, one of which was on the stand of the coachbuilder, Arthur Mulliner. This was a four-door, sports saloon with sports-type wings, rear boot and special long bonnet, painted in black and gold.

The landaulette shown on the Lanchester stand was, for the second year running, one coachbuilt by Maythorn and finished in blue picked out with white lines on mouldings and wings. The radiator, lamps and all fittings were silver-plated. Companion trays, overhead lamps, a communication system, velvet pile carpet, and cabinet work in polished walnut were among its attractions, all for sale at £1,775.

These two cars were the very last ones to be exhibited at a motor show, although other 21 h.p. cars were sold during 1931 and 1932.

Chapter Twenty-Two

DR. FREDERICK LANCHESTER

At the War's end in November 1918, Fred was commuting endlessly to Coventry and Birmingham from his London home, to fulfil his duties as Consultant Engineer and Technical Adviser to the B.S.A. group of companies, and Daimler in particular. In line with all other motor manufacturers, Daimler immediately demanded a saleable product, and Fred duly sent the Managing Director, Percy Martin, a new design of car which he had been working on. It was an incredibly radical approach to motor engineering design principles, and offered the Daimler Company the speculative risk of either a great worldwide success, or a great loss of capital investment.

In Fred's design, steel suspension springs and shock absorbers were completely dispensed with, as he offered instead a system of hydraulic suspension. (The Citroen 2CV model was the next to offer hydraulic suspension, 30 years later.) A large pressure chamber, maintained by either the car's engine or by a pump driven from a road wheel, supplied pressure to the hydraulic rams. Oil in compression was employed as the medium for supporting the weight of the vehicle, as well as absorbing the road shocks. Fred also provided for the driver an instantly variable degree of damping, merely by turning a knob. The traditional front and rear axles were abandoned and instead the front and rear wheels were mounted on short stub-axle assemblies on pillar-type, independent, suspension legs. These allowed enough movement for road inequalities and enabled the heavy Lanchester worm gear for the rear drive to be secured to the chassis, achieving the technical benefit of having no unsprung weight, other than the wheels and hubs. Other advantages of this system were an increase of 1.5ins. in ground clearance and an increase of 5ins. to the rear body width.

The rear wheels were driven by a jointed shaft, similar to that used earlier on the De Dion cars. Strangely, this futuristic design did not advocate front-wheel brakes, although Fred's rear-wheel brake design more than compensated for this omission. Instead of brake drums secured to assemblies at the road wheels, Fred arranged them one on each side of his central worm-gear housing, to again lessen the impact of their weight. (This in-board braking system was next designed for the 1923 Benz racing car, but was not used on a production vehicle until the following decade.) His initial design of rod operation was complemented by his wish that, "The hydraulic gear itself can be used as a brake and it is contemplated that in the ordinary way it will be considerably used in this manner." (The Duesenberg was the next to offer hydraulic brakes in 1921.)

Fred's "hydraulic clutch gear box" was another innovative feature of this design. Although he could not by-pass the requirement of a heavy flywheel for the engine to achieve smoothness of running, he considered how to make the flywheel a useful item rather than just a dead weight. Using the design principle of his earlier twin-cylinder Lanchester engines, where the flywheel incorporated the large, heavy magnets for the low-tension ignition, he designed the whole clutch and gearbox to be fitted inside the drum-shaped flywheel. Its three forward speeds and reverse had a control system protruding from the centre of the floor, although this could have been transferred to the driver's right if desired. Fred was pleased with the overall mechanical layout which "lends itself to a general cleanliness and absence of rods, shackles, lugs, etc.".

He adopted this principle inside the car too. "With a view to more completely clearing up the obnoxious details with which cars of the present day are encumbered I have, in the course of design, a combined instrument and switch board to be situated just beneath the steering wheel." Proposing to make the traditional dashboard with its distant gauges and switches obsolete, his 'bought out' detachable panel of essential instruments with plug-in connectors was central to the driver and easily visible. As the steering wheel spokes would have been the only obstacle to seeing this "instrument head", he designed a steering wheel with a downwards pointing single spoke, to avoid this. (This design was next used by Citroen in 1956.) The steering column had the various throttle, ignition and mixture connections located on its outside rather than inside, for ease of both installation and maintenance.

Fred acknowledged that his far-reaching proposals had ".... hardly an existing feature of the present-day motor car in the new design ....", but was experienced enough to realise that, "the whole scheme must be looked upon as including a large element of gamble ....". Fred himself obviously felt the risk worthwhile, as he continued, "If we could win all along the line the success would be gigantic, it would pulverise people like Rolls-Royce and would paralyse the components business as it has developed in the United States. It is really an immense stake to play for." Percy Martin and his team considered the bombshell of Fred's design, but became hesitant at the project's inevitably high development costs. Fred then consulted the Lanchester Motor Company, who showed a willingness, albeit restricted by their limited finances, to become involved in the project in co-operation with Daimler. Their backing obviously was not sufficient to allay the fears of Percy Martin, however. Daimler's John Milligan recalled that Fred's design looked very promising, ".....but at the last, the Company decided that they would just put out the two largest chassis of the 1914 programme, and one new one in which he (Fred) had no voice." This decision was a great disappointment to Fred after all the time and effort he had expended on the project. John

Dr. Frederick Lanchester with his wife, Dorothea, at home reading Shakespeare.

Milligan concluded, "The new one was a disaster. I was then Sales Engineer and I let him (Percy Martin) know the troubles that could have been avoided if he (Fred) had been consulted."

However, on the personal front, 1919 was to be a much happier year for Fred. At the age of fifty, he became engaged to Miss Dorothea Cooper, the daughter of a vicar. The couple had met some years earlier when Dorothea's uncle, who was the Birmingham Cathedral organist, gave Fred lessons in singing and public speaking. When the uncle moved to take up residence in an upper flat at Bedford Square, London, there were more occasions for the friendship between Fred and Dorothea to develop. After their marriage, which took place on 3rd September, they ended their honeymoon with a long stay at Frank's holiday home on the Isle of Wight. This was interrupted by a visit to Birmingham on 18th September, when Fred was awarded an honorary degree of Doctor of Laws by Birmingham University, for one who "was pre-eminent in his own field of knowledge and who had worked for the enrichment of his profession"; he was very proud to accept this honour. Indeed, further honours were bestowed upon him over the next few years, when he was elected a Fellow of the Royal Society and also awarded the Gold Medal and Honorary Fellowship from the Royal Aeronautical Society.

Such was the volume of work he undertook for the Daimler and B.S.A. Companies that Fred's tally of twenty-one patents accepted in 1919, following just eight for the preceding three years, took a further five years for him to emulate. They were, as usual, varied in their scope and included designs for tents supported only by columns of air, ship apparatus capable of landing and launching aeroplanes, exhibition buildings, hot or cool air ventilation systems in buildings, pianofortes, and one for the "propulsion of automobile vehicles of amphibious type". The latter design consisted of a pair of feathering paddle-wheels rotating at half the speed of the vehicle's driving wheels, to which they were bolted. (Sadly, no photographs of Lanchesters or Daimlers cruising the inland waterways exist.)

His no-nonsense reports were not always what the Daimler and B.S.A. management wanted to hear, however, and his ideas often fell on stony ground, although he was always much in demand elsewhere. His versatile brain allowed him to investigate and advise on vehicle details of Armstrong Siddely, Renault, Austin, Rover, Fiat, Humber, Hispano Suiza, Buick, Standard and Minerva, for example, as well as White and Pope engines and numerous vehicle components.

His most lengthy source of study, other than those for B.S.A. and Daimler, came in January 1924, when he agreed to become Consulting Engineer to Wolseley Motors Limited, his fee being set at £1,000 per annum for one day's consultancy each week.

To signify Fred's new involvement at the firm, Wolseley decided to launch an advertising exercise using photographs of the Wolseley and the Lanchester as "the two first British-built cars on the road". Vickers Limited, the parent company, wasted no time in gaining Fred's advice. "What our management at Birmingham particularly wish is to have your opinion as quickly as possible on the specification and design of the new 16 h.p. Model." Fred subsequently gave much advice on their 10 h.p., 12 h.p., 14 h.p., 16 h.p., 16/35, 20/45, 11/22 and 24/25 cars. He also studied a scheme for a three-wheel car design offered to the company and unceremoniously concluded that it was "a quite hopeless proposal" and that the inventor's description was "three parts bluff".

Fred's responsibilities for Wolseley, as well as for the B.S.A. Group, prompted him to leave London in April 1924, and return to Birmingham. In anticipation of this move, he had designed his own large and functional house, after first rejecting a layout from his architect brother, Henry. Fred and Dorothea duly moved into Oxford Road, Moseley, and named their home 'Dyott End' (now number 128) because it was at the end of George's home, 'Annersley' (now number 30) in the adjacent Dyott Road. With their rear gardens back-to-back, the two brothers now had easy access to each other on their return from long hours at their respective motor companies. This greatly facilitated their joint spare-time hobby: the construction of another car.

The Mark One, three-wheeled Lanchester wooden car.

In view of the popular craze for motorbike combinations and cyclecars, they each had ideas of producing a cheap, light, simple yet comfortable, two-seater car for the masses. It was to have a maximum weight of only 700lbs, and a selling figure of under £100. George recalled they decided that "rather than divide our effort it would be better to work together". This work had actually commenced two years earlier, when they had laid down the basic concept and started the construction in George's garden.

Much of the work on the Mark One was actually done by George. It utilised a 'V' twin-cylinder, 500 c.c. B.S.A. motorcycle engine and was chain driven. The axle width had a narrow track of 3ft.4ins., so making a differential gear unnecessary. It was all mounted in the rear of a three-wheel chassis configuration. It had simple, straight body panels secured to the light frame, with enough space for two on the single seat. The car was certainly no Lanchester 40 h.p. or Daimler 45 h.p., but showed the brothers the worthiness of their private venture.

After providing valuable experience, the car was soon abandoned in favour of a four-wheel version. The wooden wheels of the Mark Two were constructed of six flat spokes and blended perfectly with the new bodywork, constructed of mahogany. This was made to Fred and George's own design by Saunders Boatyard in Cowes. George recalled, "When the body was ready a telegram was sent saying 'body despatched', which gave the housekeeper a great shock." Known to all as the 'wooden car', not only the bodywork but also the front axle and suspension were made from wood. The car was sprung by using two parallel lengths of Sitka spruce, about 7ft. long and 7ins. wide, tapering from two inches thickness at their centres to 1in. at their ends. These did not carry the body directly, as it was mounted on a central, transversely-mounted, tubular trunnion bar. Onto this, interposed with rubber bushes, the complete body was pivoted. Its rocking movement was tightly controlled by eight cylindrical rubber buffers, two double pairs on either side of the car. Fred felt that this use of india rubber eliminated any tendencies of the car to either "pitching or bucking". The buffers acted as subsidiary springs and ensured a considerable degree of damping.

George at the wheel of the B.S.A.-engined Mark Two Lanchester wooden car.

The Mark Two Lanchester wooden car, with hood raised.

The car's exposed, rear-mounted, B.S.A. engine was unsightly and its noise for the driver and passenger must have been excessive. However, George drove the Mark Two a great deal and it generated much interest. In his words, "If I stopped at a shop, it immediately drew a small crowd. Often I was asked, 'Where can I get a car like that?' or 'It's just the sort of car I want.'"

The Mark Three wooden car probably made its debut in late 1923, and utilised most of the Mark Two components. The motorcycle engine was abandoned in favour of a new design by Fred: a compact, horizontally-opposed, 10 h.p. two-cylinder engine, able to be housed under the flat, luggage-deck bodywork.

Although Fred had patented, in 1917, a method of diesel fuel injection which squirted directly into a cylinder's combustion space, he decided on a different design of fuel injection for the Mark Three and subsequent petrol engines. His newly-patented system allowed one half of a double-acting pump to force petrol to an internal 'carburettor' which was, in fact, an internal, air-mixing chamber directly behind the inlet valve. When this was opened, the explosive mixture was forced in, any surplus petrol being removed by the fuel pump before it admitted the next quantity of petrol. (Petrol injection was next introduced on the Mercedes in 1954.) This unique device could also have lent itself to a pressure-fed air supply for super-charging.

The more flowing lines of the Mark Three wooden car, with Fred Lanchester's new, horizontally-opposed engine beneath the rear luggage-deck.

As on the earlier wooden cars, there was no clutch and only one speed gear. The transmission on the Mark Three incorporated two friction drums, which Fred later called 'spools', one driven by the engine and one secured to the tubular rear axle. Their engagement was effected by the simple but somewhat unorthodox method of moving the whole axle forwards or backwards; large securing springs ensured that the axle did not move of its own accord. The axle employed eccentric mountings, operated by the left foot pedal, which allowed any degree of pressure to be applied to the friction drums, so giving an infinitely variable drive: fully engaged, the ratio was 5:1. Actuation of the brake pedal brought the driving shaft out of contact with the friction gears and simultaneously applied a brake shoe to the friction drums.

Fred's innovative engine design, coupled with the other tried and tested features of the development programme, successfully achieved the original design parameters set out by the two brothers. This can be supported by the fact that Fred used the car as his personal, and indeed his only, transport for many years. Even the sitka spruce springs were shown to be durable and long-lasting, no doubt to the amazement of many critics, although after Fred had travelled some 10,500 miles in the car they did show a slight set of 0.625ins. (He was later involved in a collision with a lorry, after the car had covered over 30,000 miles. Tony Jones, who worked with Fred in the early 1930s, related that "the wooden car was reduced to something like a million matches" but luckily Fred was not injured.)

Rear view of the Mark Three wooden car, showing the rear clips from which the whole engine bonnet pivots.

George felt that the car would sell readily at that time, but the idea of production was resisted by Fred, whose determined views, as ever, eventually won the day. Fred's unease at selling the car showed again the trait in his character which would only countenance perfection, in this case even against the advice of his brother. Fred envisaged, as he always could, a better way of doing things and a new spiral of development work started.

A disappointed and, it seems, hurt George Lanchester commented, "How often does the genius contrive to improve and develop his design until the original conception is lost?" The private partnership was effectively disbanded from this point and Fred started alone on the Mark Four wooden car. George thought his brother's scheme a mistake and remarked that, "His ideas became more ambitious and the petrol-electric bug bit him."

Fred had always been interested in the principles of a petrol-electric combination of power and had applied these successfully to past projects, such as his K.P.L. Omnibus, the Daimler Rail Car, the Daimler-Thomas petrol-electric system and others.

His new 'Petrelect' project incorporated all the best features of his past experiences and latest theories, and he coupled a large generator to the 10 h.p., horizontal, twin-cylinder engine. A vaned cooling impeller was interposed between them. Two pairs of electrical brushes acted on a large, circular armature.

For normal running of the wooden car, the petrol engine was used. When the three-position dashboard-mounted lever was moved to its 'start' position the batteries powered the electric motor, which revolved. This rotated the directly-coupled petrol engine which, on the application of a spark at the sparking plug via the coil ignition, started the engine. If either a better acceleration than normal, or a low speed performance, was required, the 'start' position could be maintained, so that both the electric motor and the petrol engine could be employed.

If the 'reverse' position was selected with the dashboard lever, the electricity to the armature was reversed, which correspondingly changed the rotation of the armature, allowing the car to travel backwards. All this was accompanied by the opening of the exhaust valve on each cylinder, so that the petrol engine would turn over unhindered.

Normally, when the petrol engine was running, the dashboard lever was moved to the 'charge' position. When the car was travelling at more than 10 m.p.h., the rotating armature and the other electrical apparatus were able to charge the large batteries stored under the double seat. To motorists with no mechanical interest, worried about an engine failure, the system would have been most comforting, as the car could be driven using either source of power.

Regardless of the fact that the originally conceived market had been lost, due to the Mark Four's weight, size, cost and complication, and also the launch of the Austin Seven, which was beginning to dominate the small car market, so effective was his design that, later in 1924, Fred formed a syndicate called F.W.L. Propulsion Traction Limited to manufacture his Petrelect wooden car.

Various components were re-designed and many minor improvements were carried out for the Mark Five in early 1925. The wooden springs received consideration on the Mark Six, and experiments were made using ash, hickory, English poplar and spruce.

Development continued but in Fred's understated words, "Actually the work took longer than intended and ran into more money, most of which I found myself. The Chairman ..... lost interest and the syndicate, at his request, was wound up." Fred continued alone, absorbed in his spare-time project and, as time went by, the wooden car continued to grow bigger and heavier. In 1927, his Mark Seven was forced to employ double, or bi-lateral, spruce springs measuring 9.5ft. and 6ft. long by 6.5ins. wide. The suspension was kept deliberately soft, and felt to the passengers as if they were travelling in a rowing boat.

The wooden bodywork was compact yet comfortable and provided good vision, while the two-seater compartment was given the luxury of a fully-opening hood. As on previous designs, the bonnet over the petrelect mechanism and transmission was flat and incorporated a useful luggage deck. The whole bonnet, including wings, was pivoted at the rearmost part of the car and could be raised in an arc from behind the passenger compartment and supported on a prop, to allow easy access to the mechanics. As alternatives, Fred wrote that, "The motor cover may be raised on its hinges and then thrown right back out of the way on to the road to give complete access. It then rested on the luggage deck. It could be readily removed if desired, after separating the rear light wiring."

The petrol injection and the Petrelect design and transmission layout were retained, and the friction drive assembly was still operated by the left foot pedal.

The braking system was ingenious but simple. Fred realised that a driver who wished to stop quickly experienced a time delay, due to removing his foot from the accelerator pedal to depress the brake pedal. He therefore devised a system whereby the roller pedal, operated by the right foot, acted as the accelerator for the first half of its travel. Beyond this point, the accelerator was automatically released and the brakes were applied. Apart from the friction brake, Fred also incorporated front wheel brakes of the external-contracting, drum type. The spoked wire wheels, secured to aluminium rims, supported the 3.75ins by 19ins. tyres, in place of the earlier wooden wheels. Bill Sweeney, who assisted Fred with the routine mathematical design calculations for this Mark Seven car (which was later bequeathed to him in Fred's will), remembers that it took time for a driver to become accustomed to the

The underside of the Mark Seven wooden car, showing the bi-lateral suspension planks and the four pairs of body mountings.

car, as it was so radically different from any other. Its overall performance was more than adequate as, apart from an acceleration which was described as "gradual", it had a top speed of 50 m.p.h. and, more importantly for this type of car, a petrol consumption of just under 40 m.p.g.

A new wooden car, the Mark Eight, was designed as a four-seater and was a natural, although heavier, successor. It was to use a brand-new engine, of 2.75ins. bore and stroke, again designed by Fred: a 12 h.p., horizontally-opposed, four-cylinder petrol engine which, incidentally, utilised valve cams from the Lanchester Motor Company. Fred also gave this new, 50 m.p.h. engine the petrol-injection system, similar to that used on the Mark Seven model. The advanced petrol engine and electrical motor were constructed in 1927 as one compact unit and it was surely the ultimate petrol-electric combination. Unfortunately, the projected Mark Eight car was never built and the two completed and fully-tested engines were never used.

Fred did allow himself just one more flirtation with his wooden car project, when he laid out designs for a totally new concept in 1928. He rekindled the original aims of 1922 and designed his Mark Nine 'Monocar'. This single-seater car had a wheelbase of 7ft.7ins. and a track of 3ft.2ins. Its new two-cylinder engine, of 2.5ins. bore and stroke, was noted by Fred as, "Rating 5 h.p. for £5 tax". Revving at 6,000 r.p.m., it was predicted to give 80 m.p.g. and 55 m.p.h. Sadly, this exciting project was never built and indeed the whole wooden car programme was "indefinitely postponed" as Fred was experiencing difficult times with B.S.A. and Daimler, relating to his other enterprise, Lanchester's Laboratories Limited.

From the early to middle 1920s, the policy of the Daimler Company had been to develop a full range of cars, and no fewer than ten different basic models were put on the market in that time. Judging by Daimler's growing debt at the bank, the policy was ill-guided. The rush for production which, as Fred conceded, was needed "to prevent the stagnation in design and loss of market", was at the expense of research and therefore against all his instincts. To counteract this, he persuaded the firm, in the summer of 1925, to set up a small company for research and development purposes, and the partnership of Lanchester's Laboratories Limited was formed. The Daimler directors, Percy Martin and Richard Rotherham, as Chairman and Managing Director respectively, joined Fred. The Company's registered address was 38 Bailey Road, Coventry, although all meetings were held at the Daimler Works in Sandy Lane, Coventry.

In return for the input of £12,000 cash from Daimler, half the shares were allocated to Fred in exchange for the transfer of many of his British and foreign patents, mostly relating to the Petrelect wooden car programme of his existing company, F.W.L. Propulsion Traction Limited. The benefit for the latter Company would be adequate research funding, as the earlier syndicate backing the project had by this time been disbanded.

However, with the ink barely dry, Daimler "borrowed back" £11,000 on the pretext that it would be returned as and when required. It was never repaid, and the new Company was left with little working capital. Fred stubbornly carried on with his Petrelect project regardless.

In 1926, the B.S.A. and Daimler companies felt that Fred's many consultancy appointments to other motor manufacturers were not compatible with their own needs. Even though their financial resources were worsening, they contracted to pay Fred £3,000 per year to remain as their Consultant Engineer and Technical Adviser, but with the proviso that he was to give his services solely to them. Fred felt that this agreement compensated somewhat for the previous unfair treatment meted out to him at Lanchester's Laboratories Limited.

Although much varied work was successfully undertaken by Fred for Daimler and B.S.A., their attitude to him gradually began to harden. After various minor disagreements, there was another dispute in October 1927, this time regarding the size and types of tyres fitted to the Daimlers.

Fred had long been concerned with the problems of front wheel wobble, steering problems, high pressure and low pressure tyres. This, coupled with comprehensive research for his March 1928 paper to the I.A.E. on the problems of steering gear, stood him in good stead to counter certain criticisms from his Managing Director at Daimler.

Due to customer complaints about the steering, compounded by a report from the Dunlop Tyre Company, Percy Martin had instructed his drawing office to provide for larger tyres on the cars to solve the problem. Impetuously, he blamed all on Fred for the costly error of judgement. Fred retorted that, as Consulting Engineer and Technical Adviser, he had nothing whatsoever to do with tyre choice, but assumed that the problem was due to the fact that chassis weights were often altered two or three times after the initial tyre size had been fixed. This fact was confirmed by Frank Compton, Daimler's Chief Draughtsman, who stated that the Company always rushed out new models each year for the Motor Show, even though they had not been thoroughly developed. Stung by Percy Martin's inaccurate criticism, Fred made himself aware of all the facts and, on his own initiative, visited Fort Dunlop. After much debate, he "drew attention to certain observations" which he had made and, in his words, "came out of it with flying colours". By his personal intervention, Fred saved the Daimler Company between £3,000 and £4,000.

This success did not deter Percy Martin from asking Fred to accept a one-third reduction in wages for the whole of 1928. Although Fred could have insisted on his contracted salary, he reluctantly agreed to this, as he knew that Daimler's mananagement staff had had the decision imposed upon them in order to ease the Company's debt. He was thus working for the Company at just £2,000 per year, was under contract not to work for other firms, and was still running Lanchester's Laboratories Limited with no capital, as the return of the 'loan' of £11,000 was still not forthcoming.

The situation worsened for Fred in October 1928, when a letter from the Managing Director gave twelve months' notice of the ending of Fred's contract, replacing it with a one day per week consultancy. On a brighter note, the letter also detailed the Company's

Frederick William Lanchester.

desire to refund Fred's previous salary loss and to "forward the interests of Lanchester's Laboratories Limited". Fred had been able to supplement his income from his various patented inventions, with another twenty-seven being accepted since 1925, including such items as electrical warnings for low liquid levels and also electrically-controlled, hydraulic-steering dampers. However, to compensate for these changes at Daimler, it was agreed that he should be allowed to work for other firms again and, in October 1928, Fred accordingly took up the position of Consultant Engineer to the William Beardmore Group, specialising in locomotive diesel engines.

Although concerned about the future prospect of losing his full-time employment at Daimler, he nevertheless continued to advise them on a whole range of issues and technical matters. One such report to the Managing Director, in March 1929, showed that Fred was "entirely favourable to the adoption of a revised Fottinger Coupling, provided that the difficulties incidental to its adoption are properly faced and overcome". He was very concerned that the coupling, later known as the fluid flywheel, was to be put into production before proper development work had been conducted. The main instigator of this short-cut was the Chief Engineer, Laurence Pomeroy; a man with whom Fred had clashed many times over the years. After analysis, Fred concluded that he saw no reason to predict problems with the device, but "never in my whole life have I put forward such a lot of undigested experimental material as part of a pending production programme, as in the present case".

Percy Martin, and no doubt Laurence Pomeroy, was not amused, and told Fred that there would be no programme of production at all if they listened to him. In Fred's words, "He behaved like a spoiled child whose favourite new toy was being snatched from him." He surely smiled when he also recalled, "My report was ignored with the result that on the 30th March 1929 the trial, incorporating the system I had criticised, in the hands of Pomeroy himself, took charge when waiting its turn at a petrol station near Lynton, North Devon, and 'mowed down' two petrol pumps and demolished a cashier's kiosk, before it could be stopped. Then and then only was the system abandoned." (The fluid flywheel was further developed and brought into use during 1930.)

Although he was highly regarded by practically everyone else at the Company, there was obviously a clash of personalities between Fred and the hierarchy at Daimler. This was demonstrated when Fred attended a top-level meeting headed by Percy Martin. Frank Compton recalled that, with great anger, Percy Martin blazed at Fred, "Why are you reading and writing letters when we are paying for you to attend?" Fred demolished the attack by repeating everything said at the whole of the meeting thus far, in great detail.

Fred's days at Daimler were obviously numbered, and on 5th April 1929 he was informed that his "agreement with B.S.A. will therefore come entirely to an end at October 31st 1929". Fred was somewhat aggrieved, as he felt that he had saved them money far in excess of his salary over the years.

Daimlers also wanted to end their association with Lanchester's Laboratories Limited and the directors, who had been joined by Laurence Pomeroy in June 1928, effectively offered Fred an ultimatum. One course of action was for him to buy out Daimler's interest, and the other was to go to Court to resolve the issue. He accepted the Company's plea of financial worries affecting the Managing Director's decision and he felt it ".....inconceivable that after my twenty years of personal association with him, I could take up in hostile attitude and add to his difficulties." Against his own solicitor's strong advice, Fred bought out Daimler's interest for £8,000, and in April 1929 the three highly-satisfied Daimler directors resigned. Fred later admitted that he had in fact ".....made a clear gift of £5,000 to Messrs. Daimler ... to my shame as a man of business."

His determination for Lanchester's Laboratories Limited to succeed was reflected in the fact that, of his eighteen inventions patented between 1929 and 1930, thirteen related to this Company's main business of the production of microphones, loudspeakers, gramophones and sound-reproduction apparatus. He was joined at his Company by other independent directors, and business continued. Fred subsequently purchased land at the junction of Spring Road and Olton Boulevard, in Tysley, Birmingham, and a laboratory was erected.

A personal boost for Fred was the appointment of George as a director on 9th January 1931. However, the economic depression drained the Company funds further, and by May 1934 the decision was reached to sell the building, land, plant, stores, stock and patent rights. All was settled by June. Lanchester's Laboratories Limited then ceased trading 'temporarily' and though annual meetings were held until 1939, business never re-started.

Chapter Twenty-Three

THE ARMOURED CARS

The renowned qualities of the Lanchester 40 h.p. cars, together with War-time memories of the impressive performances of the 38 h.p. armoured cars, combined to win a contract from the War Office, in the autumn of 1927, to develop a new armoured car based on 40 h.p. components.

George Lanchester's subsequent design was Britain's very first of a completely new breed of fighting vehicle; that is, one that was not an adaptation from a private or commercial vehicle chassis. His design consisted of deep side-members made from 9mm. bullet-resisting armour plate, which protected all the vehicle's vital mechanical parts; a feature which had been previously neglected. Large-diameter steel tubes crossed the chassis to give torsional rigidity and the rearmost tube of 10ins. diameter served also as a reserve petrol tank. The armoured car gave 9 m.p.g., and two other fuel tanks were secured at the rear of the chassis to allow for nearly three hundred miles of travel.

The chassis of Britain's first specifically-designed armoured car: the 1927 40 h.p. Lanchester.

Twin rear axles were fitted, each one utilising double wheels. These were all driving wheels and were provided with their own brakes. The Lanchester worm gear was utilised for the rear axles in order to gain the benefit from its maximum torque capabilities at low speed. The axles were mounted on the ends of two semi-elliptic springs, the centres of which pivoted on the chassis frame mounting. This system gave an exceptional degree of flexibility to the suspension for traversing rough terrain, and gave it traction close to the performance of semi-tracked vehicles, although tracks for the rear wheels were also available if required. The front axle was 'unbraked' and was a heavy gauge, 'H' section forging, suspended on semi-elliptic springs.

The steering was of the traditional Lanchester pattern, but at its base a permanently-

The chassis of the second armoured car, showing the optional steering position at the rear.

engaged gearbox and linking shaft provided for an easily-removable additional steering column at the rear of the vehicle, able to be used for urgent rearward travel when there was no time or space to turn the vehicle.

The engines employed in the armoured cars were unchanged from the usual 40 h.p. production models, and the three-speed gearbox was also the standard item, except that no reverse gear was included. Secured to the rear of this main gearbox was a low-ratio gearbox for use in cross-country work. It provided, when needed, a 2:1 low-ratio gear and effectively gave six forward speeds. The low-ratio gearbox also contained the reversing gear and this combination cleverly gave three reverse gears; a most useful feature when travelling backwards rapidly in an emergency.

The armoured body shell, which housed a .303 machine gun, and the drum-shaped turret cupola, which carried both a .303 and a .5 machine gun, were probably designed and built at the Royal Ordnance Factory at Woolwich; although the radiator guard was certainly fitted at Armourer Mills. The vehicle's overall size was nearly 20ft. long, 7ft. wide and 9ft. high and it could carry its crew of four at a maximum of 48 m.p.h.; no mean achievement for a vehicle weighing 7.25 tons. The construction of two prototypes began in late 1927, and after completion they were subjected to stringent War Office testing and trials. Much impressed by their performance, the War Office placed an order in July 1928 for another twenty-two armoured cars to be produced.

The cavalry regiments of the 11th Hussars and the 12th Lancers, using these Lanchester armoured cars and also some from Rolls-Royce, subsequently became the very first British Army regiments to be mechanised, apart from the Royal Tank Corps. Both regiments were initially stationed in England, but were later despatched, with their Lanchesters, to Egypt. One Lancer complimented the Lanchester: "It was outstanding for its time, better than the Rolls in armament and cross country performance, except desert where its weight was a handicap." A Hussar also remarked about the armoured cars in use: "These were Lanchesters, excellent vehicles, slower but more robust than the Rolls-Royces."

A Mark Two version was put in hand when it was decided to experiment with single, larger wheels on the rear axles, instead of twin wheels. The turret was changed to the 'Bishop's Mitre' cupola and other small modifications were also installed. An order for eight of these Mark Two armoured cars was placed with the Lanchester Motor Company, and the 12th Lancers took delivery in 1930. A further order for seven armoured cars from the War Office forced the Company to seek funding from their bank, as they were caught in the severe recession affecting the whole of the country. These armoured cars were duly constructed and were delivered at the end of 1931.

The Lanchesters were subsequently used on the European front in Saarland, under the control of the League of Nations International Force. By the beginning of the Second World War, some of these armoured cars were sent to Malaya and Singapore for use by

The Mark One 40 h.p. armoured car fitted with tracks, undergoing evaluation tests.

the Argyll and Sutherland Highlanders, only to be subsequently captured by the attacking Japanese Army.

One was converted in 1940, to be used as transport for members of the Royal Family or senior politicians, to provide protection in and around London. It is not known if a Lanchester armoured car was chosen by Major Morris, the detachment's officer-in-charge, as he knew them well from his service with the 12th Lancers regiment, or whether King George VI himself requested the use of a Lanchester, again promoting the use of his favoured Company. Either way, it seems that the vehicle was not ideal for conversion to this role.

In early 1944, two Lanchester armoured cars were delivered to the only armoured division of the Home Guard in the British Isles. Ivor Howell remembers his platoon of around two dozen men being specially trained by the regular army in order to protect the vital Royal Arsenal establishment, at Woolwich in London, against saboteurs. In operation alongside two Rolls-Royces, two Crossleys and eight Beaverettes, the Lanchester armoured cars were

The Mark Two 40 h.p. armoured car chassis and test body at Armourer Mills, displaying its suspension characteristics with the rear axles criss-crossing obstacles one foot high.

Lanchester 40 h.p. armoured cars in action, with the Mark Two versions in the foreground and at the rear.

on constant patrol of the premises, which were about the size of a big town. Although they were quite old by this time, the Lanchesters were in excellent order and ready for action. At the end of the Second World War the Home Guard was disbanded and the fate of these two Lanchesters after the early months of 1946 is unknown.

Earlier, while construction of the first batch of these 40 h.p. armoured cars had been taking place in 1928, George Lanchester had also been deeply involved in further negotiations with the War Office. They had been keenly interested in his project of a light armoured car, and due to this great encouragement he had commenced his design work on the vehicle.

The engine of George's light armoured car was a modified version of the 21 h.p. It used a 3.375ins. bore by 4.5ins. stroke to develop 85 b.h.p. at 2,500 r.p.m., with a maximum engine speed of 3,150 r.p.m.

The gearbox incorporated a low-ratio unit, offering a 2.37:1 reduction. Its propeller shaft connected with the Lanchester worm gear using a 7:38 differential ratio for the driven, twin rear axles, similar to the larger 40 h.p. design.

The vehicle had a wheel base of 10ft.6ins. and a basic weight of 2.6 tons. With its armoured bodywork, armaments and other fittings, however, its finished weight was anticipated to be 4.5 tons. Unlike its 40 h.p. stable-mate, no order for production was given, as the War Office then decided they would prefer a single rear axle, four-wheel design.

George's consequent Mark Two version was a highly-advanced armoured car, smaller in size and weight than previously, and necessitated a total re-think of design. The 9ft.3ins. wheel-base and 5ft. track was based around an armour-plated hull of 8mm. plate. It was given independent suspension to both front and rear axles, and incorporated transverse overhead springs. Luvax hydraulic shock absorbers were secured to the fully-floating, universally-jointed axles, which employed bevel drive gears instead of worm and wheel.

Another departure from Lanchester practice was the steering, which utilised standard equipment from the Marles Company, for both the vehicle's forward and rearward driving positions. The turning circle was 45ft.

The four-wheel drive vehicle was equipped with Lockheed hydraulic brakes on all four wheels, and the handbrake worked on the transmission propeller shaft. The engine was that of the Mark One version, and was fed by a Zenith twin-choke carburettor. The integral gearbox incorporated a low-ratio device allowing for six forward gears and two reverse. With its low chassis weight of 2.25 tons, the vehicle was designed to give 10 m.p.g. and a maximum speed of 50 m.p.h.

This Lanchester Light Armoured Car exceeded the capabilities of any military vehicle previously produced. The Mechanised Warfare Department were so enthusiastic about the project that a wooden mock-up body was made at Woolwich in preparation for full production. It all came to nothing, however, as the British Government then adopted a policy of disarmament, and the design was left unfulfilled from 1929.

Apart from being a tragedy for the Lanchester Motor Company, who had hoped for valuable orders, the decision was felt on a personal level by George, who had put so much thought and effort into the new design concept, and had expected great things in return. He later wrote, "It was a bitter disappointment to me, as the design anticipated by ten years the 'Light Scout Armoured Cars' of the 1939-45 War."

Fabric-covered Lanchester Southport saloon 30 h.p. on 1929 chassis no. 8035. (See appendix.)

The author's 1929 30 h.p. chassis no. 8072, with Lanchester landaulette coachwork.

1930 30 h.p. Lanchester chassis no. 8077 with Weyman-style fabric coachwork. (See appendix.)

30 h.p. chassis no. 8087, fitted with Lanchester fixed-head sports-saloon coachwork. The car has been owned by the same family since 1941. (See appendix.)

Chapter Twenty-Four

THE EIGHT-CYLINDER 30 H.P.

In view of the Board's repeated rejection of George's small 16 h.p. design, he was forced to consider other options in order to produce another model to boost sales. He investigated the possibilities of a twelve-cylinder engine in 'V' formation, incorporating a valve arrangement actuated by push rods, but this never materialised.

Instead, more work was undertaken in 1927 on an ambitious design of engine incorporating eight cylinders in line. Its bore and stroke were considerably smaller than those of the current 21 h.p. six-cylinder model, being 2.56ins. and 4.25ins. respectively. It was calculated that at 3,000 r.p.m. it would develop 57 b.h.p., with a maximum of about 76 b.h.p. at 4,000 r.p.m. The proposed wheel size of 30ins. diameter, when coupled to a rear axle having a ratio of 7:34., would have given a comfortable 60 m.p.h. at 3,200 r.p.m. and a maximum of 75 m.p.h. at 4,000 r.p.m. Whether by coincidence or design, the treasury rating for the engine was the familiar sounding 21.1 h.p. On paper the project appeared to have potential and one can only conjecture as to why it did not proceed. It may have been the Board's usual reluctance to inject the necessary funding, or perhaps it seemed too similar to the existing 21 h.p., but the most likely reason seems to be that it was superseded by an entirely new proposal.

This came not from George, Frank or the Directors, but from Fred, a shareholder who had the Company's welfare very much at heart. He had analysed the situation at Lanchesters and, in 1928, he suggested a means to satisfy all parties. His plan was simple in its approach to the problem, and was to provide a car which would be relatively cheap to design and construct, would fall perfectly between the two models currently offered and would justify the claim of many experts that the Lanchester was 'the finest car in the world'. In George's words, "The car was to be a natural development from the existing 23 h.p. All components were interchangeable, excepting that the engine was increased in length by two extra cylinders and the chassis frame lengthened to accommodate it." The 30 h.p. Straight Eight Lanchester was born.

The car took relatively little time to design, but certain detailed work was necessary, and in some of this George was supported by Fred. The Works Manager, Jasper Cooper, recalled the debt that the Company owed Fred when he wrote to him, "I well remember you suggesting the Straight Eight, a model which proved very successful both financially and technically. The idea seemed to me to be a stroke of genius mainly, I must confess, because it created a new and very successful model with the least possible expenditure of money and time on jigs and tools. Its success from the sales point of view also proved that you scented the public's demand for a high class car with the smooth running and wonderful acceleration which the Straight Eight possessed in such a marked degree. Your assistance in this matter was invaluable to the Company."

Although the first manufacturer to produce a straight-eight car was Isotta Fraschini in 1920, the idea of an engine with eight in-line cylinders did not become popular for some years. It was certainly very much in vogue at the time of the Lanchester 30 h.p., however, and the Company used this to good effect in all their advertising. For the first time they changed the traditional nameplate on the radiator, producing a larger alternative which

30 h.p. Straight Eight chassis at Armourer Mills.

proclaimed the car to be a Lanchester Straight Eight. When straight-eight cars became more commonplace, however, they reverted to the original.

Either nameplate would have been unnecessary to identify the car, as the unique Lanchester radiator water-level sight-glass was retained. The depth of the radiator was given room to be increased in the same manner as on the contemporary 21 h.p., by lowering the front chassis cross-member.

The Straight Eight engine used the same 3.1ins. bore and 4.5ins. stroke as the current 21 h.p., gave 4,437 c.c. and had a treasury rating of 30.8 h.p. The engine's differences from the 21 h.p. six-cylinder were few, the main alteration being that of the newly-designed induction system.

George and his design team tried many different shapes and sizes of inlet manifold and make of carburettor, in order to achieve perfect breathing for the engine. Robert Boden, who assisted with the experimental work, recalled that Mr. Skinner was involved with George on tests which proved the latter's theory of gas flow and vacuum created in the manifolds. Eight S.U. (Skinner Units) carburettors were fitted and tested successfully at an initial trial but, due to their height, the bonnet line was too high and an alternative system was sought. The use of two Strombergs was concentrated on for a while, but eventually a

1929 30 h.p. with Lanchester two-door, drop-head coupe coachwork.

single up-draught, twin-choke carburettor from the American company, Zenith, was installed. It was modified by the Lanchester Motor Company and gave 14 m.p.g.

The inlet manifold was the subject of much experimentation and employed angled sides for deflection of the petrol vapours into each port, ensuring perfect distribution and therefore better performance and fuel economy. Tests were also carried out to determine the back pressures created in the exhaust system, and modifications were made to suit.

The engine used two ignition systems: one, a B.T.H. Polar inductor magneto similar to that in use on the current 21 h.p., and the other a coil and battery system incorporating a Delco distributor protruding, in ungainly fashion, from the top of the valve cover. The system was intended to be used on a 'one or the other' basis, but most owners ran both together to avoid one of the two separate banks of sparking plugs becoming dirty. For such a big car the actuating keys for the two systems were incongruously dainty.

The gearbox generally employed gears of the ratio of 4:1 on first, 2.4:1 on second and 1.55:1 on third, with a 1:1 direct drive on fourth speed and 4:1 on reverse; although it was advertised on this and the other two models that "ratios vary according to body style". The only difference between the 30 h.p. and the 21 h.p. gearboxes was that of the third gear ratio. The standard rear axle ratio offered on the 30 h.p. was 8:35 (4.375:1), although an option of 9:34 (3.777:1) was later available.

The steering column was located on the chassis frame by a long pivot-bolt and allowed for driver's adjustment in four positions. The patented steering gear itself was identical to that which had been employed on Lanchester cars for about twenty years, so effective was the original design, and it had also been adopted by other companies, such as Buick. The design consisted of a tubular shaft which had a left hand and a separate right hand square-section thread cut upon it. Two long half-nuts engaged on these threads and when the steering wheel was rotated, one half-nut moved down and the other up. An adjustable collar and thrust bearing were provided at the head of the device to take up any backlash or wear, in steps of 0.0018ins. Below the half-nuts, hardened pads operated against large rollers housed within a rocking shaft, and these correspondingly moved the steering drop-

Four-door, sports-saloon coachwork by Lanchester on 1929 30 h.p. chassis.

arm and linkage. Centre-lock wire wheels measuring 32ins. by 6ins. were fitted; the turning circle was 51ft.5ins. on left lock and 55ft. on right lock.

The track of the Straight Eight 30 h.p. was the same as the 21 h.p. at 4ft.8ins., although the wheelbase was 9.5ins. longer at 11ft.10.5ins. The chassis overall length was 15ft.8ins.

The chassis and mechanics underwent extensive testing before the car, complete with bodywork, was officially revealed to the public at the Southport International Motor Rally in September 1928. It was duly awarded the 150 guineas Silver Trophy for the finest car in the Concours d'Elegance, as well as the Premier Prize for the best car in its class and a Banner of Honour. That particular body design was known from then on as the Southport model. Although the event was comparatively small, the response surprised the Company, as an initial batch of chassis which had already been produced was immediately sold, and

A 30 h.p. Lanchester-bodied limousine.

similarly a second batch. Each chassis sold at £1,325, which must have provided a great boost to the Company.

The Straight Eight was next shown at the Olympia Motor Exhibition the following month. Here it was open to inspection by thousands of visitors and the Company's single 30 h.p. exhibit was described as: "A 4/5 seater owner-driver fabric saloon bodied by Lanchester, finished in Brown with cream mouldings at the waist and cream wheels."

Fabric-covered panels were the height of fashion at this time, as the Weymann or Flexible style of coachwork had captured the public's imagination. It had evolved from the need for a flexible body on most other chassis, which did not have the Lanchester's rigidity, in order to avoid cracking and distortion of the bodywork. Lanchesters followed the trend by making their new car appear the same, although this was for cosmetic purposes only. The fabric adhered to 20 gauge aluminium coachwork panels, 0.036ins. thick, and the effect gained general approval. It was not a cheap car, however, as the complete Southport model was advertised at £1,830.

The praise given to the new car by the public and motoring press alike was tremendous and future sales prospects looked promising, in spite of the recession. Following numerous road tests on the 30 h.p., each and every magazine article was couched in glowing terms. They contained statements such as "..... a dazzlingly impressive motor car ..... the most interesting, fascinating English motor car that I have ever driven ..... American, French and Italian straight-eights; none of them has been so good ..... an excellent example of the best English motor design ..... without exaggeration a beautiful piece of workmanship ..... as a performer there is little to be desired ..... to drive it is sheer pleasure ..... indubitably one of the finest examples of advanced automobile engineering ever produced in Great Britain ..... the nearest approach to Atlantic liner smoothness that I have ever experienced on a motor vehicle ..... honest-to-goodness luxury." The words of the Automotor Journal summarised for them all: "I have never sat behind a silkier, stealthier motor, and the rest of the car is pure, unadulterated and therefore impeccable Lanchester."

Despite this fulsome praise, some employees were worried that sales of Lanchester cars

A 30 h.p. four-door sports saloon.

were too low and left the Company for better prospects elsewhere. A few key members of staff thought that there was more wrong with the Company than just the national and worldwide recession. They felt that the Company policy was at fault for lack of investment in the products and in the staff, whose wages and conditions were poor compared to those offered at other companies.

Fred himself felt unappreciated, yet again, by the Board, who had never thanked him for his efforts. Jasper Cooper took umbrage with his Directors about the cavalier treatment meted out to Fred with regard to his contribution to the Straight Eight design. In a letter to him, Jasper Cooper protested, "I recall with astonishment, in fact a stronger word would be more suited to the facts, that the Directors of the Company at that time made no acknowledgement to you for your idea nor for your practical assistance and advice in carrying it out." The letter continued, "It naturally followed that from about that time the Directors, as a Board, lost your goodwill and assistance. It seems to me that there followed a slow but continuous decline in the fortunes of this fine old pioneer Company." Jasper Cooper had also become completely disillusioned with the Board and left in 1929. "As time passed I lost confidence in those who were directing the policy of the Company and resigned my position." He was replaced as Works Manager by the popular Arthur Bird. Eric Sisman, who had faithfully served the Company since 1921, was another who resigned in protest at the lack of direction of the Company.

Lanchester fabric-roof saloon coachwork on a 30 h.p. chassis.

George and Frank's constructive proposals were continually thwarted, and the Company Secretary was "a continual thorn in George's side", blocking practically all his ideas for the Company's advancement.

In spite of all this, the 30 h.p. car sold at the steady average rate of one per week which, in those days of recession, was not quite as bad as it may appear now. There were too many luxurious and expensive cars chasing too few customers; indeed, a highly-skilled worker at the Company would have had to work for ten years or more before he would have earned enough money to buy even the cheapest Straight Eight Lanchester.

The Company tried to find wealthy customers at the 1929 Motor Exhibition. One of their two chassis on display was clothed in brown coachwork by Hooper with a sedanca folding-extension roof over the driver, and with a rear compartment of the highest standard, including pile carpet, silver and ivory fittings and two electric lamps; all for £2,465. An identical chassis was adorned with a blue five-seater Windover body, blue interior and even a blue steering wheel, dashboard, brake and gear levers. This was slightly cheaper at £2,100.

The Windover stand displayed a 30 h.p. Lanchester with their own sedanca coupe-deville coachwork. It was painted grey with black wings and was finished inside and out in the most luxurious way. No price was given on this car, but a 30 h.p. landaulette, bodied by Rock, Thorpe and Watson Coachbuilders, was for sale at the very competitive price of £1,912. This car was finished in crimson-lake with black wings, and included every conceivable luxurious extra.

After making such an effort to promote Straight Eight sales at the motor shows, it seems strange that potential customers from all over Great Britain and abroad had to travel to Birmingham to actually view a 30 h.p. car. When he was asked by The Auto magazine for such an inspection, Frank himself answered at his Bond Street headquarters, explaining that they were unable to keep a chassis in London; no doubt this was the case for the Manchester depot also. A visit to Birmingham would, however, include a conducted tour of the main works, as well as the coachbuilding works, before the potential customer enjoyed a thirty-mile return trip in a Lanchester. In fairness to the Company, many motoring correspondents insisted that it would be hard for anyone to resist a car after such treatment.

Ways to improve the Straight Eight were continually being considered, and a number of minor changes took place in 1929 and 1930. One such difference was the introduction of a new exhaust system. In order to reduce back pressure and obtain a higher piston speed and

Modified exhaust and bulkhead layout of the 1930 Lanchester 30 h.p., also showing the engine's oil filler (D) and the automatic by-pass oil-cleaner (E).

power, the original system of two central, vertical down-pipes, similar to the 21 h.p. engine, was abandoned in favour of eight, separate pipes converging into each other at different intervals to obtain a better flow of exhaust gases.

The engine compartment bulkhead was altered from its angled design to being completely upright. The fuse box and electrical connections were bolted to this, in preference to the previous aluminium box housed on a horizontal shelf, on top of the bulkhead. A standard canister was used with the Autovac instead of being encased within Lanchester's own aluminium housing, as had been used previously on the 21 h.p.

The frontal aspect of the car was given a face-lift, by covering the 1in.-higher radiator with a chromium-plated shroud. This incorporated the new thermostatically controlled, louvred radiator shutters; the water temperature was able to be checked by a dashboard temperature gauge. A ribbed dumb-iron fairing was added, as well as a horizontal chrome-plated bar supporting the new Lucas P100 twin-bulb headlamps.

Windover-bodied limousine with dummy hood-irons and a sun-roof, on a 30 h.p. chassis.

Dunlop centre-lock wheels were included in the specification, as opposed to the earlier Rudge-Whitworth type, and 6.00ins. by 20ins. well-base tyres were standard fitting; reports showed an improved turning circle of 47ft. Connected to the massive chassis were new spring mountings and on the front were Luvax shock absorbers; Lanchester cars were never fitted with rear shock absorbers. A belt-and-braces job was applied to the chassis by including underhung struts and tie-rods, helping to resist any twisting motion.

Many engine, transmission and chassis fittings previously produced in aluminium were made lighter by the introduction of Elektron metal. It weighed about half that of aluminium and was used extensively throughout the car, including the imposing valve-rocker cover with LANCHESTER emblazoned on each side. Even so, the 30 h.p. car weighed between 38 and 44 cwt, depending on the style of coachwork used. The engine, which had a compression ratio of 4.8:1, would idle at 150 r.p.m. At 1,000 r.p.m. it produced 35 b.h.p., at 3,000 r.p.m. it produced 90 b.h.p. and at its maximum of 3,400 r.p.m. it produced nearly 100 b.h.p. The maximum road speeds on each gear were 21 m.p.h. on first, 33 m.p.h. on second and 55 m.p.h. on third; George's personal straight eight tourer was timed at 89 m.p.h. on top gear. A car's performance was checked at the Brooklands Motor Course and on the 154ft. long Test Hill, whose slope averaged 1 in 5 and climaxed in a 1 in 4 incline, the climb was made easily in second gear in 17.4 seconds. In first gear this was reduced to

THE EIGHT-CYLINDER 30 H.P.

George Lanchester's daughter, Nancy, in her 30 h.p. Lanchester-bodied sports-tourer of the late 1930s.

15 seconds. Both times compared favourably to an early 40 h.p. car which averaged 17.92 seconds.

In 1930, Captain Malcolm Campbell, holder of various world speed records, drove a Straight Eight Lanchester which had already covered 20,000 miles and issued a factual report. His comments included: "It is certainly the quietest motor I have ever sat behind that has been equipped with this type of overhead gear ... the engine is vibrationless ... a remarkable degree of flexibility, in fact almost astonishing ... the acceleration is exceedingly good ... the gearbox and back axle are dead quiet ... the clutch is extremely nice to handle ... the suspension could not be bettered ..." However, in his opinion, "the brake and change speed levers are rather close to each other, and one is inclined to rub ones knuckles between the two ..." He also complained, "A carriage key has to be used, and it takes longer than usual to open the bonnet." The Lanchester Motor Company deliberately used a key, uncommon to other makes, for reasons of security, and the aluminium bonnet top did indeed take marginally longer to remove than one of the top-hinged type. This minor delay was more than offset by the facility to remove all three bonnet sections to allow complete access to the engine; something that could not be readily achieved with the top-hinged type.

Other small items of complaint had already been amended on the newer 30 h.p. version and Captain Campbell conceded that the "slight tendency to wag its tail when cornering fast over wet and slippery roads" was due to the "disappearing treads of the back tyres".

The last criticism he made referred to the lack of a one-shot chassis lubrication system. Lanchesters preferred their bearing surfaces to be greased manually, via extension pipes from an easily accessible grouping of nipples. Rolls-Royce, who adopted the method espoused by Captain Campbell, actually later abandoned the system. Apart from these few criticisms, Captain Campbell's report on the 30 h.p. Straight Eight echoed the theme of praise and confidence in the car set by a score of other correspondents.

Pride in the Company's product was evident at the 1930 Motor Show, where no fewer than eight Lanchesters were on display, including four Straight Eights. Of the two on the Lanchester stand, neither had Lanchester coachwork, showing the Company's acceptance of customers wishing to order outside coachwork. The Hooper-bodied limousine was "painted blue, picked out with yellow line" and complemented the Windover sedanca coupe-de-ville finished in two shades of brown. Prices were £2,300 and £2,435 respectively, and were inclusive of the very finest of fittings. The same applied to the 30 h.p. on Windover's own stand, which was a landaulette finished in yellow and black, as well as the Maythorn stand's ivory and black limousine, which was offered for sale at £2,045.

1931 30 h.p. chassis number 8096 fitted with sedanca-de-ville coachwork by Barker. (See appendix.)

Two other Lanchesters on display were part of the Society of Motor Manufacturers and Traders' historical collection of ten early motor vehicles. In pride of place as Exhibit Number One was Lanchester's 1895 car, with its five-seater Stanhope Phaeton body, and Exhibit Number Ten was a 12 h.p. Lanchester twin-cylinder tonneau.

These veterans demonstrated to everybody their durability and quality; attributes which were proudly continued on the current models and should have assured the future success of the Company. Sadly, this was not to be the case, and this magnificent display of Lanchesters old and new proved to be the Company's swan song.

LANCHESTER
PRE-EMINENT AMONG THE WORLD'S FINEST CARS

Chapter Twenty-Five

THE END OF THE ROAD

The economic climate was worsening by the day. Despite the fact that present historians of such matters regard the collapse of world trade at this time as something that should never have happened, it was nevertheless reputed to be the worst trade depression since that following the Napoleonic Wars.

The motor trade had suffered badly: the one hundred and five British car manufacturers in 1920 had dwindled to just forty-two by late 1929, and this was a year or so before the full impact of the depression was felt. Most of the bankrupt companies were of small and medium size and had enjoyed a large customer base, compared to the luxury motor trade where the demand for high-priced cars diminished to a mere trickle.

The 30 h.p. limousine owned by Sir George Bernard Shaw, one of the many famous Lanchester customers.

The continuing glowing reports of the Lanchester cars, as well as of their coachwork, had kept the majority of the workforce fairly confident of the future, and this was encouraged by news of the War Office order, in 1930, of the seven additional armoured cars. Due to the ill-health of the Lanchester Motor Company's Managing Director and Chairman, Hamilton Barnsley, in June Frank Lanchester was appointed Acting Managing Director. Frank immediately persuaded the Company's bank to supply extra funding for the order and, in return, he began a cost-cutting exercise to lessen the Company's overdraft.

In spite of the resulting improvements over the months, the bank called for urgent meetings to discuss the position. Frank, still as Acting Managing Director, and other Board members, would have stressed again the world acclaim for their products and the fact that the Company was still favoured by Royalty, the rich and the famous, both at home and abroad.

Lanchester devotees, the Duke and Duchess of York, visiting Kings College Hospital in their 40 h.p. landaulette, complete with chromium-plated radiator surround; their other 40 h.p. follows.

The bank pointed out the inescapable fact that the Company's overdraft still amounted to £28,000. Even though the Company was trading with material assets many times in excess of the overdraft, the bank was swept along in the maelstrom of financial short-sightedness and gave the Company just over a fortnight to clear its debt.

The intervening period was frantic as options were considered by the Board. Of their wealthy customers, it seemed that the time-limit precluded any of them from injecting money into the Company, although at least one was known to be very enthusiastic. It was reported that the Napier Company, which had ceased production of their own cars in 1924 to concentrate on aircraft engine design, were keen to acquire part or all of the Company. Frank Lanchester recalled that Rolls-Royce were also more than interested, but the drastic deadline beat all concerned.

There was no doubt among the Lanchester employees of the time that the nearby Birmingham Small Arms engineering group, who had a massive account at the same bank as Lanchester, applied persuasion to the bank to withdraw its support of their beleaguered neighbour, in order to make a swift take-over bid themselves. The degree of truth in this rumour will remain unknown, but the fact is that, on the 28th December 1930, Hamilton Barnsley signed over the Company to B.S.A. lock, stock and barrel, pending shareholders' approval. Hamilton Barnsley, who had been a Director at the Lanchester Engine Company, had led the Lanchester Motor Company as Managing Director since its formation and had also been Chairman since 1924, took the matter badly and was, according to the foreman, Sid Dangerfield, "a shattered man". The next day he died of a heart attack.

The 27th Ordinary General Meeting of the Lanchester Motor Company was held on the 31st December. George, Frank, and Fred were all present to pay their respects to Hamilton

Loyal Indian Maharajahs continued to purchase Lanchesters. This 21 h.p. tourer was fitted with Venetian blinds at the windows to offer complete privacy.

Barnsley. Fred, as a shareholder, and his brothers in their capacity as Directors, discussed the prevailing situation. The Board called on Frank Berresford to become Chairman and Frank Lanchester to take over as Managing Director forthwith. Accordingly, Frank personally interviewed each of the shareholders and advised them of B.S.A.'s progressive plans for the future. He asked their opinion on the proposed take-over, which would involve B.S.A. purchasing all of the Lanchester Motor Company's shares. In addition, Frank and the Board of Directors had been informed that "financial support would be given to allow them to continue in charge at the Montgomery Street Works in Birmingham". Faced with the bank's inflexible demands, the shareholders, including Fred, realised that financially the arrangement was better than the alternative of bankruptcy. Consequently, in order to save the name of Lanchester, they agreed unanimously to the sale of their shares to the B.S.A. Company. Although deeply saddened, the Board also accepted the take-over as inevitable. The B.S.A. board meeting minutes showed that Percy Martin, one of the directors on the board of B.S.A., made the arrangements for the purchase of the whole of the issued share capital. This consisted of 32,167 ordinary shares at £1 each and 32,154, 6% cumulative preference shares of £1 each. All these shares were purchased for only 8 shillings and 3 pence each, merely 41.25% of their face value, costing the B.S.A. Company £26,532.8s.3d. Coupled with sundry small expenses and the Lanchester overdraft, of £28,000, the total sum of £55,000 was reached.

On 9th January 1931, the Autocar wrote: "A Lanchester Development - One of the oldest concerns in the British automobile industry, the Lanchester Motor Co. Ltd., has been acquired by the Birmingham Small Arms Co. Ltd. This development is in the nature of an act of industrial rationalisation, the acquiring company being what is known as a parent, or holding, company, and the transaction just concluded not implying any particularly important changes, nor any amalgamation in the ordinary sense of that term. In its new status the Lanchester Motor Co. now becomes a member of the group of companies under the aegis of the B.S.A. Co. Ltd."

The B.S.A. firm had been founded in 1861 and had become very wealthy through the sale of armaments. The Company had diversified and also produced bicycles, machine tools and a great many other products. Their motor car production had begun in 1907, and two years later had gained greater status when they acquired the Daimler Motor Company, and coincidentally the services of Fred Lanchester as Consulting Engineer and Technical Adviser.

There is no doubt that their experience of Fred's work over two decades would have

The last coachwork design from Lanchester before the Works was disbanded: the two-door, four-seater sporting-saloon on the 30 h.p. chassis. It was displayed at the Scottish Automobile Show in November 1930, priced at £2,025.

encouraged B.S.A. to mount a take-over bid. There was also the consideration that it offered a face-saving way of discontinuing the sleeve-valve engines, which B.S.A. and Daimler both used, and replacing them with the Lanchester-style, poppet-valve engines instead.

Inevitably, the acquisition resulted in comparisons being drawn between B.S.A.'s Daimler and Lanchester companies. They were both pioneering British companies, with Lanchester having produced the first four-wheel, petrol-driven, all-British car in 1895 and Daimler having been the first registered motor company in Britain in 1896, although their Coventry works did not produce a car of their own design until 1897. Both companies' cars had been produced as quality items and had been designed on rather conservative lines. Furthermore, both the Daimler and Lanchester companies had been struggling financially since the middle of the 1920's. Ironically, Lanchester's operating costs were actually lower than those of Daimler, as was the amount of their overdraft at the bank. The crucial difference between the companies, however, was that Daimler had the protection of the giant B.S.A. Group behind it, whereas Lanchester had no such support and had been operating on a shoestring since its inception. To all then, it appeared quite logical and appropriate that, if Lanchester had to submit to 'industrial rationalisation', there could be no better parent company to co-exist with than B.S.A.

The first statements from the B.S.A. Company after the take-over offered much optimism for the future. A weakness in the Lanchester sales operation had always been the lack of outlets and in view of this it was announced that Stratton-Instone Limited, with their national network, would be made sole distributors for Lanchester cars, along with their existing B.S.A. and Daimler vehicles. Frank Lanchester was "invited to join their Board of Directors" and assured all the Lanchester customers shortly afterwards, by letter, that the Lanchester staff in London, Birmingham and Manchester would continue "to give you the closest personal attention". He also confidently wrote that the changes made "cannot fail to be beneficial to you and to all users of Lanchester cars". He had scarcely finished writing, it seems, before the new bosses began to alter their plans. The "comfortable marriage" was not to be; the Birmingham operation would not continue as before. The coachbuilding works would not be building any more fine carriages. Armourer Mills would not see any more Lanchester precision engineering. The positions of George and Frank would be devalued. George and Frank's previously confident words, "this purchase will in no way mean the sinking of the Lanchester identity and interests", began to look rather misplaced.

A further statement was issued by B.S.A. stating that, "It has been decided to transfer the offices and head works of the former concern to Sandy Lane, Coventry, the headquarters of the Daimler Company Limited, which is also controlled by B.S.A. Limited." No doubt trying to alleviate the worries of the Lanchester customers, it continued, "Mr. G.H.Lanchester will be in charge of affairs at Coventry. Lanchester cars will retain their own identity; in no sense will there be any fusion of the Lanchester and Daimler cars." To maintain some semblance of continuity, it was decided that the factory's name of Armourer Mills should front the new Coventry address, although this was a short-term measure.

Gradually realisation dawned: the fine words had meant little. As the full force of the transaction took effect, George and Frank were devastated. Neither could face their dedicated and loyal workforce of whom it had been said more than once, "They were like one large happy family."

Percy Martin, the Managing Director of Daimler, sent his Works Manager over to Armourer Mills. Bill Hancock had already worked with Fred at Daimler, and had the greatest respect for him. He also knew the proud record of the Lanchester Company and fully understood the reluctance of the brothers "to discharge staff and men with whom they had worked for so many years"; indeed, one elderly labourer had been with the firm since 1895. Bill Hancock carried out his morbid duty, "the worst job in my industrial life", and addressed the workforce: the old Lanchester Motor Company was effectively dead. Each employee received a signed letter from the old Company, on Armourer Mills, Birmingham, headed paper, thanking them for their loyal service. They also received two weeks' wages as a goodwill gesture from the new owners, but desperate times must have awaited the Lanchester craftsmen in those days of massive unemployment. Only a handful were transferred to Daimler's Coventry works, with a few others moving to B.S.A.'s Birmingham factory. Repeated mass sackings took place at Armourer Mills, as well as at the Fallows Road coachbuilding works. The Birmingham operation was shut down by 2nd February 1931.

The Lanchester service department at Fallows Road was reprieved for a short while and utilised by the Daimler agents, Stratton and Instone Limited.

The B.S.A. Company directors were certainly more astute businessmen than their

A 30 h.p. four-door sports-tourer of 1931, which was George Lanchester's personal car.

counterparts at Lanchester had ever been; they knew the true value of their purchase and profited handsomely from the deal. After B.S.A. and Daimler had had the pick of the precision machinery, tooling and equipment throughout the Birmingham works, the surplus was sold off.

The 'old' Lanchester Motor Company's proud boast in all its literature, referring to earlier models dating back to the previous century, was that they "had been produced in quantities on the interchangeable system, and we still manufacture and stock replacement parts". Of these replacement parts, those relating to the 40 h.p., 21 h.p. and 30 h.p. cars were removed to Daimler's stores in Coventry. The large stock of parts for all previously-built models, which kept a great many cars in regular use, were unceremoniously thrown from the first floor store to smash on the ground beneath, before being carted away and sold as scrap metal.

Regarding the Fallows Road coachbuilding operation, there were about a dozen bodies under construction at the time, some nearing completion on their chassis. All were taken away to be finished at the Daimler Works. There were also numerous car chassis and armoured car chassis at Armourer Mills, at different stages of construction, and these too were removed to Coventry. With everything gone, the surplus empty buildings were then sold (although Armourer Mills was retained and used for a while as a store for B.S.A. bicycles). For these tangible assets alone, it was estimated that the B.S.A. Company received double the total amount they had paid for Lanchester.

That was not the end of it. As part of the take-over arrangements, B.S.A. had stipulated that George and Frank remain with the Company, to ensure continuity. B.S.A. received great commercial advantage from employing George as Chief Engineer of the new Lanchester Motor Company, and Frank Lanchester as Sales Director of their subsidiary, Stratton and Instone Limited, although their annual salaries were immediately slashed from £2,000 to £1,000, causing further resentment. Another enormous benefit for B.S.A. was that the famed name of Lanchester could be applied to a new range of engines and cars designed by Daimler's Laurence Pomeroy, albeit with a small input by George. The whole operation by B.S.A. was ruthlessly executed and demonstrated beyond doubt that there was no room for

> **THE LANCHESTER MOTOR Cº LTD**
>
> All Cars and other goods sent to us for Repairs, Storage, etc., are held only at Owner's risk and we accept no responsibility for damage or loss to same arising from fire or any unforeseen causes.
>
> All communications to be addressed to the Company.
>
> Customer's Cars are driven by our representatives entirely at Customer's own risk.
>
> CONTRACTORS TO THE ADMIRALTY, WAR OFFICE & AIR MINISTRY.
>
> HEAD OFFICES & WORKS, MONTGOMERY STREET, BIRMINGHAM.
> BIRMINGHAM REPAIR DEPT., MONTGOMERY STREET.
> LONDON DEPOT & SHOW ROOMS, 95, NEW BOND STREET, W.1
> LONDON REPAIR DEPT., STAR ROAD, WEST KENSINGTON, W.14
> MANCHESTER DEPOT & REPAIR DEPT., 88, DEANSGATE.
> TELEGRAPHIC ADDRESSES, MOTIVITY, BIRMINGHAM. REBALANCE, WESDO, LONDON. MOTIVITY, MANCHESTER.
>
> TELEPHONES, 0526 (4 LINES) VICTORIA.
> 6138 (2 LINES) MAYFAIR.
> 5296 FULHAM.
> 1763 CENTRAL.
> CODES, A.B.C. 5TH EDITION. WESTERN UNION. PRIVATE. MARCONI.
>
> ARMOURER MILLS, MONTGOMERY STREET, BIRMINGHAM.
>
> 4th February, 1931.
>
> Mr. S. Dangerfield,
> 133, Warwick Road,
> SPARKHILL.
>
> Dear Sir,
>
> The Directors of this Company wish to place on record their appreciation of your loyal services given over the past 17 years, and regret that a change in our manufacturing plans renders it necessary for you to sever your connection with us.
>
> If required, we shall be pleased to assist you in any way possible to obtain employment with another Company.
>
> Yours faithfully,
> For THE LANCHESTER MOTOR CO. LTD.
>
> Secretary.

One of the last letters sent out on Lanchester's Montgomery Street paper, informing one of their foremen that he was no longer required.

sentiment in business. (Indeed, a short time later the Rolls-Royce acquisition of cash-starved Bentley Motors resulted in that marque also losing its identity, their 8-litre spares being smashed and scrapped, and W.O. Bentley himself being unwanted.)

The Lanchester 30 h.p. and 21 h.p. chassis and bodies which had been taken to Coventry were finished and made ready for sale. Daimler's own service department started maintaining the old Lanchesters for guarantee, repair and servicing work and the Fallows Road service department was then sold.

The October 1931 Motor Exhibition was the very last chance to inspect the features of a 30 h.p. Straight Eight Lanchester, when the 'new' Lanchester Motor Company offered for sale a beige and brown Windover coupe-de-ville for £1,975. It was displayed alongside four examples of a Daimler-built Lanchester 18 h.p. model, which was ironically just the type

> **CHOICE of a motor car by name at once takes the buyer to the makers of the first British car—the Lanchester.**
>
> **Selection on performance, with refinement, arrives at the selfsame decision—the modern Lanchester.**
>
> # Lanchester
> *The Pioneers of the British Motor Industry*
>
> THE LANCHESTER MOTOR Co., Ltd.,
> ARMOURER MILLS, BIRMINGHAM.
> London: 95, Bond Street, W.1.
> Manchester: 88, Deansgate.

The last proud advertisement from The Lanchester Motor Company of Birmingham before the B.S.A. take-over.

of car George and Frank had been advocating since 1926. Although the 21 h.p. and 30 h.p. chassis were still advertised, the 'new' Company was eager to sell 'old' stock and the 30 h.p. chassis price was dropped by £275 to £1,050 accordingly.

The last of the pre-B.S.A. Lanchesters left the Daimler works in 1932, taking with it thirty-seven years of pioneering achievement and excellence, leaving the road clear for a new chapter to be forged in the Lanchester legacy.

Appendix

SURVIVING LANCHESTERS

The first Lanchester car produced, in 1895, was classified as Number One. All subsequent twin-cylinder cars continued with consecutive chassis numbering, although an exact quantity of those produced is not known. A Lanchester Engine Company advertisement of August 1904 stated: "300 Lanchester cars are now running in England, Scotland and Ireland." Many twin-cylinder cars had been exported before this date, however, and with four different models being produced until 1906, plus a few afterwards built to special order, this number was obviously exceeded. Although one historian later wrote that only 350 twin-cylinder cars were actually built, a Lanchester Motor Company advertisement of 1930 stated: "Some 500 cars of this type were manufactured." The details of the surviving twin-cylinder cars are believed to be as follows:-

	Model	Chassis	Wheelbase	Engine	Year	Reg.No.	Coachbuilder/Body
1	8 hp	2	5ft.10ins.	3	1897	A 259	Lanchester phaeton
	The National Museum of Science and Industry, London, England.
2	10 hp	88	7ft.9ins.	88	1901	O 166	tonneau
	G.Pilmore-Bedford, England.
3	10/12	91	7ft.10in.	91	1901	FRW 766	Lanchester tonneau
	Jaguar-Daimler Heritage Trust, Coventry, England.
4	10/12	151	7ft.10ins.	151	1902		
	G.Pilmore-Bedford, England.
5	12 hp	241	7ft.10ins.	248	1903	AR 621	Lanchester tonneau
	J.Hutton-Stott, Berkshire, England.
6	12 hp	321	7ft.10ins.	268	1904	O 778	Lanchester tonneau
	Unrecorded owner
7	12 hp	339	7ft.10ins.	274	1904	O 928	Lanchester tonneau
	C.Matthews, Buckinghamshire, England.

From the first experimental 20 h.p. cars of 1904, whose numbers may have started at 501, all chassis were numbered consecutively, regardless of model type. No production figures are known for the various models, but collectively the number of chassis produced up to August 1919 was 1,632. The details of the remaining cars are believed to be as follows:-

	Model	Chassis	Wheelbase	Engine	Year	Reg.No	Coachbuilder/body
8	20 hp	578	9ft.5ins.	576	1904	O 1230	tourer
	Jaguar-Daimler Heritage Trust, Coventry, England.
9	20 hp	727	9ft.5ins.	501	1907	M 2096	tourer
	R.Brown, Warwickshire, England.
10	20 hp	747	9ft.5ins.	746	1907	LN 7815	Lanchester landaulette
	The Yorkshire Car Collection, Keighley, West Yorkshire, England.
11	20 hp	771	9ft.5ins.	766	1908	D 4227	Lanchester landaulette
	G.Pilmore-Bedford, England.
12	20 hp	829	9ft.5ins.	869	1908	O 4498	Lanchester phaeton
	Hull City Museums and Art Gallery, Yorkshire, England.

	Model	Chassis	Wheelbase	Engine	Year	Reg.No	Coachbuilder/body
13	28 hp	970	11ft.5ins.	1015	1909	O 5339	Lanchester landaulette

G.Milligen, Norfolk, England.

	Model	Chassis	Wheelbase	Engine	Year	Reg.No	Coachbuilder/body
14	20 hp	979	9ft.5ins.	974	1910	E 1061	tourer

J.Zimbler, England.

	Model	Chassis	Wheelbase	Engine	Year	Reg.No	Coachbuilder/body
15	28 hp	1013	11ft.5ins.	1053	1910	EC 626	Lanchester landaulette

Beyring Motor Museum, California, U.S.A.

	Model	Chassis	Wheelbase	Engine	Year	Reg.No	Coachbuilder/body
16	28 hp	1047	11ft.5ins.	1084	1910	E 6103	Lanchester limousine

W.Matthews, Buckinghamshire, England.

	Model	Chassis	Wheelbase	Engine	Year	Reg.No	Coachbuilder/body
17	38 hp	1154	11ft.7ins.	1197	1912	AC 2746	Lanchester limousine

J.Matthews, Buckinghamshire, England.

	Model	Chassis	Wheelbase	Engine	Year	Reg.No	Coachbuilder/body
18	38 hp	1367	11ft.7ins.	1415	1912	LB 7923	Lanchester limousine

The National Museum of Science and Industry, London, England.

	Model	Chassis	Wheelbase	Engine	Year	Reg.No	Coachbuilder/body
19	25 hp	1434	10ft.7ins.	1497	1912	BJ 1950	Lanchester tourer

G.Pilmore-Bedford, England.

	Model	Chassis	Wheelbase	Engine	Year	Reg.No	Coachbuilder/body
20	38 hp	1460	11ft.7ins.	1501	1913	K 6731	Lanchester tourer

W.K.Haines, Ohio, U.S.A.

	Model	Chassis	Wheelbase	Engine	Year	Reg.No	
21	Sport 40	–	12ft.8ins.	1569	1915	O 7574	

M.Wurst, Australia.

The 40 h.p. range continued the consecutive sequence of chassis numbers, the first one being number 1633. The last one produced was thought to be number 2025, making a total of 392. The details of the remaining cars are believed to be as follows:-

	Model	Chassis	Wheelbase	Engine	Year	Reg.No	Coachbuilder/body
22	40 hp	1634	11ft.9ins.	1635	1919	XC 5629	Lanchester saloon

L.Aas, Oslo, Norway.

	Model	Chassis	Wheelbase	Engine	Year	Reg.No	Coachbuilder/body
23	40 hp	1755	11ft.9ins.	1755	1921	HN 69	Lanchester tourer

J.Tattershall, Lancashire, England.

	Model	Chassis	Wheelbase	Engine	Year	Reg.No	Coachbuilder/body
24	40 hp	1762	12ft.6ins.	1768	1921	KE 5050	Lanchester tourer

G.Pilmore-Bedford, England.

	Model	Chassis	Wheelbase	Engine	Year	Reg.No	Coachbuilder/body
25	40 hp	1768	11ft.9ins.	1771	1921	SD 4336	

M.Gray, New South Wales, Australia.

	Model	Chassis	Wheelbase	Engine	Year	Reg.No	Coachbuilder/body
26	40 hp	1774	12ft.6ins.	1746	1921	XF 7024	Gill saloon

J.Oates, Hampshire, England.

	Model	Chassis	Wheelbase	Engine	Year		
27	40 hp	1783	11ft.9ins.	1774	1921		

J.Winter-Irving, New South Wales, Australia.

	Model	Chassis	Wheelbase	Engine	Year	Reg.No	Coachbuilder/body
28	40 hp	1789	11ft.9ins.	1778	1921	S 2455	Lanchester tourer

P.Harwin, New South Wales, Australia.

	Model	Chassis	Wheelbase	Engine	Year	Reg.No	Coachbuilder/body
29	40 hp	–	–	–	1921	Gondal	Lanchester tourer

H.H. The Maharajah of Gondal, India.

	Model	Chassis	Wheelbase	Engine	Year		
30	40 hp	1804	12ft.6ins.	1880	1921		

A.Falstein, New South Wales, Australia.

	Model	Chassis	Wheelbase	Engine	Year	Reg.No	Coachbuilder/body
31	40 hp	1843	11ft.9ins.	1836	1921	Q 52	Lanchester tourer

E.Eby, Michigan, U.S.A.

	Model	Chassis	Wheelbase	Engine	Year		
32	40 hp	1844	11ft.9ins.	1844	1921		

M.Gray, New South Wales, South Australia.

	Model	Chassis	Wheelbase	Engine	Year	Reg.No	Coachbuilder/body
33	40 hp	1915	12ft.6ins.	1872	1924	MPA4598	Lanchester limousine

M.Dollershall, Germany.

	Model	Chassis	Wheelbase	Engine	Year	Reg.No	Coachbuilder/body
34	40 hp	1925	15ft.9ins.	1919	1924	ALWAR 1	Lanchester landau

H.H. The Maharajah of Alwar, India.

	Model	Chassis	Wheelbase	Engine	Year	Reg.No	Coachbuilder/body
35	40 hp	1936	12ft.6ins.	1916	1924	TDP 5	Lanchester limousine

E.Ryan, Dublin, Eire.

	Model	Chassis	Wheelbase	Engine	Year	Reg.No	Coachbuilder/body
36	40 hp	1953	12ft.6ins.	1953	1925	MB 9999	Windover limousine

T.Jefferies, England.

SURVIVING LANCHESTERS

	Model	Chassis	Wheelbase	Engine	Year	Reg.No	Coachbuilder/body
37	40 hp	1965	12ft.6ins.	1945	1925	CF 5051	Barker saloon
				J.Clouston, New Zealand.			
38	40 hp	1968	11ft.9ins.	1897	1925	DS 8015	Lanchester tourer
				P.Noble, Essex, England.			
39	40 hp	1984	12ft.6ins.	1963	1926	MK 2428	Lanchester limousine
				A.Simms, Yorkshire, England.			
40	40 hp	1999	12ft.6ins.	1971	1927	YF 8147	Lanchester saloon
				D. and F. Simms, Kent, England.			
41	40 hp	2009	12ft.6ins.	1994	1928		Lanchester tourer
				A.Falstein, New South Wales, Australia.			
42	40 hp	2010	11ft.9ins.	1922	1928	KO 9985	Lanchester tourer
				D.Bonhomme, Jersey, U.K.			

The 40 h.p. based Lanchester Armoured Cars were, it is believed, numbered from LAC 1. Only one remains of 39 built:-

43	40 hp	LAC34	14 feet	LAC17	1931	MT 9755	Armoured Car
				The Tank Museum, Wareham, Dorset, England.			

The 21/23 h.p. range used their own numbering sequence starting with number 3001. It is thought that 735 were produced. The details of the remaining cars are believed to be as follows:-

	Model	Chassis	Wheelbase	Engine	Year	Reg.No	Coachbuilder/body
44	21 hp	3057	10ft.9ins.	3060	1924	I 486	Lanchester f/h coupe
				T.Wilson, Victoria, Australia.			
45	21 hp	3077	10ft.9ins.	3077	1924	NP 4954	Lanchester limousine
				Author, Gloucestershire, England.			
46	21 hp	3081	10ft.9ins.	3078	1924	XW 5524	Lanchester landaulette
				T.Hutton-Stott, Berkshire, England.			
47	21 hp	3086	10ft.9ins.	3091	1924		tourer
				P. and R.Thian, Christchurch, New Zealand.			
48	21 hp	3087	10ft.9ins.	3090	1924		Lanchester tourer
				A.Falstein, New South Wales, Australia.			
49	21 hp	3108	10ft.9ins.	3114	1924		Lanchester d/h coupe
				P. and R.Thian, Christchurch, New Zealand.			
50	21 hp	3123	10ft.9ins.	3129	1924	VP 9860	Penman landaulette
				B.Willis, Cork, Eire.			
51	21 hp	3133	10ft.9ins.	3128	1924	NP 5384	Lanchester tourer
				D.Russell, England.			
52	21 hp	3153	10ft.9ins.	3156	1924		
				Author, Gloucestershire, England.			
53	21 hp	3180	10ft.9ins.	3183	1924	SB 27	Lanchester d/h coupe
				M.Stanton, West Sussex, England.			
54	21 hp	3188	10ft.9ins.	3188	1924	OM 7357	Lanchester limousine
				M.Shapland, Avon, England.			
55	21 hp	3230	10ft.9ins.	3222	1925	NE 3337	Lanchester tourer
				D.Shapland, Avon, England.			
56	21 hp	3242	11ft.1in.	3241	1925		Lanchester limousine
				P. and R.Thian, Christchurch, New Zealand.			
57	21 hp	3272	11ft.1in.	3271	1925	YF 6785	
				Author, Gloucestershire, England.			
58	21 hp	3284	10ft.9ins.	3292	1925	HC 6001	Caffyn saloon
				G.Allen, West Sussex, England.			

	Model	Chassis	Wheelbase	Engine	Year	Reg.No	Coachbuilder/body
59	21 hp	3320	10ft.9ins.	3304	1926	PE 8988	
	J.Weber, Pennsylvania, U.S.A.							
60	21 hp	3334	11ft.1ins.	3333	1926	TD 5549	Lanchester tourer
	G.Rutter, Surrey, England.							
61	21 hp	3355	11ft.1in.	3345	1926	IK 9849	Hooper landaulette
	L.Lawlor, Dublin, Eire.							
62	21 hp	3390	Shortened	3390	1927	RJB 848	Moyle tourer of 1942
	J.Whittaker, South Australia.							
63	21 hp	3401	10ft.9ins.	3399	1927	5840 PK	Lanchester d/h coupe
	H.Shilden, Derbyshire, England.							
64	21 hp	3402	10ft.9ins.	3398	1927	DR 1159	Incomplete Lanchester tourer
	S.Danaher, Suffolk, England.							
65	21 hp	3437	11ft.1ins.	3428	1927	TE 2459	Lanchester tourer
	J.Weber, Pennsylvania, U.S.A.							
66	21 hp	3445	10ft.9ins.	3443	1927	TS 6539	Connaught saloon
	D.Abbey, Essex, England.							
67	21 hp	3474	11ft.1in.	3460	1927	OX 1844	Lanchester limousine
	P.Harwin, New South Wales, Australia.							
68	21 hp	3497	11ft.1in.	3501	1927	OA 2000	Maythorn limousine
	R.Halliwell, Yorkshire, England.							
69	21 hp	3513	11ft.1in.	3488	1927	RX 1128	Lanchester limousine
	D.Bonhomme, Jersey, United Kingdom.							
70	21 hp	3535	11ft.1in.	3519	1928	–	tourer
	A.Falstein, New South Wales, Australia.							
71	21 hp	3550	11ft.1in.	3533	1928	EAF 932	Lanchester limousine
	D.Offler, Nottinghamshire, England.							
72	21 hp	3552	10ft.9ins.	3540	1928	TU 8518	Lanchester tourer
	I.Maxwell, Berkshire, England.							
73	21 hp	3558	11ft.1in.	3551	1928	RF 4350	Lanchester limousine
	E.Baptiste, West Midlands, England.							
74	21 hp	3574	11ft.1in.	3596	1928	MW 3440	Hooper landaulette
	R. Boxley, West Midlands, England.							
75	21 hp	3583	11ft.1in.	3550	1928	EN 1925	Lanchester tourer
	G. McConnachie, Perthshire, Scotland.							
76	21 hp	3616	11ft.1in.	3602	1929	OF 4213	Lanchester limousine
	J.Grindlay, Perthshire, Scotland.							
77	21 hp	3618	11ft.1in.	3616	1929	GC 4447	Corsica tourer
	Author, Gloucestershire, England.							
78	21 hp	3642	11ft.1in.	3619	1930	UJ 2666	Maythorn landaulette
	D.Widgery, Hertfordshire, England.							

The 30 h.p. Straight Eight range used its own numbering sequence, starting with number 8001. It is thought that 126 were produced. The details of the remaining cars are believed to be as follows:-

	Model	Chassis	Wheelbase	Engine	Year	Reg.No	Coachbuilder/body
79	30 hp	8010	11ft.10.5	8003	1928		tourer
	C.R.Kumar, Bangalore, India.							
80	30 hp	8014	11ft.10.5	8004	1928	UU 9065	
	N.Jarvis, Shropshire, England.							
81	30 hp	8019	11ft.10.5	8008	1929	VR 36	Lanchester fabric-saloon
	B.Hills, England.							
82	30 hp	8035	11ft.10.5	8027	1929	OF 7366	Lanchester fabric-saloon
	J.Logue, Orkney Islands, United Kingdom.							

	Model	Chassis	Wheelbase	Engine	Year	Reg.No	Coachbuilder/body
83	30 hp	8046	11ft.10.5	8077	1929	UW 7666	sports-tourer

Sparkford Motor Museum, Dorset, England.

	Model	Chassis	Wheelbase	Engine	Year	Reg.No	Coachbuilder/body
84	30 hp	8072	11ft.10.5	8072	1929	KJ 5595	Lanchester landaulette

Author, Gloucestershire, England.

| 85 | 30 hp | 8077 | 11ft.10.5 | 8063 | 1930 | PL 1493 | Weymann fabric-limousine |

C.Sherwood, Worcestershire, England.

| 86 | 30 hp | 8079 | 11ft.10.5 | 8080 | 1930 | TK 8587 | sports-tourer |

M.Whitehouse, Bedfordshire, England.

| 87 | 30 hp | 8087 | 11ft.10.5 | 8073 | 1930 | DG 1746 | Lanchester saloon |

W.& J.Best, Dorset, England.

| 88 | 30 hp | 8094 | 11ft.10.5 | 8094 | 1931 | GW 3088 | tourer |

B.Yates, Cambridgeshire, England.

| 89 | 30 hp | 8096 | 11ft.10.5 | – | 1931 | – | Barker coupe-de-ville |

G.Gluckstein, Normand Garage?

SUMMARY

Excluding the privately constructed Petrelect and wooden cars from the 1920s, the total number of Lanchester cars produced to the pre-1931 Lanchester designs was between 2,882 and 3,033.

Including the only surviving wooden car, owned by the Birmingham Science and Industry Museum, the total number of Lanchesters believed remaining in the Centenary Year of 1995 is 90.

THIS 40 H.P. IS RUMOURED TO STILL EXIST IN INDIA.
MAY OTHER LANCHESTERS BE FOUND IN FUTURE YEARS!

TABLE OF ABBREVIATIONS AND COMPARISONS

Conversion of Imperial Measurements to metric

1 in.	=	1 inch	=	2.54 centimetres (cms.)
1 ft.	=	1 foot	=	12 inches = 30.48 cms.
1 yard	=	3 feet	=	91.44 cms.
1 mile	=	1,760 yards	=	1,609.3 metres
1 sq.ft.	=	1 square foot	=	0.0929 square metres
1 cubic inch	=	16.387 c.c.	=	16.387 cubic centimetres
		1,000 c.c.	=	1 litre
1 lb.	=	1 pound	=	453.6 grams
1 stone	=	14 lbs	=	6.3503 kilograms
1 cwt	=	1 hundredweight	=	50.802 kilograms
1 ton	=	20 cwt	=	1.0161 tonnes
1 gallon	=	8 pints	=	4.546 litres
1 p.s.i.	=	1 lb per square inch	=	0.0703 kg/sq.cms.

32 degrees Fahrenheit   =   0 degrees Centigrade or Celsius
212 degrees Fahrenheit   =   100 degrees Centigrade or Celsius

BIBLIOGRAPHY

Personal papers and notes of Dr. Frederick, Frank, and George Lanchester, Lanchester Motor Cars by Anthony Bird and Francis Hutton-Stott, The Life of an Engineer by P.W. Kingsford, Aerial Flight by F.W. Lanchester, Aircraft in Warfare by F.W. Lanchester, The Complete Motorist by A.B. Filson Young, The Age of Motoring Adventure by T.R. Nicholson, The Best Twin by Jeff Clewes, Daimler 1896–1946 by St.John C. Nixon, The Czar's British Squadron by Bryan Perret and Anthony Lord, British Armoured Cars at War by Charles Messenger, The History of Brooklands by W. Boddy, Rolls-Royce and Bentleys in Princely India by John Fasal.

INDEX

In four parts, compiled by C.S.Clark.

PART ONE – index of 'firsts' by Lanchester.

1890   gas engine starter: 3, 4
1892   engine-dynamo installation: 4, 5
1894   all-British motor boat: 8
        fundamentals of flight: 82
1895   all-British, four-wheel, petrol car: 16
        oil scraper piston ring: 13
        balanced engines: 14
        epicyclic gearbox: 13
        swivel-pin castor-angle: 12
        'live' rear axle: 12
        1:1 direct transmission drive: 13
        torsionally rigid chassis: 14
        tangentially set wheel spokes: 12
        car specifically designed to take pneumatic tyres: 12
1896   high speed engine: 19
        magneto ignition: 14
1897   automatic lubrication of engine.
        inertia governor: 22
        mechanically-operated inlet valves: 22
        fan-cooling for engines: 26
        rotating magnet, magneto ignition: 14
        accelerator foot-pedal: 22
        precision roller-bearings: 25
        cantilever suspension springs: 26
        centre-less grinding machines: 25
        splined ends on shafts: 66
        worm-drive rear axle: 22
        forerunner of B.S.F. threads: 30
        full interchangeability of parts: 37, 39, 49
        'go' & 'no-go' gauges: 39
1898   rack & pinion steering: 22
1901   removeable rear wheels: 41
        pre-selector gear-change: 45
        compounding epicyclic gears: 67
        metal to metal clutch: 46
        founding of the Midland Automobile Club: 80
1902   detachable brougham hard-top: 53
        turbo-charging: 76
        disc-brake: 75, 76
        power-steering: 79
        founding of the S.M.M.T.: 80
        outboard boat engine: 81
1903   the word 'streamlined': 55
1904   four-wheel drive: 81
        high-pressure engine lubrication: 96
        box-section chassis members: 97, 98
        unit construction of engine, clutch & gearbox for production cars: 99
        car featured in first motoring film: 40
1905   first car up Shelsley Walsh Hill Climb: 80
        engine completely free of torsional vibration: 114
1907   chauffeur's seat: 117
1908   separate sub-frame used for body: 121
1909   crankshaft vibration damper: 82
        monocoque construction for commercial vehicles: 85
        aluminium-panelled omnibus coachwork: 85
        forward-facing omnibus seats: 85
        four-wheel brakes on commercial vehicles: 86
1910   aluminium-clad aeroplane: 87
1919   welded aluminium body panels: 194
        adoption of duralumin body frames: 193
1921   export of left-hand-drive cars: 175
1923   fuel injection: 237
1925   first European company to use cellulose paint: 191
1927   purpose-built armoured cars: 244

PART TWO – index of cars.

5 h.p. Stanhope Phaeton: 11–19, 21, 22, 24, 66, 126, 258
8 h.p. Stanhope Phaeton: 19–25, 43–45
8 h.p. Gold Medal Phaeton: 24–28, 30, 34–37, 40, 43–45, 47, 64
8 h.p. Mail Phaeton: 30, 34–37, 40, 43–45, 64
10 h.p.: 39–57, 59–62, 64, 65, 67–72, 75, 76, 78–80, 100
16 h.p.: 58–64, 75–79
12 h.p.: 64–79, 99–101, 258
18 h.p.: 75–79, 92, 101
20 h.p.: 92–114, 120, 130, 140, 201, 217
28 h.p.: 114–124, 130, 138–40, 150
50 h.p.: 81, 82
38 h.p.: 130–40, 142–46, 153, 155, 157, 159–63, 178, 201, 205
25 h.p.: 140–46, 153–56, 158–60, 163, 205

Sporting 40: 147–52, 163, 164, 171, 176
40 h.p.: 163–89, 208, 213, 216–18, 220, 222, 224, 230, 260, 264, 271
Racers: 139, 205–16
21 h.p.: 176–80, 191, 217–31, 249, 261, 264, 266
Armoured cars: 139, 153–58, 244–48
Wooden cars: 235–41
30 h.p.: 189, 249–59, 262–264, 266

PART THREE – *index of the Lanchester brothers.*

(1) *Frederick William Lanchester*
youth: 1
first patent: 2
becomes Works Manager of Forward Gas Engine Co.: 3
starts bicycle company: 5
resigns from Forward Gas Engine Co.: 7
studies flight: 5, 8
designs high-speed engines: 8
launches first Lanchester boat: 8
inspects continental car designs: 11
begins first Lanchester car: 11
road tests first Lanchester car: 16
re-designs 5 h.p. Lanchester: 19, 21, 23
buys new workshop: 20
first road trial in 1897 Stanhope Phaeton: 21
designs roller bearings and machinery: 25
first road trial of Spirit Passenger Phaeton: 26
designs water-cooled engine: 28
second boat launched: 29
constructs Mail Phaeton: 30
invents screw threads: 30
founds Lanchester Engine Company: 33
becomes General Manager & Chief Designer: 34
delays production model: 37
buys Armourer Mills: 38
designs 10 h.p.: 40
designs bodywork system: 48
argues with Directors over prices: 53
designs small runabout; rejected: 55
abducts a policeman: 57
designs 16 h.p.: 58
designs 12 h.p.: 64
details driving skills: 67
introduces disc brakes: 75, 76
designs turbo charging system: 75, 76
designs power steering: 79
designs four-wheel drive: 81
investigates relationship between bore and stroke: 81
designs 50 h.p.: 81
writes Aerodynamics and Aerodonetics: 82
joins Daimler Motor Co.: 82
designs crankshaft vibration damper: 82
challenges Rolls-Royce: 84, 85
improves Daimler's Knight engine: 85
designs K.P.L. omnibus: 85
becomes Consultant & Technical Adviser to B.S.A.: 87
joins Government's Aeronautics Committee: 87
joins White & Thompson aircraft business: 87
reduced responsibility under Lanchester Motor Co.: 91
designs 20 h.p.: 92
between seats engine: 97
resigns as Chief Engineer: 125
resigns completely from Lanchester Motor Co.: 129
advises on 21 h.p.: 217
designs radical car: 232
marries: 234
Becomes Doctor Of Laws: 234
designs wooden cars: 235
starts Lanchester's Laboratories Ltd.: 240
advises on 30 h.p.: 249
advises on fluid flywheel: 242

(2) *George Herbert Lanchester*
youth: 1
joins Forward Gas Engine Co.: 5
becomes Works Manager: 5
deputises for Fred: 5
helps build river launch: 8
works on 5 h.p. Lanchester: 14
first road trial of 5 h.p.: 16
resigns from Forward Gas Engine Co.: 20
becomes assistant to Fred at Lanchester Engine Co.: 34
drives on 1,000 Mile Trial: 35
drives on Glasgow Trial: 46
becomes engaged to pioneer woman driver: 80
retains position at new Lanchester Co: 92
designs single-lever gear control: 110
modifies coachwork with bonnet: 123
becomes Chief Designer: 125
designs Sporting 40: 147
welds aluminium panels: 194
adopts duralumin body frames: 193
designs 21 h.p.: 217
designs 30 h.p.: 249
joins Fred at Lanchester's Laboratories: 243
joins B.S.A./Lanchester as Chief Engineer: 264

(3) *Francis Lanchester*
youth: 1
joins Fred's bicycle company: 5
Secretary and Captain of Speedwell Cycle Club: 7
sells Fred's patent to Harry Lawson: 7
appointed cashier at Forward Gas Engine Co.: 7
helps build river launch: 8
helps with first Lanchester: 16
graduated as a demonstrator: 30
appointed Secretary of Lanchester Engine Co.: 34
organises sales policy: 50
competes in Phoenix Park trials: 55
founds Midland Automobile Club: 80
founds S.M.M.T.: 80
marries: 80
demoted to Chief of Sales: 92
gains War Office orders: 153
promotes sales to India: 180
secures orders from Duke & Duchess of York: 183
becomes Managing Director: 261
becomes Director of Stratton-Instone Ltd.: 263

INDEX

PART FOUR – general index.

A.C. cars: 76
accelerator pedal: 22, 26
A.C.G.B.I. (later R.A.C.): 27, 35, 37, 52, 80, 109
agents: *see* sales agents
aircraft design: *see* flight and aeronautics
Alfa Romeo cars: 216
Alldays & Onions cars: 191
apprentices: 202
Ariel cars: 83, 90
Argyll cars: 224
Armstrong Siddely cars: 234
Arnold car: 143
Aston Martin cars: 207, 212
Astor Engine Co.: 141
Austin, Herbert & cars: 16, 80, 86, 126, 159, 196, 204, 207, 224, 234
Australia: 179
Austro-Daimler cars: 228
axles, front,
   pre-1914: 12, 15, 25, 26, 40–42, 76, 81, 93, 96, 98, 100, 101, 112, 130
   post-1914: 149, 164, 177, 224, 228
axles, rear, (*see also* worm gear)
   pre-1914: 12–14, 21, 22, 25, 26, 32, 40 , 41, 60, 65, 66, 76, 81, 86, 98, 101, 107, 112, 126, 127, 131
   post-1914: 164, 165, 222, 224, 230, 251

Barker, Tom Birkett: 3–5, 7, 8, 11, 16, 18, 33, 101
Barnsley, Godfrey: *see* Directors, Lanchester Co.
Barnsley, Hamilton: *see* Directors, Lanchester Co.
Barter, Joseph: 37
Bateman, Arthur: 16
Beardmore Co. & cars: 154, 242
Bentley, W.O. & cars: 84, 216, 265
Benz vehicles: 11, 232
Berresford, Frank: *see* Directors, Lanchester Co.
Bersey, Walter: 16
boats, ships & marine: 8, 9, 14, 28, 29, 64, 81, 91, 126
bodywork, (*see also* painting, bodywork)
   pre-1914: 15, 16, 18, 24, 30, 48, 49, 54, 59, 64, 74, 77, 94, 99, 109, 110, 120, 121, 123, 124, 138, 145
   post-1914: 150, 151, 163, 181, 190, 191, 193, 194, 196, 198, 219, 253
brakes, four-wheel: 86, 224
brakes, systems,
   pre-1914: 13, 22, 25, 26, 55, 65, 66, 68, 70, 75–77, 99–101, 110, 124, 131, 145
   post-1914: 149, 164–166, 176–178, 188, 189, 224, 227, 230
Bremer, Frederick: 16
Brewster & Co. Inc.: 175, 176
British Standards: 30, 39, 159, 173, 219
Brooke motor cars: 51, 83
Brooklands Motor Course: 142, 188, 206–13, 256
Brown, Samuel: 16
B.S.A. = Birmingham Small Arms: 38, 85, 87, 232, 234, 235, 237, 240, 243, 260–64, 266
Bugatti cars: 207, 212
Buick cars: 234, 251
Butler, Edward: 16

Cadillac cars: 49, 54, 143
Canal, Warwick & Birmingham: 38, 39, 58, 190, 202
carburettors: (*see also* wick vaporiser) 149, 174, 202, 211, 213, 222, 248, 250, 251
Chamberlain, Joseph: 105, 106
chassis,
   pre-1914: 12, 24–26, 40, 41, 43, 44, 59, 64, 77, 97, 98, 101, 108, 130, 131, 140, 142
   post 1914: 147–149, 164, 165, 181, 217, 224, 227–29, 231, 251
chauffeurs' seats: 117, 118, 139
Chrysler cars: 76, 142
Churchill, Winston: 154, 155
Citroen cars: 232, 233
Clerk, Dugald: 3–5, 8, 88
clutches,
   pre-1914: 13, 22, 26, 45, 46, 47, 65, 66, 68, 99, 100, 110, 114
   post-1914: 166, 222, 227
coachwork companies,
   Barker: 136, 137, 258
   Blake: 173
   Brainsby: 230
   Cann: 142
   Gurney Nutting: 230
   Hooper: 186, 187, 198, 225, 227, 230, 255, 257
   Kellner: 231
   Lacre: 92, 93
   Maythorne: 231, 257
   Mulliner: 30, 35, 198, 223, 231
   Penman: 230
   Rock, Thorpe & Watson: 255
   Rothschild et Cie: 49
   Thrupp & Maberley: 175
   Vanden Plas: 221
   Weymann: 230, 253
   Windover: 198, 255–57, 265
Colt cars: 142
crankshafts,
   pre-1914: 13, 14, 19–21, 43, 59, 65, 66, 114
   post-1914: 149, 182, 218, 220
crankshaft vibration & Lanchester Damper: 82–85, 114, 132, 150, 168, 183, 226
Crossley, W.J. & cars: 4, 246
Crystal Palace, The: 40 (*see also* exhibitions)
cycles,
   Lanchester Brothers, The: 5–7, 20
   Rudge Cycle Co.: 51
   Speedwell Cycle Club: 7, 30
   Whitworth Cycle Co.: 9, 34, 51

Daimler Motor Co.,
   pre-1914: 11, 75, 82–87, 89, 90, 127, 132
   post-1914: 141, 159–161, 175, 179, 181, 183, 193, 196, 208, 218–20, 224, 227, 232–35, 240–43, 261–64, 266
David Brown Co.: 128
Davis, Sammy: 16
De Dion & Bouton: 19, 45, 49, 232
Delage cars: 96
Delco Co. & electrics: 143–145
diesel engines: 87

Directors, Board of: (*see also* financiers) 48, 49, 54–56, 61, 63, 64, 81, 82, 90, 91, 119, 125, 128, 147, 150, 163, 188, 217, 225, 228, 249, 254, 259
Directors, Lanchester Co.,
  Barnsley, Godfrey: 225, 231
  Barnsley, Hamilton: 33, 55, 80, 90, 91, 105, 129, 217–20, 225, 259–61
  Beresford, Frank: 225, 261
  Gibson, Arthur: 90, 91, 129, 225
  Lanchester, Frank: 225, 261
  Lanchester, George: 225
  Pugh, Charles Vernon: 33, 49, 54–56, 61, 90, 91, 108, 109, 129, 225
  Pugh, John: 33, 55, 58, 61, 90, 129
  Taylor, Joseph: 33, 90, 129, 225
  Thornley, Noel: 92
  Whitfield, Allen: 58, 129
  Whitfield, James: 33, 90, 129, 225
disc brakes: 75–77, 101
Dodge cars: 142
Douglas motor cycles: 37
driving skills: 68–72, 98, 102, 199
Duesenberg cars: 232
Dunlop Co.: *see* tyres
Durkopp cars: 83
dwellings, Lanchester,
  Frank: 1, 2, 9, 80, 120, 234
  Fred: 1, 2, 8–10, 27, 91, 161, 234
  George: 1, 2, 5, 8, 9, 80, 81, 234

Edge, S.F.: (*see also* racing drivers) 50, 82, 178
Emancipation Run 1896, The: 23
employees, Lanchester, (*see also* Millership, Archie)
  pre-1914: 20, 31, 49, 57, 60, 92, 112
  post-1914: 158, 169, 178, 188, 190, 191, 196, 201–4, 208, 209, 212, 213, 214, 219, 225, 249, 250, 254, 260, 265
engines, Lanchester, (*see also* performance, *see also* tests, car acceptance)
  5 h.p.: 13, 14, 17
  8 h.p.: 19, 20, 21, 23, 26, 28, 36
  10 h.p.: 40, 43–45, 53, 65
  16 h.p.: 59, 60, 63
  12 h.p. 64, 65, 66
  18 h.p.: 75
  20 h.p. 93–98, 102, 105, 106
  28 h.p.: 114, 116
  38 h.p.: 132, 133, 160
  25 h.p.: 140, 142
  Sporting 40: 147, 148, 150
  40 h.p.: 164, 167, 168, 178, 181, 182, 185, 188
  21 h.p. 217–22, 225–27, 230
  30 h.p.: 250, 255, 256
exhibitions & shows,
  pre-1914: 35–37, 40, 54, 55, 74, 79, 82, 92, 94, 100, 102, 107, 110, 111, 116, 117, 123, 126, 145, 146
  post-1914: 173, 174, 178, 185, 186, 188, 224, 227, 230, 231, 252, 254, 262, 265

factories, workshops & showrooms,
  Alliance Works: 190, 191
  Alpha Works: 58, 59, 105, 120, 122, 125, 158, 190, 191
  Armourer Mills: 34, 38, 45, 48, 56, 58, 59, 64, 76, 77, 92, 98, 102, 107, 108–11, 113, 119, 120, 122, 125, 158, 159, 165, 173, 178, 191, 194, 201, 208, 211, 213, 250, 263, 264–66
  Fallows Road: 191, 199, 201, 263, 264
  Forward Gas Engine Co.: 11, 17, 18, 20
  Ladywood Road: 20, 33, 38
  London: 1, 120, 121, 199, 255
  Manchester: 126, 199, 255, 266
  Radix Works: 59, 173
  St. Paul's Square: 5, 6
family, Lanchester: 1, 2, 16, 46, 80, 87, 110, 117, 233, 234
F.I.A.T. cars: 234
finances, business,
  1894 syndicate: 11, 16, 18, 20, 33
  Lanchester American Patents syndicate: 58
  Lanchester Brothers, The: 5
  Lanchester Engine Co.: 33, 34, 48, 49, 51, 56, 58, 61, 90, 91, 92, 103
  Lanchester Motor Co.: 91, 92, 103, 125, 129, 163, 188, 227, 228, 260, 261
financiers, 1894 syndicate: 11, 16, 18, 29
flight & aeronautics: 5, 7–9, 81, 82, 87, 92, 126, 160–162
fluid flywheel, Daimler: 75, 242, 243
Ford, Henry & cars: 13, 49, 113, 142, 175
Forward Gas Engine Co., The: 2–5, 7–9, 11, 14, 16, 20
four-wheel drive: 81
Frazer Nash cars: 207
front-wheel drive: 33

Gardner engines: 84
gas engines: 3, 4, 19, 20, 38
Gauge Limits, Unilateral System of: 39, 173
gears & gearing,
  pre-1914: 13, 21, 22, 26, 45, 65, 66, 67, 69, 93, 99, 102, 110, 111–14, 120, 123, 124, 128, 129, 131, 137
  post-1914: 147, 149, 164, 166, 167, 172, 183, 218, 219, 222, 224, 225, 227, 251
Georges Richard car: 61
Gibson, Arthur: 90 (*see also* Directors, Lanchester Co.)
Great Horseless Carriage Co.: 7
guarantees: 103, 104, 125, 135

Hancock, Bill: 263
Harmonic Balancer, Lanchester: 141, 142
Hispano-Suiza cars: 179, 234
homes: *see* dwellings, Lanchester
horse-power ratings, engine: 81, 88, 89
Humber cars: 84, 234

ignition system,
  pre-1914: 8, 14, 17, 22, 35, 45, 55, 59, 60, 68, 95, 96, 105, 107, 110, 116, 132
  post-1914: 149, 169, 174, 178, 188, 191, 220, 225, 228, 232, 251
Institution of Automobile Engineers: 16, 81, 82, 88, 126, 128, 218, 241
Institution of Civil Engineers: 88

INDEX

Institution of Naval Architects: 126
Isotta Fraschini cars: 249

Jensen cars: 81
Johnson, Claude: *see* Rolls-Royce Ltd.

Kidner, Percy: 16
Kipling, Rudyard: 50, 51, 57, 71, 79
Knight, Charles: 85, 86, 87
Knight, John: 16
K.P.L. omnibus: 85, 86, 238

Lacre Motor Car Co.: *see* sales agents
Lanchester businesses,
   F.W.L. Propulsion Traction Ltd.: 239
   Lanchester American Patents Syndicate: 57
   Lanchester Brothers, The: *see* cycles
   Lanchester Engine Co.: 33–92, 260
   Lanchester's Laboratories Ltd.: 240–43
   Lanchester Motor Co.: 91 to end
Lawson, Harry J.: 7, 23
legislation: (*see also* police)
   8, 11, 17, 23, 25, 32, 35, 36, 56, 57, 71, 85
Lenoir motor: 11
Levassor & cars: 12, 37, 42, 94
Leyland motor cars: 175
Lifu car: 28
Linford, Charles: 3
Locker-Lampson, Commander: 157
Locomobile cars: 30, 50
London Motor Garage Company Limited: 50
Lorraine Dietrich cars: 208
lubrication & oils,
   pre-1914: 45, 51, 96, 97, 125, 126, 131, 133, 137, 145
   post-1914: 149, 168, 181, 182, 202, 221, 222, 226, 230
Lucas, Joseph, Ltd.: 191

machines & machining: 26, 59, 98, 202–4
Martin, Percy: 232, 234, 241, 243, 261, 263
Maudsley cars: 83
Mercedes cars: 237
Michelin tyres: *see* tyres
Midland Automobile Club: 56, 60, 73, 74, 80
military: 50, 51, 56, 57, 87, 153–62, 202, 226, 244–48, 259
Millership, Archie: 9, 31, 35, 36, 39, 40, 45, 46, 52, 58, 62, 63, 67, 80, 92, 100, 126, 172, 183, 210, 211
Milligan, John L.: 32, 39, 87, 233, 234
Minerva cars: 32, 234
Mors cars: 30, 94

Napier cars: 50, 82, 90, 119, 159, 175, 178, 179, 260
National Physical Laboratory: 81, 128, 161
Newell limits system: 39
New Engine Co.: 123

oils: *see* lubrication & oils
Oldsmobile car: 56
omnibus, K.P.L.: 85, 86, 238
Otto & Langham engine: 4

owners, Lanchester car: (*see also* testimonials, *see also* Royalty) 46, 47, 51, 52, 61–63, 67, 72, 109

Packard cars: 84
painting, bodywork: 121, 122, 191, 196
Panhard cars: 11, 37, 50, 94
Parfitt, George: 16
performance of cars, (*see also* testimonials)
   5 h.p.: 17
   8 h.p.: 19, 23, 26–28, 34, 35
   10 h.p.: 43, 54
   16 h.p.: 59, 60, 63
   12 h.p.: 64–66, 73
   18 h.p.: 75
   20 h.p.: 94, 95
   28 h.p.: 114, 116, 124
   38 h.p.: 133, 135, 138
   25 h.p.: 142
   Sporting 40: 150, 152
   40 h.p.: 171, 172, 183
   21 h.p.: 224–26
   30 h.p.: 256
   Racers: 205–216
Perkins engines: 142
petrol-electric systems: 85, 86, 238–240
petrol supply, (*see also* wick vaporiser, *see also* carburettors)
   pre-1914: 14, 26, 44, 45, 68, 69, 73, 98, 99, 105, 110, 132
   post-1914: 149, 171, 174, 222, 237, 256
Petter, Percival: 16
Peugeot cars: 11, 12, 62
Physical Society, The: 1, 82
piston design: 13, 19, 20, 95, 132, 168, 181, 220, 230
police: (*see also* legislation) 17, 31–33, 56, 57, 80
Pomeroy, Laurence: 88, 89, 242, 243, 264
Porsche cars: 142
power-assisted steering: 79
price of cars,
   pre-1914: 34, 54, 55, 62, 63, 74, 79, 107, 108, 116, 123, 139, 145
   post-1914: 152, 173–76, 186, 188, 224, 227, 230, 231, 253, 255, 257, 265, 266
Pugh, Charles: (*see also* Directors, Lanchester Co.) 9, 51, 80, 108, 109
Pugh, John: (*see also* Directors, Lanchester Co.) 9, 96, 101

R.A.C. = Royal Automobile Club: 81, 88, 89, 109
racing cars, Lanchester,
   Hoieh-Wayaryeh-Gointoo: 205, 206
   Lord Ridley's 38 h.p.: 139
   Rapson 40 h.p.: 211–16
   Softly-Catch-Monkey: 206, 207
   Winni-Praps-Praps: 208, 209
   works 40 h.p.: 210, 211
racing drivers, Lanchester,
   Bellingham-Smith, Tony: 207
   Bird, C.A.: 208–11
   Campbell, Malcolm: 207, 257
   Duller, George: 213–16
   Edge, S.F.: 208–11
   Gedge, D.R.W.: 216

racing drivers, Lanchester—*cont*
   Hann, Tommy: 142, 205–7
   Millership, Archie: 211
   Rapson, Lionel: 177, 212, 213, 215, 216
   Thomas, J.P. Parry: 212, 213, 215, 216
radiators,
   pre-1914: 64, 67, 76, 93, 96, 101, 111, 121, 139, 143, 145
   post-1914: 150, 152, 171, 174, 181, 222, 231, 256
Railcar, Daimler: 86, 87, 238
Ravenglass & Eskdale Railway Co.: 139
Renault cars: 13, 234
Ricardo, Harry R: 25, 40, 83, 85, 87, 128
R.N.A.S. = Royal Naval Air Service: 154
Roadtrain, Renard: 87
Rolls, Charles: 50
Rolls-Royce Ltd. & cars: 84, 85, 123, 136–138, 163, 171, 175–77, 179, 183, 188, 189, 190, 196, 211, 220, 224, 227, 231, 232, 245, 246, 257, 260, 265
Roots & Venables: 16
Rover cars: 128, 234
Royal Aircraft Factory: 158, 160, 161
Royalty,
   British: 109, 173, 179, 183, 191, 192, 221, 230, 259, 260
   Indian: 72, 116, 117, 138, 139, 143, 179, 180, 182, 183, 195, 196, 230, 261
   Japanese: 188, 189, 195
Royce, Henry: 84, 123, 137, 138
Rudge Cycle Co.: *see* cycles
Rudge-Whitworth Co.: (*see also* wheels) 51, 119, 120, 125, 131, 149, 165, 190, 229, 256

Saab cars: 142
sales agents: 50, 54, 78, 90, 92, 93, 105, 263, 264
sales of Lanchesters, (*see also* owners, *see also* Royalty)
   pre-1914: 34, 49, 50, 51, 53, 59, 63, 72, 79, 90, 91, 92, 101, 103, 105
   post-1914: 152, 174, 188, 189, 200, 217, 254
Santler, Thomas and Walter: 16
Serck Radiators: 181
Sheffield Simplex cars: 127, 175
Shelsley Walsh Hill Climb: 80, 211
ships & designs: *see* boats, ships & marine
shows, motor: *see* exhibitions & shows
Siddely Deasey cars: 127
Siddely, John: 80
sleeve-valve engines: 85, 86, 87
S.M.M.T. = Society of Motor Manufacturers & Traders: 80, 173, 258
Speedwell Cycle Club: *see* cycles
splines: 66, 119
sprag: 26 (*see also* brake, systems)
Spyker cars: 83
Standard cars: 128, 234
Stanley cars: 56
Star cars: 227
starter/generator set: 141–145
steering:
   pre-1914: 12, 22, 26, 34, 42, 43, 48, 79, 100, 111, 120, 123
   post-1914: 149, 185, 186, 251,
Sunbeam cars: 83, 159, 196, 208

suspension,
   pre-1914: 12, 14, 15, 24, 26, 40–42, 64, 98, 110, 123, 130, 138
   post-1914: 147, 164, 172, 181, 222, 224, 229

Talbot Co.: 154
Taylor, Joseph: *see* Directors, Lanchester Co.
testimonials,
   pre-1914: 27, 28, 40, 43, 45–47, 52, 63, 67, 74, 79, 82, 94, 102, 104, 106, 107, 109, 110, 112, 113, 115–17, 122, 135, 136, 138
   post-1914: 152, 172, 173, 178, 180, 208, 209, 224, 228, 253, 257
tests, car acceptance: 104, 105, 201, 202, 251
Thornley, Noel: *see* Directors, Lanchester Co.
threads, Lanchester 'M': 30, 159, 173
track, front wheel,
   pre-1914: 12, 25, 48, 59, 64, 75, 108
   post-1914: 165, 230, 252
Trevethick, Richard: 16
trials & tests,
   pre-1914: 16, 17, 27, 28, 35–37, 40, 46, 47, 52, 53, 55, 56, 60–62, 65, 66, 72–74, 80, 86, 87, 109, 124
   post-1914: 172, 177, 185, 188, 201, 219, 245, 252
trimming: 15, 24, 112, 117, 119, 122, 134, 171, 173, 178, 188, 189
tuition & training: *see* driving skills
turbo-charging: 75, 76, 143
tyres,
   pre-1914: 12, 25, 40, 51, 55, 113, 119, 123, 131, 142
   post-1914: 149, 165, 174, 183–86, 213, 224, 227, 229, 231, 241, 256

Unholy Trinity, The: 1, 8, 9
United States of America: 5, 33, 38, 56, 58, 83, 85, 113, 142, 143, 175, 179, 191
Upholstery: *see* trimming

valves & gear,
   pre-1914: 22, 44, 64, 81, 82, 96, 133, 221
   post-1914: 150, 169
Vauxhall cars: 89, 141, 212

Ward, Ratcliffe: 16
Warner Gear Co.: 83, 84, 127
warning alarms, car: 15, 16, 69
War Office: *see* military
weight of cars,
   pre-1914: 17, 26, 54, 59, 63, 65, 76, 101, 116, 142
   post-1914: 183, 224, 256
welding & brazing: 14, 15, 194
wheelbase,
   pre-1914: 12, 25, 41, 59, 64, 75, 116, 130, 142
   post-1914: 148, 172, 174, 176, 181, 224, 252
wheels,
   pre-1914: 12, 25, 41, 59, 64, 76, 100, 101, 119, 120, 131
   post-1914: 149, 165, 224, 227, 229, 230, 252, 256
White & Thompson aircraft: 87, 88, 161
Whitfield, Allen: (*see also* Directors, Lanchester Co.) 11, 18, 29, 30, 33
Whitfield, James: (*see also* Directors, Lanchester Co.) 11, 18, 22, 28–30, 33

Whitworth Cycle Co.: *see* Rudge-Whitworth, *see* cycles
wick vaporiser: 8, 14, 21, 22, 26, 44, 68, 69, 73, 98, 99, 131, 147, 148
Willys motors: 84, 142
Wilson gearbox: 67
Wilson, Montague S.: 67
Wilson-Pilcher cars: 67, 83

Wolseley cars: 16, 35, 51, 54, 90, 92, 126, 207, 234
workshops: *see* factories
worm gear drive,
    Lanchester: 22, 25, 65, 81, 86, 112, 126–29, 137, 147, 164, 222
    parallel: 128
Wrigley Transmission Company: 128